T0323323

Measurements-Based Radar Signature Modeling

MIT Lincoln Laboratory Series

James Ward, Series Editor

Perspectives on Defense Systems Analysis: The What, the Why, and the Who, but Mostly the How of Broad Systems Defense Analysis, William P. Delaney, 2015

Ultrawideband Phased Array Antenna Technology for Sensing and Communications Systems, Alan J. Fenn and Peter T. Hurst, 2015

Decision Making Under Uncertainty: Theory and Application, Mykel J. Kochenderfer, 2015

Applied State Estimation and Association, Chaw-Bing Chang and Keh-Ping Dunn, 2016

Perspectives in Space Surveillance, Ramaswamy Sridharan and Antonio F. Pensa, eds., 2017

Mathematics of Big Data: Spreadsheets, Databases, Matrices, and Graphs, Jeremy Kepner and Hayden Jananthan, 2018

Modern HF Signal Detection and Direction-Finding, Jay R. Sklar, 2018

Measurements-Based Radar Signature Modeling: An Analysis Framework, Joseph T. Mayhan and John A. Tabaczynski, 2024

MIT Lincoln Laboratory is a federally funded research and development center that applies advanced technology to problems of national security. The books in the MIT Lincoln Laboratory Series cover a broad range of technology areas in which Lincoln Laboratory has made leading contributions. The books listed above and future volumes in the series renew the knowledge-sharing tradition established by the seminal MIT Radiation Laboratory Series published between 1947 and 1953.

Measurements-Based Radar Signature Modeling

An Analysis Framework

Joseph T. Mayhan and John A. Tabaczynski

The MIT Press
Cambridge, Massachusetts
London, England

The MIT Press would like to thank the anonymous peer reviewers who provided comments on drafts of this book. The generous work of academic experts is essential for establishing the authority and quality of our publications. We acknowledge with gratitude the contributions of these otherwise uncredited readers.

This book was set in Adbode Garamond Pro by Toppan Best-set Premedia Limited. Printed and bound in the United States of America.

Library of Congress Cataloging-in-Publication Data

Names: Mayhan, Joseph T., author. | Tabaczynski, John A., author.
Title: Measurements-based radar signature modeling : an analysis framework / Joseph T. Mayhan, John A. Tabaczynski.
Description: Cambridge, Massachusetts : The MIT Press, [2024] | Series: MIT Lincoln laboratory series | Includes bibliographical references and index.
Identifiers: LCCN 2023010453 (print) | LCCN 2023010454 (ebook) | ISBN 9780262048118 | ISBN 9780262374545 (epub) | ISBN 9780262374538 (pdf)
Subjects: LCSH: Radar cross sections—Computer simulation. | Radar simulators. | Signal detection—Mathematical models.
Classification: LCC TK6578 .M39 2024 (print) | LCC TK6578 (ebook) | DDC 621.3848—dc23 /eng/20230414
LC record available at https://lccn.loc.gov/2023010453
LC ebook record available at https://lccn.loc.gov/2023010454

10 9 8 7 6 5 4 3 2 1

John Tabaczynski
1939–2021

This text is dedicated to Dr. John Tabaczynski, who passed unexpectedly on June 25, 2021. "Tab" was an innovative leader, educator, mentor, and friend to so many at Lincoln Laboratory. His mentorship, teaching expertise, and knowledge of radar signal processing was instrumental in developing and elucidating the framework presented in this text. His constant persistence that the material presented needed more clarification, background material, and technical justification was invaluable.

Tab was blessed over many years to have the companionship and support of his wife of 56 years, Glenda, as well as his children and grandchildren, and the friendship of his many colleagues at the Laboratory.

Contents

Preface   ix

Introduction   1

**PART I   BACKGROUND, SIGNAL MODELS, AND
             PROCESSING TOOLS**

1   Background   13

2   Target Signature Modeling and the Role of Sequential
    Estimation Processing   43

3   Acceleration Estimation: Extending the All-Pole Model   73

4   Autocorrelation Measures and Signal Coherence   99

PART II   THE JOINT TARGET-MOTION ESTIMATION PROBLEM

5   A Solution Framework for the Joint Target-Motion
    Estimation Problem   127

6   Narrowband Signature Modeling Techniques   159

7   Interferrometric ISAR   195

8   Motion Estimation Techniques   217

9   Joint Motion and 2D/3D Characterization of Tumbling Targets   233

10 Joint Target-Motion Solution from Range-Only Data   265

PART III   DATA EXTRAPOLATION AND SENSOR FUSION PROCESSING

11 Multisensor Fusion and Mutual Coherence   279

12 Data Extrapolation and the Composite Target Space Mapping   305

13 Colocated Sensors: Sparse Frequency Band Processing   329

14 Signature Modeling Using Sparse Angle Data   345

PART IV   MEASUREMENTS-BASED RCS SIGNATURE MODELING

15 An Integrated Predictive/Measurements-Based RCS
    Signature Model   363

16 Component Modeling Using Measurement Data   393

Acknowledgments   405
Appendix A: Characterization of Torque-Free Euler Rotational Motion   407
Appendix B: 2D Spectral Estimation: A State-Space Approach   413
Appendix C: 2D Spectral Estimation: An ESPRIT Approach   425
Appendix D: Location Estimation Using Target Space Filters
    for $\left\{R_n, \dot{R}_n\right\}$ Observables   433
Appendix E: Acceleration Estimation MATLAB Code and
    Input Parameters   447
Appendix F: Integrating Static Range and Field Test Measurements
    into a Computational, Measurements-Based Signature Model   457
Appendix G: A Polynomial Filter Estimate of Scattering
    Center Acceleration   477
References   483
Contributors   493
About the Authors   495
Index   497

Preface

This book brings together for the first time the results of research papers and technical reports published in several diverse technical areas with the goal of achieving improved resolution and precision in the area of radar target signature modeling, and target-motion solutions. Information about objects (targets) interrogated by a radar is typically characterized by the target radar cross section (RCS), commonly referred to as the radar signature of the target. The radar transmits a waveform, which is reflected from the target, and information about the target is obtained by processing the received waveform through the radar's receiver chain. Typical radar observables that are extracted include the location (position), motion, and radar signature of the target. The radar signature provides a unique characterization of the target. Depending on the radar transmitting parameters and target characterization objective, the radar signature might be characterized by a single number (e.g., the amplitude of the RCS) or more detail about the target (i.e., its scattering component locations and amplitudes). To better understand and interpret the signature measurement, estimates of the measured signature can be made using a computational signature code based on electromagnetic scattering theory foundations using estimated target model parameters as input to the code. Comparing simulated signature data generated using computational codes to the measurement data is important for obtaining a proper understanding of the measurement data.

Radar signature modeling using a variety of electromagnetic codes is a well-established field. However, the incorporation of measurement data into the computational signature model has not been addressed in book form. The analysis techniques developed in the book provide the framework for a novel approach called measurements-based modeling (MBM) to model target signatures by incorporating measurement data into the signature model.

Characterizing the radar signature of targets that are in motion is composed of two basic operations: estimating the target motion and estimating the target model parameters. Radar signature measurement data can typically be divided into two categories: static range data, for which the motion of the target is well controlled; and field measurements, for which the target motion must be estimated. Application of the techniques developed in the book are developed relative to each measurement scenario.

The analysis framework introduced in this book aimed at achieving the MBM objective covers several diverse technical areas and addresses the following topics:

An introduction to basic radar concepts, signal representation, and radar measurements.

The development of advanced analysis tools essential for high-resolution signature modeling.

The development of novel wideband and narrowband radar imaging techniques.

The application of 2D spectral estimation theory to wideband signal processing.

Ultrawideband scattering phenomenology and sparse-band sensor data fusion.

Integration of field measurements into the radar signature modeling process.

These topics are integrated into a unified analysis framework and focus on the general radar target signature modeling problem. Throughout the book, examples are presented that compare the performance of the new techniques with that using conventional analysis techniques. The book begins with the development of some advanced processing tools essential to the framework and then branches to consider three closely related subject areas: the joint target-motion estimation problem; data extrapolation and sensor fusion; and measurements-based RCS signature modeling.

The book is organized into four sections appropriate to the MBM modeling objective:

Part I: Background, Signal Models, and Processing Tools

Part II: The Joint Target-Motion Estimation Problem

Part III: Data Extrapolation and Sensor Fusion Processing

Part IV: Measurements-Based RCS Signature Modeling

The book presents the mathematical development of the various techniques with references to a number of original papers, technical reports, and books. Discussion of the differences between conventional radar signal processing and the advanced approaches are included throughout.

This book is intended for graduate students and advanced undergraduates as well as established professionals with the desire to learn and understand the techniques for processing radar observables in order to characterize an observed target and its detailed motion.

Introduction

This text brings together for the first time the results of previously published research papers and technical reports on the subject of obtaining improved precision for radar target signature modeling and target-motion solutions. It provides an analysis framework that incorporates a new class of signal and data processing techniques capable of extracting a high level of detail characterizing the target radar signature. The performance of the new approach exceeds that of conventional Fourier-based processing techniques. The enhanced performance provided by the new framework is made possible by recent advances in radar signal processing technology. The material is drawn from research dedicated to developing high-fidelity signature models of targets that are difficult to model, either because they are very complicated and standard computational codes cannot be applied, or because the data comes from field tests using the target where the motion parameters are unknown and must be estimated. The techniques extract target scattering center locations and their associated diffraction coefficient behavior and integrate the scattering center model into a fast-running signature generation code. The framework assumes that the radar data are recorded and processed collectively to achieve a comprehensive signature model of the target.

An important element of the new framework is the target model used to represent the radar signature. This model along with the advanced spectral estimation techniques developed in the text allow one to extract high-quality information to characterize the dominant scattering centers on the target. A major aspect of the framework addresses the problem of jointly estimating the target component scattering center locations and its motion solution. This step is necessary to convert the time sequence of signature measurements into an unambiguous target signature model as a function of observation angle. A natural consequence of the modeling and estimation methodology is the development of a high-fidelity signature simulation tool called measurements-based modeling (MBM) that incorporates target measurements directly into the signature model to efficiently generate radar cross section (RCS) signature simulations. Although the emphasis in this text is on extracting the signature model from field measurements, the techniques are equally applicable to static range RCS measurements.

The framework development begins by considering the single radar case. The techniques are then extended to allow integrating the returns from multiple radars operating in different frequency bands into the signature model. To achieve the enhanced capability, recent advances in radar technology, analysis techniques, and spectral estimation theory are brought together. The primary contributors are as follows:

The use of a linear all-pole model that is directly related to the geometrical theory of diffraction (GTD) scattering theory to characterize scattering centers on the target.

The application of recent advances in two-dimensional (2D) spectral estimation theory to process 2D (frequency-time) blocks of uncompressed received radar waveform data.

The use of sequential estimation techniques to process successive blocks of processed radar data to extract motion and detailed target information.

The application of novel concepts and techniques to the general problem of estimating the target motion and RCS signature model.

Techniques for processing radar data to characterize the target signature, and extract target-motion solutions are demonstrated using simulation and static range examples. A number of recently published results from diverse technical areas are brought together and interpreted in terms of radar applications.

The text begins with an introduction to the basic concepts of radar systems and the necessary elements of signal representation. Proper understanding of these concepts is essential for separating the target radar signature from other artifacts inherently present in the radar receive processing chain. The text provides tutorial material describing the conventional Fourier-based approach for forming range-Doppler images using 2D blocks of compressed radar pulses, and compares that to the new methodology. A detailed examination of the traditional radar imaging signal processing chain is presented to show how the techniques of the new framework can be used to provide more robust characterization of the target's radar signature.

The conventional approach to extracting target information achieves its high point with the implementation of wideband waveforms, and the generation of the range-Doppler image [1, 2]. The new analysis techniques achieve improved resolution, and more accurate characterization for each scattering center on the target. In the new framework, radar target characterization means not only the identification of each scattering center on the target in terms of estimating its location and complex scattering properties, but also the general topology of the target, its motion, and potentially its material composition. All of these attributes are important in the formation of a measurements-based signature model. In terms of target characterization, the new framework achieves a level of precision and detail surpassing that possible with conventional analysis techniques. The new approach is enabled by recent advances in 2D spectral estimation theory. The conventional range-Doppler image process [3] is replaced by a technique that extracts the information from the corresponding block of uncompressed received radar data, to yield not only enhanced resolution, but also information about the nature of the scattering centers located on the target. (Note, throughout the text, the terms Doppler frequency, f_D, and range-rate, $\dot{R}$, are used interchangeably because they are related by the simple scale factor, $f_D = 2\dot{R}/\lambda$. Thus, a range-Doppler image readily converts to a range, range-rate image by simple scaling of the Doppler axis. In this text, the use of range-rate in lieu of Doppler is preferred.) A key to enhanced performance is the ability to combine the processed outputs of sequential *blocks* of radar data that are widely separated in time, and to incorporate highly sophisticated motion estimation and target characterization schemes.

The framework integrates a number of modern signal processing and spectral estimation tools into a unified theory for target characterization and motion estimation. These tools have been treated piecemeal in a variety of references including, but not limited to: range-Doppler imaging [1, 2, 3, 4], three-dimensional (3D) and bistatic imaging [1, 5, 24], signal autocorrelation [6], multiple sensor fusion [7, 8], and RCS signature modeling [9, 10, 11]. The framework incorporates enhanced versions of these tools specifically formulated for the target modeling objective.

A key aspect of the framework is the use of sequences of blocks of uncompressed received radar waveform data from which the target information is extracted. Each data block is organized into a 2D array in *frequency-time* space and consists of received radar pulses sampled in frequency and ordered by transmit time. Conventional processing often uses similar blocks of data arranged in order of transmit time, but the waveforms are pulse compressed and sampled in range. The two approaches are illustrated and discussed in detail in section 1.3.3.

A parametric linear *all-pole* model is used to represent the target signature and governs the quality of information extraction that can be achieved. It is based on a GTD scattering model [10, 12, 13], and its parameters are estimated using the newly developed high-resolution spectral estimation techniques.

The interplay of target signature modeling and the use of advanced information extraction techniques are major themes in this book. An important collateral benefit of the approach is that the target models provide the underpinning for an efficient means to generate high-fidelity, computationally robust signature representations of complex targets.

A number of the most important of the processing techniques and concepts developed in the text are listed as follows:

1. Two-dimensional spectral estimation lies at the heart of the new framework. Two independent, two-dimensional high resolution spectral estimation (HRSE) techniques [14, 15, 26], each based on a different mathematical formulation, are developed in Appendices B and C, and used throughout the text. Comparing the development of the two HRSE estimators using a common signal model and notation provides much insight, not only into the similarities of the techniques, but to the contrasting underlying assumptions which lead to each formulation. MATLAB analysis code for each technique is included. Applying either of these techniques to a two-dimensional *frequency versus time radar data block* results in estimates for the range, range-rate, and amplitude observables $(R_n, \dot{R}_n, D_n), n = 1, \ldots, N$, characterizing the instantaneous motion of each scattering center located on the target relative to the line-of-sight (LoS) between the radar and the target. These estimates are referred to in the text as the *extracted observables*.

2. Improved estimates of the scattering center observables are obtained by extending the number of pulses contained in the data block beyond that consistent with a linear signal model. A more robust nonlinear signal model is applied to the data block as the block size is incrementally increased. The enhanced model yields a new set of extracted observables $(R_n, \dot{R}_n, \ddot{R}_n, D_n), n = 1, \ldots, N$ that includes the effects of the acceleration associated with each scattering center. The addition of the acceleration variable $\ddot{R}_n$, exploited in references [16, 17, 18] using sequences of range-Doppler images, adds considerable robustness to the techniques used to estimate the motion and target solutions, and is a critical element of the framework.

3. The time evolution of the extracted observables introduced in 1. and 2. over multiple data blocks centered at times $t = t_b$, $b = 1, \ldots, B$ provides the highest quality target characterization. This sequence of data blocks allows the motion of the target to be determined relative to the radar line of sight over multiple view angles to the target. Obtaining the time evolution of the extracted observable

sequence is referred to as *sequential estimation processing*, and forms the basis for a large number of techniques developed in the text.

4. By extending the temporal length of the frequency-time data block well beyond the limits appropriate for linear imaging, a measure of the temporal coherence of the data associated with each scattering center, indicative of the target motion, can be obtained. This measure is quantified by introducing the concept of the *wideband autocorrelation filter* in chapter 4 and provides complementary information when compared to the conventional range, range-rate image (see figure 4.15). It can also be used to align the receive radar data relative to the target center of rotation (CoR).

5. In order to characterize targets in motion, two basic elements that comprise the signature must be decomposed: the target motion, and the target's scattering center properties. A unique matrix coupling that relates the target's rotational motion and the target's scattering center locations (see also the development in [19, 33]) is exploited using the extracted observable sequence introduced previously. This matrix can be decomposed into a structure that isolates the motion effects from the target scattering center attributes. Solutions addressing a variety of sensor, motion, and target constraints are considered in chapters 5–10.

6. The use of a target-centered reference frame, defined relative to the target CoR, to organize the sequence of 3D scattering center location estimates into a common coordinate system. This allows the processing of many data blocks observed over wide viewing angles. Mapping the extracted observable sequence to a target-centered coordinate system referenced to the target-motion CoR (using a known motion solution) as developed in [20] is essential to obtaining a comprehensive characterization of the target incorporating differing viewing angles. The composite of scattering centers viewed in this reference system is referred to in the text as the *composite target space mapping*.

7. The single-pulse synthetic bandwidth extrapolation (BWE) technique developed by [21, 22] is extended using a 2D signal model, and is referred to as *enhanced BWE (EBWE)*. The 2D signal model is used to represent the data over the frequency-time data block, and the analytic nature of the signal model provides robust extrapolation simultaneously in both dimensions of frequency-time space. Additionally, single-pulse performance versus signal to noise is enhanced due to the coherent integration gain realized from processing the 2D data block.

8. Although the formation of radar images using Fourier-based processing of narrowband data has been discussed by various authors (e.g., [20, 23]), a

narrowband radar is not traditionally considered an imaging sensor. However, the enhanced resolution that can be obtained using the framework achieves a new level of performance using narrowband processing techniques. Techniques for generating a high-resolution, composite target space mapping from narrowband data are developed in chapter 6. These techniques exploit the properties of the scattering center range-rate, range-acceleration sequence $(\dot{R}_n, \ddot{R}_n)_b$, where b indicates the bth data block in the sequence, and eliminates many artifacts that can occur in a conventional range-Doppler image.

9. A methodology is developed for generating computationally efficient, high-fidelity RCS signature models of the target that directly incorporate measurement data into the signature model. Extensive discussion of this approach is found in chapters 14, 15, and 16, as well as appendix F.

The framework is extensive and covers a broad range of applications. It is difficult to describe the framework in a linear fashion, so a roadmap is provided in this introduction to help the student negotiate this complex subject. The book is organized into four major parts.

Part I: Background, Signal Models, and Processing Tools

Chapter 1 presents an overview of the basic radar concepts helpful to understanding the techniques discussed in the text. It provides a brief tutorial overview of the radar signal processing chain, and the radar calibration techniques required to remove system artifacts from the target radar cross section and motion attributes contained in the received waveform. The concepts of frequency-time data blocks, matched filters [28, 29], and high-resolution spectral estimation are also introduced. A detailed examination of the traditional signal processing chain resulting in a compressed radar pulse is presented and expanded to portray a conceptual next generation radar signal processing architecture. In chapter 2 the electromagnetic scattering theory foundation for characterizing the received waveform in terms of an all-pole signature model is developed. This all-pole model is exploited to extract key information contained in the frequency-time data block.

The conventional approach to characterizing the target using such a data block would be to use Fourier processing [1, 2]. Instead, the new approach models the received signal using a 2D frequency-time all-pole signal model, each pole pair corresponding to the (range, range-rate) pair for each of the target's scattering centers contained in the data block. The 2D HRSE technique is applied to the data block to obtain estimates of the all-pole model parameters. A critical aspect of the new spectral estimation technique is the pole-pairing technique used to associate the range and range-rate observables extracted from the data block unique to each scattering center. By processing a single data block, centered at $t = t_b$ and applying simple scale factors, the range, range-rate and amplitude observables $(R_n, \dot{R}_n, D_n)_b$, $n = 1, \ldots, N$ are obtained. The sequential evolution of these estimates, where b ranges over a set of B data blocks, is referred to as sequential estimation processing and provides information about the target's motion relative to its CoR as well as the potential for identifying specific properties associated with each scattering component [20]. The exploitation of this information for characterizing individual scattering centers is an important element of the framework. The first half of chapter 2 develops the 2D all-pole model using the GTD [10, 12, 13]. The second half of chapter 2 compares performance of the new spectral estimation techniques to Fourier-based processing.

The sequential processing techniques initially introduced in chapter 2 extract only the range, range-rate and amplitude information for each scattering center present in the data block. In chapter 3 a more robust nonlinear model is introduced that allows the estimation of the acceleration properties for each scattering center, yielding the sequence $(R_n, \dot{R}_n, \ddot{R}_n, D_n)_b$ for $n = 1, \ldots, N$, $b = 1, \ldots, B$, where N denotes the number of scattering centers extracted from the data block and B is the number of data blocks This sequential information is then incorporated into candidate scattering center location estimation techniques used in later chapters to improve the quality of target characterization. The $(\ddot{R}_n)$ extraction enhancement applies to both narrowband and wideband data blocks, and adds considerable robustness to the target-motion estimation solution. An additional feature of the $(\ddot{R}_n)$ extraction enhancement technique using the nonlinear model is that it acts as a smoothing filter when compared to the sequential observables obtained using the linear model.

Chapter 4 introduces a set of tools to provide a quantitative measure of coherency for the data contained within a given data block. It also examines fundamental limits on the type of processing (e.g., coherent versus noncoherent) that can be applied to the data block. A *coherence measure* characterizing the data block is defined and used to set the processing parameters applicable to either coherent or noncoherent processing. This coherency measure is extended to introduce the concept of the autocorrelation filter that can be applied to either narrowband or wideband data blocks.

Part II: The Joint Target-Motion Estimation Problem

Chapters 5–10 concentrate on developing solutions for the joint target-motion estimation problem. Radar signatures incorporate both the target's scattering characteristics as well as the motion that the target is undergoing. An important facet of target characterization is the process of separating the effects of target motion from the observables to yield an uncorrupted picture of the target alone. The framework for decomposing the motion from the target information embedded in the received signal is developed in chapter 5. A unique motion-target coupling matrix is developed (analogous to the development in [19]) that is directly related to the range, range-rate, range-acceleration observable sequence, and allows the target motion to be isolated from the target attributes. Three important solution constraints are considered in detail. Chapters 5, 6, and 7 consider the constraint where the target motion is known, such as could be obtained by independent analysis. Using the sequential processing concepts developed in chapters 2 and 3, significantly improved 3D composite scattering center location estimation techniques are developed that achieve higher resolution compared to conventional Fourier processing [20]. Chapter 8 examines the case where the target is known, and the motion is to be determined. Such a case might occur in a field experiment containing a known target, and the dynamics of the target are to be inferred. Finally, in chapters 9 and 10 two joint solution techniques are developed that are applicable to the case where neither the target nor the motion is known, and a joint solution is required.

Because the nature of the motion-target matrix coupling depends on the particular properties of the sensor waveform used, it is necessary to consider a variety of sensor constraints including bandwidth, pulse repetition frequency limitations, and target viewing aspects. Chapters 6 and 7 address sensor constraints applied to narrowband and bistatic sensors, respectively. The application to bistatic sensing to obtain 3D target scattering center location estimates using interferometric processing [1, 25] is incorporated into the framework, and provides a high-resolution extension for the conventional Fourier-based solution.

Part III: Data Extrapolation and Sensor Fusion Processing

Chapters 11–14 address the general problem of data extrapolation in frequency-time space, and the fusion of data from colocated as well as noncolocated sensors [7, 8, 25]. These techniques are particularly valuable in developing a comprehensive model for each GTD scattering center diffraction coefficient using parametric models. Chapter 11 introduces the concept of *synthetic data* and develops the sensor processing corrections that must be applied to noncolocated sensors to cohere the data received by each sensor as if it were obtained from a single sensor. This process is referred to in this text as *sensor mutual coherence*. In chapter 12 the extracted poles and complex amplitudes of the all-pole model are used to reconstruct the data within the data block as the first step in developing techniques for data extrapolation outside the data interval, both in frequency and time space. Because the all-pole model is analytic in nature, it is particularly robust in extrapolating the data over very wide bandwidths and angles. Because this data reconstruction uses data over the entire 2D frequency-time data block, the resultant extrapolation techniques are considerably more robust in performance relative to signal-to-noise limitations when compared to conventional single-pulse frequency space only extrapolation techniques [21, 22]. These 2D data extrapolation techniques are applied in chapter 13 to develop techniques for fusing data between colocated, sparse-band sensors, and again in chapter 14 to develop techniques for cohering and fusing data from two noncolocated sensors. The former allows for the fusion of data from separate, sparse-band sensors by coherently combining their waveforms, in effect joining the two sparse bands to achieve ultrawideband performance. Finally, in chapter 14, the techniques and limitations associated with fusing data over widely diverse viewing angles are developed.

Part IV: Measurements-Based RCS Signature Modeling

In chapters 15 and 16 a methodology for developing a computationally efficient RCS signature model that incorporates measurement data is presented. This model is referred to as the MBM. The objective of the MBM modeling technique is to integrate both the static range and field measurement data into a common signature model, referenced to a particular origin located on target. A detailed discussion of the data alignment required to accomplish this fusion is developed in appendix F. Comparing simulated signature data (using computational codes applied to the estimated target model) to the measurements is important for interpreting the signature measurement, as well as assessing the fidelity of the target model.

A common element of the signal-processing framework is the use of appropriate target models. A GTD-based electromagnetic scattering theory formulation [12, 13] is used to model the scattering cross section of the target, and is shown to be directly related to the HRSE linear all-pole signal model that characterizes the received waveform. This relationship facilitates the incorporation of measured diffraction coefficients directly into a computationally efficient RCS signature code. Using this approach, a complex target that is difficult to model using existing electromagnetic codes [9, 10] is decomposed into two separate pieces. One contains the attributes that are simple to model (e.g., a symmetric body of revolution), while the other uses the estimated diffraction coefficients to represent the components that are difficult to model using standard RCS codes. The data exploitation techniques introduced in the signal-processing framework are precisely those tools required to extract the component signature models from the measurement data. The relationship of the HRSE model amplitudes to the GTD-based diffraction coefficients is validated using simulation and static range data. Because the measurements-based signature model has the attribute of a fast-running computational code, it is appropriate for use in providing high-fidelity signature inputs for simulation studies. The GTD-based signature model is augmented with a physical theory of diffraction signature model as an alternate mechanism for characterizing the target topology and simple-to-model components. The high-fidelity signatures generated using the resulting MBM signature model, and their representation of the measured data, are the best measure of the power of the approach to extract information about the target from the received radar signal.

I

Background, Signal Models, and Processing Tools

1

Background

1.1 Historical Background

Before proceeding to the development of the framework it is useful to understand the historical timeline and the advances in both hardware and software technologies that have been exploited to achieve the improved capability for target feature extraction. These technological advances set the baseline for the advanced signal-processing framework developed in this text. Figure 1.1, adapted from [34], illustrates an approximate timeline of radar technology advances starting with narrowband instruments, and continues to the modern wideband, high-frequency, multifunction sensors of today.

Table 1.1 lists the corresponding acronyms used in the figure.

The early high-power radars were primarily narrowband, used high-power transmitter tube technology, and operated at relatively low pulse repetition frequencies (PRFs). Advances in radio frequency (RF) hardware technology made high pulse-repetition rates, wider instantaneous bandwidths, and higher operating frequencies a reality. Concomitant with the hardware development, a significant effort was taking place in the analysis community to develop techniques to extract and exploit information embedded in the observations, eventually leading to the analysis framework presented in this text.

A key to the improvements in information extraction is the transition to higher operating frequencies, higher PRFs, and wide bandwidths. The resolution achievable for isolating scattering centers located on the target improves in proportion to decreasing wavelength. The scattering center location resolution in range is given by λ/FBW where λ denotes the center frequency wavelength and FBW the fractional bandwidth. Assuming a PRF consistent with changes in the target motion, a number of pulses can be processed as the target rotates over an angle $\Delta\theta$ and a two-dimensional (2D) range,

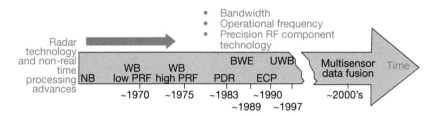

Figure 1.1 Evolution of radar technology capability and target characterization. *Source:* Adapted from [34].

Table 1.1 Acronyms for figure 1.1

NB	Narrowband (Having fractional bandwidths less than 0.1 percent)
WB	Wideband (Having factional bandwidths ~5–10 percent)
PRF	Pulse repetition frequency
PDR	Phase derived range (Extracting differential range information, ΔR, from phase changes $\sim 4\pi\Delta R/\lambda$)
BWE	Bandwidth extrapolation (Synthetic bandwidth obtained by extrapolating the target response outside the measurement band)
ECP	Extended coherent processing (Extended processing of range-Doppler data blocks over wide angles)
UWB	Ultrawideband (Having fractional bandwidths ~50 percent)
RCS	Radar cross section

cross-range image of the target can be formed having cross-range resolution given by $\lambda/(\Delta\theta)$. Thus, as higher frequency, wide bandwidth hardware technology advanced, analysis techniques for target feature extraction evolved, starting with the estimation of only narrowband target position, velocity, and radar cross section, and progressing to high-resolution target feature extraction. To achieve wide bandwidth, high-power wideband radars utilize coded waveforms having large time-bandwidth products. Wideband chirp waveforms discussed in section 1.3.2, are commonly used because of their ease of generation and processing in the radar receiver. Figure 1.2 contrasts a typical narrowband radar and wideband radar target response using the compressed waveform. The narrowband response provides the location of the target with the peak corresponding to the narrowband radar cross section (RCS) response and a target location estimate that may have an uncertainty on the order of the target size. The wideband

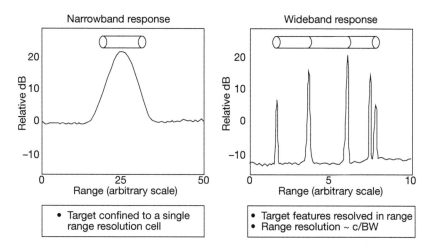

Figure 1.2   Typical narrowband and wideband radar target responses using the compressed waveform.

response provides resolution that can be a small fraction of the target's range profile. Individual scattering centers are isolated into small range resolution cells that provide a more direct measurement of the target scattering center locations. A time sequence of wideband pulses can be processed to form range-Doppler images, and the exploitation of precise phase information (extracting range information from phase changes $\sim 4\pi\Delta R/\lambda$, commonly referred to as phase derived range) from specific scattering centers isolated on the target enables estimates of changes in target motion. Signature models could now be constructed and validated based on the signature and relative locations of specific target components rather than on a simple narrowband RCS estimate.

Even with these advances, the desire for more resolution led to the development of synthetic bandwidth techniques, commonly referred to in figure 1.1 as bandwidth extrapoloation (BWE), which provided even better isolation between scattering centers. The synthetic bandwidth is obtained by extrapolating the target response outside the measurement band, and is further considered in chapters 11 and 12. The recorded data can be processed collectively, leading to a processing technique referred to as *extended coherent processing* (ECP) [35], which consists of coherently processing sequences of range-Doppler data blocks over very wide viewing angles. Finally, the break in the technology evolution arrow depicted in figure 1.1 represents the leap in capability enabled by the analysis framework developed in the text. Data from multiple sensors viewing the target from differing aspects can be processed and integrated into a robust signature model of the target. Advances in spectral estimation theory and ultra-wideband processing can be used to provide enhanced precision using the frequency-time data blocks that are traditionally used to form range-Doppler images. Sequential processing of the information extracted from these data blocks, incremented in time sequence, provides a new framework to support comprehensive motion and target feature estimation.

1.2   Basic Radar Concepts

This section introduces the fundamental concepts of radar signal processing that are essential to understanding the development of the new framework. Figure 1.3 illustrates a simple radar transmit/receive scenario, and introduces the nomenclature used in this text. A clean environment, without clutter, is assumed.

The figure includes the multistatic case in order to illustrate monostatic as well as bistatic receiver modes, both of which are addressed in the framework. The simplest case is monostatic, where the transmitting sensor receives its own waveform reflected from the target. The term bistatic generally refers to a separate receive only sensor,

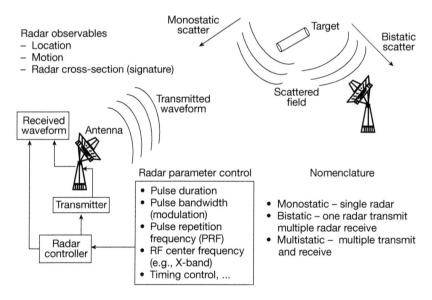

Figure 1.3   Basic radar concepts and nomenclature.

whereas the most general multistatic case assumes all sensors both transmit and receive. The general multistatic case requires each sensor ID encode its transmitted waveform so that the multiple receive signals at each sensor can be separated and processed appropriately. Typically, each transmitter is controlled by a radar controller that governs the waveform transmitted, pulse duration, PRF, RF center frequency, and keeps accurate account of the timing. The transmitted waveform is scattered off the target, and received by the sensors either in the monostatic direction (backscatter) or in the direction of the bistatic receiver (bistatic scatter). Target information is obtained by processing the received waveform through the receiver chain, and typical radar observables that are extracted include the location (position), motion, and RCS of the target.

Figure 1.4 depicts a conceptual modern radar showing the major functions such as timing and control, as well as transmit and receive subsystems. It also illustrates an important decomposition of the modern signal processor into a *front-end* processor, and a *back-end* processor. The front-end processor operates on each radar return in serial fashion and must keep up with the transmit schedule, while the back-end processor is not so constrained, because it has the capability to sort and buffer data over a limited, but multipulse time interval.

In discussing modern radar signal and data processing systems it is convenient to define three different time scales. The three-time scales of interest are the transmit pulse

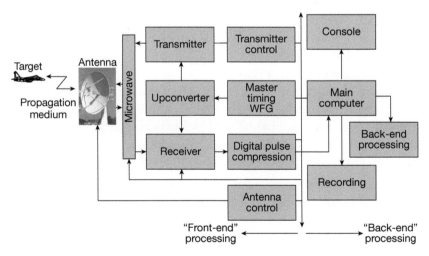

Figure 1.4   Block diagram illustrating the major functions of a conceptual modern radar.

duration time (*fast time*), the pulse repetition interval time (*slow time*), and the back-end processing time (*block time*). These are characterized by the following:

1. Transmit pulse duration time. This time scale is typically very short in duration, on the order of microseconds, and defines the time span of the transmitted pulse. For example, for a *chirped* waveform, the transmitted frequency changes in a linear sweep covering the waveform bandwidth over the pulse duration time. This time scale is often referred to in the literature as fast time. Typically, the transmit pulse duration time is very short relative to the changes in the motion of the target. Sometimes however, for very large time-bandwidth waveforms, the target moves over this transmit duration time leading to an effect referred to as range-Doppler coupling, where the motion of the target significantly affects the estimate of target range and must be removed by the processor.

2. The pulse repetition interval (PRI) time. This interval (on the order of milliseconds) denotes the time between waveform transmissions. Upon reception, the front-end processor must complete its operation on the waveform within the PRI duration in order to be ready to operate on the next transmission. This operation usually consists of matched filtering, resulting in a compressed pulse analogous to those depicted in figure 1.2. The processed pulses are then buffered and sent to the back-end processor. In the more general phased array radar case, the radar can track more than one target, and these targets may be located in different radar beams. However, the transmitter can illuminate only a single beam at a time. Hence, the transmitter PRI can be much shorter than that of the PRI allocated to any single target.

3. The back-end processing time, or block time. This time scale refers to an extended time interval over which multiple received radar pulses on a single target are buffered and processed as a single block of data. In this text, the block time is referenced to the center time of a buffered block of radar pulses. The concept of a data block is discussed in section 1.3.2. The techniques developed and discussed in this text rely on processing a sequence of data blocks.

In order to prevent confusion as to which time scale is under discussion, the book adheres to a consistent notation to differentiate among them. Because processing takes place in the time sampled domain, the distinction is identified in the sample indices as either t_k, t_q, or t_b.

The subscript k represents the time samples for fast time and corresponds to the use of f_k for frequency samples. This is consistent with $s(t)$ representing the received time waveform, and $S(\omega)$ its frequency spectrum where $s(t_k)$ and $S(\omega_k)$ are discrete Fourier transform (DFT) pairs, and $\omega_k = 2\pi f_k$.

The notation t_q denotes the single waveform transmission time. A sequence of received waveforms is buffered to form a data block consisting of Q_0 pulses, so that q ranges over the interval $q = 1, \ldots, Q_0$. The notation t_b denotes the block time referenced to the center of the data block, and the range of the subscript b is consistently labeled B, denoting the total number of data blocks contained in the block sequence.

A radar performs many functions in addition to target characterization, such as search, target acquisition, and track, so it is important that the target characterization waveforms are time multiplexed in such a manner that they are transmitted at a sufficiently high rate to capture the details in the target motion. Thus, the sequence of sensor transmissions versus time is a complex choice of multifunction waveforms organized by the radar resource allocation function. At a higher level of complexity, modern phased array radars are capable of simultaneously handling multiple targets, each of which may reside in separate radar beams. It is the job of the main computer to separate the individual target data streams into a consistent database for each target.

While the basic building blocks of the radar system are depicted in figure 1.4 at a level adequate to discuss the signal-processing framework, it is also important to keep in mind the wide variety of engineering disciplines required to actually design, develop, and field a complete radar system. Some of these are listed here for completeness:

- Engineering disciplines for modern radar design
  - RF technology
    - Antenna theory and design
    - Microwave excitation and circuitry
    - Precision RF component technology
  - Digital signal processing
    - Digital circuit design and signal distribution
    - Digital signal processing theory
    - Waveform design and control
  - Mathematical prerequisites for back-end processing
    - Radar signal processing fundamentals
    - Electromagnetic propagation and scattering theory
    - Theoretical mechanics and rigid body dynamics
    - Spectral estimation concepts and theory

- ○ Real-time software design
- ○ Engineering disciplines
  - Mechanical engineering
  - Power generation and distribution
  - Human interface engineering
  - Structural engineering
  - Control systems engineering

A simple example of the importance of understanding the complete set of engineering disciplines is illustrated by the structural changes that occur with a very large steerable dish antenna as its pointing angle changes. The structural changes introduce errors in pointing estimation and must be calibrated and removed by the back-end processor. Such calibration models require an understanding of the structural mechanics of the antenna motion. The back-end processor also compensates for pointing errors introduced by atmospheric and ionospheric propagation environments that require knowledge of the interaction of electromagnetic waves with the propagation media. The list is extensive and the texts by Skolnik [36, 37] provide a more complete set of examples.

The design, development, and fielding of a radar system requires a broad set of engineering disciplines, and the understanding of the effect of the physical limitations on the signal processing requirements cannot be overemphasized.

1.3   Radar Signal Representation

This section introduces the mathematical tools required for understanding the new framework. Additional tutorial material is presented as required throughout the text.

1.3.1   Received Waveform and Sensor Calibration

The objective of this section is to identify the corrections that must be applied to the received signal so that it represents an accurate representation of the target RCS, a process commonly referred to as sensor calibration. This step is critically important in order to separate the target's signature contributions to the receiver output from artifacts introduced by the antenna, transmitter, receiver, and other system components. It is instructive to trace the signal flow from transmit to receive using a simple frequency domain representation of a canonical radar system. This representation is illustrated in figure 1.5 and is adequate to represent the essential calibration corrections required. A more detailed characterization of the various stages of processing the received waveform is developed in section 1.3.2.

The generation of a single pulse begins with the waveform generator (WFG) at the *baseband* signal; that is, the frequency spectrum of the waveform, $T_1(\omega)$, is centered about zero frequency with bandwidth BW. In this context (ω) denotes the angular baseband frequency $\omega = 2\pi f$, where $-BW/2 < f < BW/2$. Note that in figure 1.5, a critical variable, the angle of incidence of the wavefront on the target relative to a coordinate system fixed on the target, is notionally indicated by ϕ. This variable is in fact dependent on two angle variables, the aspect (θ) and roll (φ) angles defined relative to a coordinate system referenced to the target center of rotation (CoR). The aspect and roll angles change with time as the radar viewing angle to the target changes relative to the target's CoR as a consequence of the targets rotational motion. A detailed definition and discussion of the importance of referencing the received signal relative to this target-centered coordinate system is found in section 2.2 and appendix F.

Let F_1 represent the antenna frequency response, and note that because the antenna response is tuned to a high RF frequency, $\omega_1 = 2\pi f_1$, the baseband waveform must be upconverted to the RF frequency as indicated by the mixing symbol. The transmitted signal radiated from the antenna illuminates a plane wave incident on the target at range R_0, and is reflected with complex amplitude σ_c, the complex RCS of the target, defined over the RF frequency band $f_1 - BW/2 < RF\_frequency < f_1 + BW/2$. In this text, the complex RCS of the target[1], σ_c, is defined such that

$$\sigma = |\sigma_c|^2 \tag{1.1}$$

where σ represents the conventional RCS of the target. The reflected wavefront is received by the antenna, again weighted by the antenna frequency response F_1. Denote the power received at the output of the antenna as P_0 indicated in figure 1.5. P_0 is given by

$$P_0 = P_T \left(\frac{A_e}{\lambda} \right)^2 \frac{1}{4\pi R_0^4} \sigma$$

where P_T denotes the transmitted power, and A_e is the antenna effective receiving area

$$A_e = \lambda^2 G / 4\pi$$

where G denotes the antenna gain, and R_0 is the range to the target. The signal-to-noise ratio at this point of the receiver chain is set by the noise level at the antenna port

---

1. Targets demonstrate polarimetric behavior [42]. The radar cross section is, in general, a function of the antenna transmit and receiver polarization. The scaler quantity σ_c is assumed relative to this polarization match.

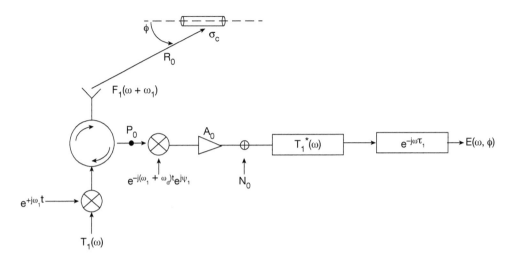

Figure 1.5 Frequency domain signal flow representation of a canonical radar system.

output, resulting from thermal noise sources present in the antenna environment. The received signal is subsequently down converted to baseband (or an intermediate RF frequency), where $\omega_d = 2\pi f_d$ compensates for the Doppler shift introduced by the target velocity. Typically, for targets at long ranges, the signal at this point of the receive chain is very small due to the $1/R_0^4$ attenuation. The signal is then amplified by introducing amplification, denoted as A_0. As a consequence, the amplifier introduces additional noise (as well as the amplified thermal noise) into the receive chain. This is modeled in figure 1.5 by the injection of noise having average noise power $\bar{N}_0^2$. The signal-to-noise ratio after amplification is given by

$$\frac{S}{N} = P_T \left(\frac{A_e}{\lambda}\right)^2 \frac{1}{4\pi R_0^4} \sigma / \bar{N}_0^2$$

The signal-to-noise ratio sets the sensitivity of the radar relative to targets having RCS σ, at a range R_0 for an antenna having receive area A_e transmit power P_T and noise power $\bar{N}_0^2$.

At this point in the processor, the signal is weighted by the conjugate of the transmitted waveform spectrum, $T_1^*(\omega)$, the matched filter [36, 37]. As the signal transits the receiver chain, additional phase and delay are introduced, denoted as ψ_1 and τ_1, respectively. Combining the various frequency response functions introduced by the receive chain, the composite received signal, represented in the baseband frequency domain, is expressed in the form

$$E(\omega, \phi) = \left(\frac{1}{4\pi R_0^4}\right)^{1/2} A_0 |T_1(\omega)|^2 F_1^2(\omega + \omega_1)\sigma_c(\omega + \omega_1, \phi)e^{-j2\left(\frac{\omega + \omega_1}{c}\right)R_0}$$
$$\times e^{j\psi_1}e^{-j\omega\tau_1} + N_0 \qquad (1.2)$$

where $-\dfrac{BW}{2} \le f \le \dfrac{BW}{2}$. The $\left(\dfrac{1}{4\pi R_0^4}\right)^{1/2}$ attenuation factor can be removed using the measured range to the target. Because A_0, T_1, F_1, ψ_1, and τ_1 are unique for each radar system, it is possible to calibrate for their contribution to the received signal output by adjusting the measured range and RCS to known values determined using a calibration target positioned at known distance from the sensor and having a known RCS. The remaining factors, in order to emphasize the decomposition of the signal into its target signature and range to the target, can be isolated by decomposing equation (1.2) into two parts as follows:

$$E(\omega, \phi) = \sigma_c(\omega + \omega_1, \phi) \times \left\{ e^{-j2\left(\frac{\omega + \omega_1}{c}\right)R_0} \right\} + N_0 \qquad (1.3)$$

The term to the left of the multiplication symbol in equation (1.3), $\sigma_c(\omega + \omega_1, \phi)$, is unique to the target. It represents truth in terms of the target's signature, and is the ultimate error-free measurement objective. The quantity within the brackets to the right of the multiplication symbol contains information about contributions made by the translational motion of the target. Corrections must be made to account for this motion and if done poorly would corrupt the interpretation of the target's true signature, σ_c.

Ideally, for signature modeling, the RCS data would be collected using a highly calibrated static RCS range. In such an environment, the target's location and position are extremely well known. The radar is operating in a well-controlled environment, uses simple waveforms, and is calibrated to high precision. This enables one to remove from equation (1.2) all the effects that might corrupt the estimate of the target radar signature. The framework developed in the text is intended for use on field test measurements as well as static range data. In the case of field measurements the additional complexities discussed previously must be addressed before the data can be handled in a manner consistent with the static range data. The ability to remove the effects of target translational motion and radar artifacts are critical to achieving good target signature models.

Target motion can always be decomposed into two components: translational and rotational [30, 31]; this is discussed further in detail in appendixes A and F. The translational motion is governed by the action of external forces referenced to the target's center of mass and is represented by the range, R_0, between the radar and the target's center of mass. The rotational motion is governed by torques acting on the target referenced to a coordinate system origin located at the center of mass, and defines the CoR. The notional incidence angle ϕ for the measured RCS characterizes the rotational motion and is defined relative to the target's rotation axis.

Correctly removing the effects of the translational motion and hardware contributions from the right side of equation (1.2), in effect, presents the field measurement data to the analyst as if it were received from a target situated in a static range measurement system positioned on a pylon having rotational motion referenced to the target CoR. Comparing simulated signature data generated using computational codes to either static range and/or field measurements is important for assessing the quality of the target model and a proper understanding of the measured data. Equation (1.2) can be used to illustrate the inherent link between RCS static range measurements, field measurements, and electromagnetic signature modeling. Static range measurements are taken using a standard RCS range having controlled known motion and range, and the calibrated RCS, σ_c, is a direct result of the measurement. The static range measurements are referenced to a CoR having origin defined relative to a fixed point on the target (typically determined by the symmetry of the target geometry) as positioned on the pylon. Field measurements are referenced to a CoR having origin determined by the target's mass distribution and dynamic equations of motion, as discussed previously, and may differ from the location of the CoR used in the static range measurement. Corrections to the received signal to remove the effects introduced to the field data by the right side of equation (1.2) are required before extracting the cross section σ_c. Signature model estimates of the target RCS are made using a computational code for which the target model dimensions input to the code are referenced to a CoR fixed on the target model, chosen by the signature analyst, having origin that might possibly differ from either the static range or field measurement. Constructing a target model based on fusing static range and field data into the computational model thus requires data alignment to account for the differences in measurement coordinate system registration, and is discussed in appendix F.

The text assumes that the radar data are recorded and processed collectively in order to develop a comprehensive signature model of the target. Using the complete time history of the data collection, the translational motion, apart from a range bias that must be estimated, can be removed from the received signal as discussed in appendix

F, and the rotational motion is used to characterize the viewing angles to the target. These effects are further discussed in chapter 9 and appendix F. Residual artifacts of the processing due to inaccuracies in completely removing the translational motion from field data applications are not addressed in this text.

For the target signature modeling techniques developed in chapters 5–10, the appropriate calibration and signal processing corrections are assumed to have been made. The data used to develop the signature model, recorded over the data collection time interval, is assumed to be directly proportional to the complex radar signature of the target as defined in equation (1.1). The incidence angle ϕ referenced to the CoR is a function of time dictated by the target motion. Typically, a number of data blocks obtained over a time interval are processed, and estimating the radar look angle to the target versus time becomes critical to interpreting the changes in the target RCS. In this text, the mapping between time and angle is termed the *motion solution*. As mentioned previously, it is the rotational motion that is the focus of the signature modeling techniques developed in this text. For exoatmospheric targets the rotational motion is characterized by Euler's equations of motion [38, 39]. Analytic expressions characterizing the temporal variation of the spherical coordinate angles (θ, φ) for a target undergoing torque-free Euler motion are developed in appendix A, and used throughout the text. For targets subjected to various atmospheric forces or torque, equation (1.3). remains valid; however, the motion solution may be difficult to extract.

1.3.2   Data Blocks, Matched Filters, and High-Resolution Spectral Estimation

In order to illustrate how the tools introduced in the new analysis framework correlate to conventional methods used for processing the received signal, a more detailed interpretation of figure 1.5 is instructive.

The signal flow depicted in figure 1.5 is represented in the frequency domain, and is useful in describing the calibration techniques required to extract the radar signature from the received signal. However, to properly model the signal at the receiver output, a time domain model representing the signal flow is required and developed in this section. The basic elements of this signal flow model, emphasizing the contrast between the signal developed by the WFG (referred to here as the transmit waveform) at the low baseband frequency, and scattering from the target at the RF frequency, are illustrated in figure 1.6. The transmit waveform bandwidth is set by the WFG at baseband, and the operational frequency is set by the transmitter feeding the antenna.

For simplicity the calibration factors discussed in section 1.3.1 are omitted. The time domain representation consists of three basic components that model (1) the transmit waveform, (2) the target signature, and (3) the interaction between the transmit waveform and the target's scattering characteristics. The time domain model for the WFG producing the transmit waveform is straightforward as it is defined in this domain. However, target signatures are typically modeled in the frequency domain using either measurement data or computational codes. The time domain model of the signature is represented by the Fourier transform of the frequency domain signature model defined by $\sigma_c(\omega + \omega_1, \phi)$, as introduced in section 1.3.1; in this case $\omega = 2\pi f$ and $\omega_1 = 2\pi f_1$, where f_1 is the operational frequency. The time domain signature model is given by $\tilde{\sigma}(t, \phi)$ where $\tilde{\sigma}(t, \phi)$ and $\sigma_c(\omega + \omega_1, \phi)$ are Fourier transform pairs. The interaction of the transmit waveform and target is determined by the convolution of $e_{T_c}(t)$ and $\tilde{\sigma}_c(t, \phi)$:

$$s(t) = \tilde{\sigma}_c(t, \phi) \otimes e_{T_c}(t)$$

To illustrate the time domain signal flow, consider a detailed representation of the signal flow indicated in figure 1.7. To model $\sigma_c(f + f_1, \phi)$ two different scenarios for target and motion are considered: multiple targets in the radar transmit/receive beam, each having independent motion; and a single rigid body target in the radar transmit/receive beam consisting of multiple scattering centers having correlated rotational motion. The target signature represented by $\sigma_c(f + f_1, \phi)$ as depicted in figure 1.6 is representative of either scenario. First consider the multitarget scenario. Consider a

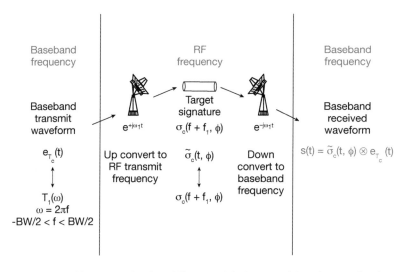

Figure 1.6   Time domain signal flow model characterizing the received waveform.

sensor operating at S-band (2.5 GHz) having a bandwidth of 10 MHz, representative of a search mode waveform. Assume a multitarget complex consisting of three targets mutually separated by 1 km. For this case, as considered later in chapter 4, $\sigma_c(f+f_1, \phi)$ is determined by the coherent sum of each of the three individual target returns defined over the 10 MHz bandwidth (e.g., a simple extension of the multitarget scenario illustrated in figure 4.9). The frequency response of $\sigma_c(f+f_1, \phi)$ over the 10 MHz bandwidth for the multitarget case is illustrated in the left side of figure 1.7. For the purposes of this example, $\sigma_c(f+f_1, \phi)$ characterizes the three targets as the composite target signature. Assume a transmitted linear chirp waveform of duration $T_c = 500$ μs. The time-based representation of the transmitted signal, denoted as $e_{T_c}(t)$, is given by

$$e_{T_c}(t) = e^{j2\pi\xi(t)} = e^{j2\pi(\alpha t^2/2)} \tag{1.4}$$

Figures 1.7 and 1.8 illustrate the signal processing flow for a single received pulse for two cases: no noise, and 25 dB compressed pulse signal-to-noise ratio.

Consider first the noise-free case. A typical frequency spectrum for $T(\omega)$ is illustrated in the left side of figure 1.7. The continuous frequency sweep associated with the chirp is given by $f(t) = \dfrac{d\xi}{dt} = \alpha t$ where the phase of the transmitted pulse $2\pi\xi(t)$ is characterized by $\xi(t) = (\alpha t^2/2)$ and $\alpha = BW/T_c$. The chirp frequency sweep is bounded by $0 < f(t) < BW$. For large time-bandwidth products ($T_c BW = 5,000$ for this example), the spectrum associated with $T(\omega)$ is flat over the bandwidth and drops off sharply as illustrated in the left side of figure 1.7. The transmitted waveform interrogates the three-target complex at continuous frequencies covering the entire bandwidth. This transmitted analog waveform is convolved with the target complex temporal response function $\tilde{\sigma}_c(t, \phi)$, where $\tilde{\sigma}_c(t, \phi)$ and $\sigma_c(\omega + \omega_1, \phi)$ are Fourier transform pairs, and the sensor receives the backscattered waveform $s(t)$. The Fourier transform pair relationship between $\tilde{\sigma}_c(t, \phi)$ and $\sigma_c(\omega + \omega_1, \phi)$ characterizing the target signature is illustrated in figure 1.7. The received signal $s(t)$ is sampled at a rate slightly greater than the Nyquist rate determined by sample spacings $\Delta t_{Nyquist} = 1/BW$. The sampling rate, r, used in the simulation is $r = 1.2BW$, so that the samples are $\Delta t = 1/r = 83$ ns. Note that the time samples t_k are associated with time samples over the duration of the transmitted pulse, and thus correspond to fast-time samples. There are $rT_c = 6,000$ samples characterizing $s(t_k)$.

The sampled frequency response of the received waveform, denoted as $S(\omega_k)$, is recovered by performing a DFT on the time samples $s(t_k)$ of the received waveform. These samples are match filtered by multiplication by $T^*(\omega_k)$, the conjugate of the transmitted frequency response, and then corrected for sensor calibration. The resulting rT_c

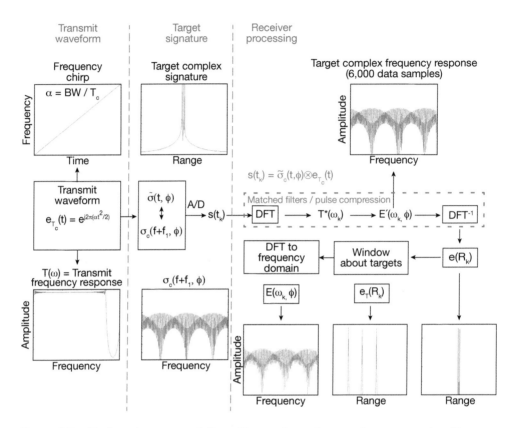

Figure 1.7   Single-pulse representation of transmit–receive waveform processing. No noise. Simulation parameters: three targets separated by 1.0 km; $BW = 10$ MHz; sampling rate $r = 12$ MHz; transmitted pulse duration $T_c = 500$ μs; signal-to-noise ratio (no noise); $rT_c = 6{,}000$ data samples for received pulse. Axis scales are notional.

frequency samples characterize the frequency response $E'(\omega_k,\ \phi)$ of the three-target complex:

$$E'(\omega_k,\ \phi) = \sigma_c(\omega_k + \omega_1,\ \phi) \tag{1.5}$$

Note that for the noise-free case, the target signature $\sigma_c(\omega_k + \omega_1,\ \phi)$ and measured signature $E'(\omega_k,\ \phi)$ are identical.

Using conventional pulse compression processing, the rT_c samples characterizing $E'(\omega_k,\ \phi)$ are weighted by a side-lobe reduction window function $SLW(\omega_k)$ and are Fourier processed using a N_{fft} point DFT, where $N_{fft} = rT_c$ is given by the number of data samples in the received pulse. The resultant N_{fft} time samples are scaled to provide

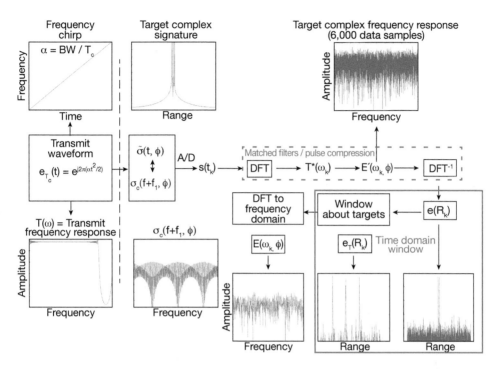

Figure 1.8   Single-pulse representation of transmit–receive waveform processing. Simulation parameters: $BW = 10$ MHz; sampling rate $r = 12$ MHz; transmitted pulse duration $T_c = 500\mu s$; signal-to-noise ratio ~25 dB (compressed pulse); $rT_c = 6{,}000$ data samples for received pulse.

a two-way transit range profile across the target. The range scaling $e(R_k)$, $k = 1, \ldots, N_{fft}$, $(R)_{scale}$ is determined by

$$(R)_{scale} = \frac{c}{2}[-(N_{fft} - 1)/2/BW : 1/BW : (N_{fft}t - 1)/2/BW] \qquad (1.6)$$

The resultant compressed pulse defined by $e(R_k)$, $k = 1, \ldots, N_{fft}$ corresponds to the compressed pulse range profile, and the number of range gates corresponds to the N_{fft} sample size. The range profile $e(R_k)$ is illustrated in figure 1.7 as the output of the N_{fft} DFT pulse compression operation. Note the three targets, spaced apart by 1 km, are contained in only a very small portion of the range window. The range extent of the compressed waveform is given by $c/2\Delta f$, where Δf is the frequency sample spacing given by $\Delta f = BW/6{,}000$. Thus, for this example, the range window extent of the compressed waveform is 90 km.

On examination of the compressed pulse, it is clear that most data samples in $e(R_k)$, $k=1,\ldots,N_{fft}$, for the noise-free case are well below a low threshold centered about the three targets present in the compressed data. The dimension of $E'(\omega_k,\phi)$ can be reduced by forming a window about the three targets excluding data below this threshold, and Fourier-transforming the windowed response function, $e_T(R_k)$, $k=1,\ldots,N_{fftW}$, into the frequency domain using N_{fftW} range samples contained in the window centered about the targets. Depending on the size of the window, zeros may be added outside the target window. Using $N_{fftW}=512$ range samples results in N_{fftW} frequency samples for the resulting frequency spectrum. The resultant spectrum, denoted by $E(\omega_k,\phi)$ in figure 1.7 is effectively down-sampled by N_{fft}/N_{fftW} samples relative to $E'(\omega_k,\phi)$. Comparing $E(\omega_k,\phi)$ to $E'(\omega_k,\phi)$ in figure 1.7, for the noise-free case, the structure of the multitarget frequency response is preserved. The $E(\omega_k,\phi)$ is simply a down-sampled version of $E'(\omega_k,\phi)$. However, for the more realistic case of targets in the presence of noise, windowing the target response represents an essential step in reducing the size of the sample set characterizing the target frequency response.

Now consider the single-pulse processing chain for the more realistic 25 dB compressed signal-to-noise ratio case illustrated in figure 1.8.

Examination of figure 1.8 shows that the N_{fft} data samples characterizing $E'(\omega_k,\phi)$ are very noise-like in nature and the structure of the frequency response no longer provides a good estimate of the signature $\sigma_c(\omega_k+\omega_1,\phi)$, as the uncompressed waveform is overwhelmed by the noise. Each sample of $E'(\omega_k,\phi)$ has a sample/noise ratio of -13 dB. Pulse compression of the $rT_c=6,000$ data samples characterizing $E'(\omega_k,\phi)$ results in 38 dB coherent integration gain using DFT processing, resulting in the 25 dB signal-to-noise ratio observed in the compressed data represented by $e(R_k)$, $k=1,\ldots,N_{fft}$, in figure 1.8. Direct down-sampling of $E'(\omega_k,\phi)$ to 512 samples results in a signal-to-noise degradation of 6,000/512, or approximately 11 dB. This degradation in signal-noise ratio can be avoided by first compressing the entire 6,000 frequency data samples, windowing the resultant compressed pulse about the targets, and then transforming the windowed response function back to the frequency domain as described for the noise-free case. This results in a version of $E'(\omega_k,\phi)$ down-sampled by N_{fft}/N_{fftW}, without any loss in signal-to-noise ratio.

Because of the large DFT processing gain (rT_c) realized from pulse compression, most applications result in data that is signal-to-noise ratio limited for the Nyquist sampled uncompressed pulse, and the window processing scheme in figure 1.8 is most often employed. For this reason it is useful to introduce a *time-domain window filter* characterized by the time domain target window processing chain indicated in figure 1.8 as being applied to the highly sampled data when extracting either the time (range) or frequency domain target response. Processing associated with this time-domain window filter is outlined in red in figure 1.8.

It becomes clear from the example shown previously that there are two versions of the uncompressed waveform: a finely sampled version $E'(\omega_k, \phi)$, and a down-sampled version $E(\omega_k, \phi)$. Computational considerations using HRSE processing dictate that the down-sampled version is most desirable for practical implementation. The frequency response $E'(\omega_k, \phi)$ contains rT_c data samples, a large number which significantly increase the run time of the HRSE technique. Windowing the compressed time domain range profile data about the multitarget structure to extract $E(\omega_k, \phi)$ provides a down-sampled frequency response having much reduced sample space dimension.

The new processing architecture advanced in this text is based on applying HRSE techniques to the uncompressed waveform. In the text the new process is referred to as the HRSE technique, and discussed in relation to figure 1.10. A more complete discussion of its application is provided in chapter 2. In the HRSE context, the uncompressed waveform is represented by a linear signal model having parameters estimated using HRSE processing. Once these parameters are determined, an estimate of the actual target frequency response $E(\omega_k, \phi)$ [or $E'(\omega_k, \phi)$] can be made. It is useful at this point to compare the linear signal model frequency response estimate to the data defined by $E(\omega_k, \phi)$ [or $E'(\omega_k, \phi)$] relative to the conventional single-pulse processing chain.

Consider now the integration of the HRSE technique into the single-pulse processing chain. The results are illustrated in figure 1.9 using the same simulation parameters as figure 1.8. The one-dimensional (1D) HRSE processing technique, discussed later in chapter 2, is applied directly to the received waveform frequency samples of $E(\omega_k, \phi)$ [or $E'(\omega_k, \phi)$], and an estimate of the frequency response is determined using a linear 1D all-pole signal model. The results are illustrated in red. Note that although the frequency data samples of $E(\omega_k, \phi)$ [or $E'(\omega_k, \phi)$] are embedded in noise, the estimate of $E(\omega_k, \phi)$ [or $E'(\omega_k, \phi)$] using the linear signal model presents a noiseless estimate of the true frequency response of the multitarget complex. The coherent processing gain associated with the HRSE technique is identical to that realized by the DFT pulse compression.

Explicit knowledge of the target frequency response, particularly as applied to the single rigid body scenario, is important when comparing the measured radar signature to computational signature models that operate in the frequency domain. Also indicated in red is the compressed waveform obtained using the linear signal model, which compares quite well with the compressed waveform determined directly using the noisy data samples.

The results of the multitarget simulation illustrated in figures 1.7–1.9 are directly scalable to the single target wideband scenario. When using a wideband waveform for characterizing a single target, the transmitted waveform would still have a large time-bandwidth product but a much larger bandwidth, chosen such that the range profile of the waveform received from a single target is capable of identifying various components on the target (see figure 1.2).

Interpretation of the results of figures 1.7–1.9 can also be used to illustrate the comparison between the signal received in a low noise environment, such as on a static RCS range versus the signal received in a noisy environment, such as with field data. For example, assume the measurement objective is to estimate the frequency response of the target at a given incidence angle ϕ. Figure 1.7 compares the measured frequency response $E(\omega_k, \phi)$, to the actual frequency response of the target in the noise-free environment. The results are identical. Figure 1.8 illustrates the measured frequency response $E(\omega_k, \phi)$ in the noisy environment. Although the compressed signal is well above the noise level, the uncompressed signal $E(\omega_k, \phi)$ characterizing the measured target frequency response is dominated by the noise. Figure 1.9 illustrates the frequency response (in red) estimated from the data using the HRSE linear signal model. Because the linear signal model is obtained using HRSE processing, the frequency response estimate is obtained realizing an integration gain equivalent to the pulse compression integration gain. The HRSE model frequency response estimate is a very close representation of the noise-free case.

1.3.3   Multipulse Block Processing

Now consider the extension of the single-pulse processing to a block of Q_0 pulses. This extension is depicted in figure 1.10. Because the single pulse received data are recorded and processed collectively, it can be organized in a manner consistent with the 2D block processing discussed earlier.

Denote the time associated with each received pulse as $t = t_q$ where $q = 1, \ldots, Q_0$, for a total of Q_0 pulses. Consistent with the notation discussed in section 1.2, each received signal in the processing chain is indexed with the label q, indicating it is the qth pulse. Thus, the fast-time samples $s(t_k)$ in figure 1.10 are now labeled as $s(t_k, t_q)$ emphasizing the interplay of fast-time index k, denoting the sampling time of the received waveform, to the slow-time index q, denoting the reference time of the qth received pulse. Analogously, $E(\omega_k, \phi)$ is now labeled as $E(\omega_k, t_q)$, where the change in incidence angle ϕ versus pulse time t_q is now referenced to the receive pulse time.

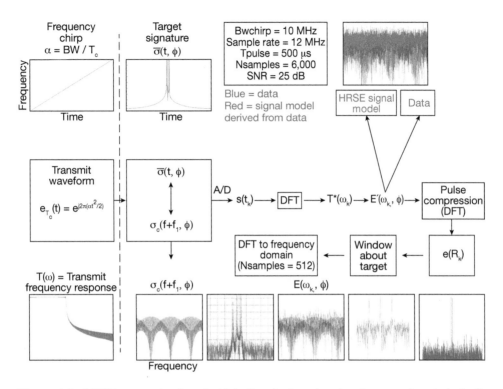

Figure 1.9 HRSE processing inserted into the single-pulse signal processing chain (red) compared to the results of figure 1.8. Fourier processing in blue. Simulation parameters: *BW* = 10 MHz; sampling rate 12 MHz; transmitted pulse duration T_c = 500µs; signal-to-noise ratio ~25 dB (compressed pulse); 6,000 data samples for received pulse.

Figure 1.10 illustrates the key functions and branch points in the 2D processing architecture important to understanding the mathematical structure of the analysis framework as well as the interplay of fast time and slow time. For branch A, each pulse is processed as described in figure 1.8, including the time-domain window filter, and buffered into a 2D data block denoted as $E(\omega_k, t_q)$. For branch B, the windowed time domain compressed pulse $e_T(R_k, t_q)$ is the direct output of the time-domain window filter.

Throughout the text it is important to keep track of the differences between Fourier-based processing, and that of the new framework using HRSE processing. The first, branch A, corresponds to the new framework developed in the text, where a block of

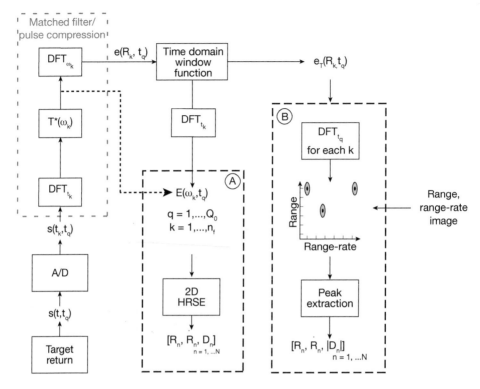

Figure 1.10   Signal flow applied to a block of Q_0 pulses (A) HRSE versus (B) conventional Fourier-based processing.

uncompressed pulses, each pulse analogous to the single-pulse example described in figure 1.8, is input directly to a 2D spectral estimation processor to obtain the HRSE estimates of the observables $(R_n, \dot{R}_n, D_n)$, $n = 1, \ldots, N$. The second, branch B, utilizes the conventional range, range-rate image processing chain, where a 2D block of range-windowed compressed pulses as described in the following section is processed to form a range, range-rate image, from which the observable set $(R_n, \dot{R}_n, |D_n|)$, $n = 1, \ldots, N$ is extracted using a peak extraction filter applied to the image. Obtaining the observable set from the range, range-rate image is referred to as Fourier-based spectral estimation as opposed to the HRSE technique that obtains the observable set (including the phase for the scattering center amplitude) without producing an image. The implications of knowing the phase properties of each scattering center are discussed later in this section.

Each of these processing branches is discussed in detail in the next few sections.

Fourier-based range, range-rate block processing

In processing the sequence of received pulses to form a conventional range, range-rate image, the compressed pulses denoted by $e_T(R_k, t_q)$, $q = 1, \ldots, Q_0$ are scaled in terms of the range transit across the target using equation (1.6), and buffered into a 2D data matrix. The transmitter $\Delta t = PRI$ spacing between the Q_0 pulses is set appropriately to observe the RCS changes due to the target motion. The length along the target of each received pulse is divided into *range gates*, with the gate width determined by the resolution of the pulse compression filter, as described in the single-pulse processing illustrated in figure 1.7. For Q_0 pulses and N_{ffiW} range gates, the data block forms a 2D matrix of size $[N_{ffiW} \times Q_0]$ consisting of complex signature samples. To obtain the range, range-rate image a weighted 1D Fourier transform is applied across the Q_0 received pulses in each range gate in the 2D data matrix. For each range gate, a larger time window is introduced over which samples of zero value are added symmetrically about the Q_0 data samples in order to achieve finer interpolation of the resultant Fourier transformed output relative to a transform of smaller size Q_0. Assume this results in N_{ffiD} samples for each range gate. The range-rate scaling, $(R_d)_{scale}$, is the dual of the range scaling equations defined in equation (1.6).

$$T_p = (N_{ffiD} - 1)\Delta t$$

$$(R_d)_{scale} = \frac{\lambda}{2}[-(N_{ffiD})/2/T_p : 1/T_p : (N_{ffiD} - 1)/2/T_p] \tag{1.7}$$

where T_p is the expanded time scale including zero padding, $\Delta t = PRI$ denotes the spacing of the time samples, and the vector notation used in equation (1.7) is consistent with MATLAB analysis software. In the case of forming a range, range-rate image using either HRSE or static range data, the initial data matrix consists of a 2D block of n_f frequency samples and Q_0 pulses.

Figure 1.11 depicts the evolution in data block size beginning with the $n_f \times Q_0$ data matrix $E(\omega_k, t_q)$ at various stages of the range, range-rate image formation process, where $k = 1, \ldots, n_f$ and n_f the number of frequency samples. Zero padding in both the frequency and time dimensions is employed to achieve fine interpolation in image quality.

An example of a conventional 2D range, range-rate image is illustrated in figure 1.12, where the image pixel color is scaled to represent the RCS magnitude of each scattering center. For the example illustrated, the target is a canonical spinning cylinder as described in chapter 2 with the sensor aspect angle $\theta = 30°$. The end portions of the

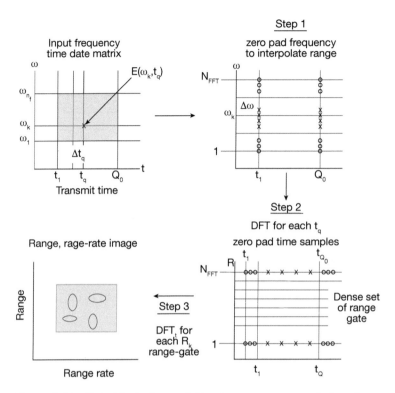

Figure 1.11   Data block size evolution for range, range-rate imaging.

cylinder each consists of four fixed point equally spaced scattering centers. As the cylinder spins, the range, range-rate image characterizes the instantaneous motion of the scattering centers relative to the LoS of the radar. Clearly evident from the image is the spinning nature of the motion as evident by the spreading of the Doppler (range-rate) at each end of the target. This example target is used in chapter 2 to compare the performance of Fourier and high-resolution spectral estimation techniques, and used throughout the text to illustrate various processing techniques.

2D high-resolution block processing

This section contrasts the differences between the two types of data blocks that are used in the new framework vice traditional processing, and the advantages provided by the new framework. Figure 1.13 depicts examples of the two types of data blocks that are highlighted on the signal processing chains A and B illustrated in figure 1.10.

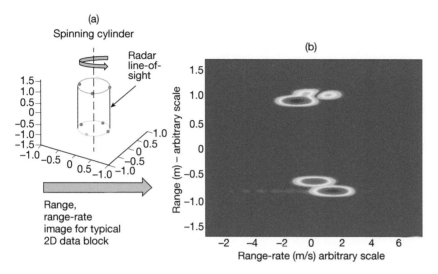

Figure 1.12 A typical range, range-rate image of a spinning cylinder.

Figure 1.13(a) depicts the frequency-time data block that is central to the new framework developed in this text. It is composed of a block of samples of 16 uncompressed pulses, $E(\omega_k, t_q)$, applicable to signal processing chain A in figure 1.10. The range, range-rate and complex observables $(R_n, \dot{R}_n, D_n)$, $n = 1, \ldots, N$, are obtained as a direct output of applying the 2D HRSE processing techniques developed in appendixes B and C to the $E(\omega_k, t_q)$ data block matrix. Figure 1.13(b) depicts the more conventional data block applicable to signal processing chain B in figure 1.10 containing compressed pulses sampled after applying the time-domain window filter. It is composed of the block of compressed pulses, $e_T(R_k, t_q)$ described in figure 1.10. The range, range-rate and amplitude observables $(R_n, \dot{R}_n, |D_n|)$, $n = 1, \ldots, N$, are extracted using a peak extraction operation on the range, range-rate image. It is important to note that in this case the phase associated with D_n is ambiguous when referenced to the range, range-rate image. It is only the relative phase between scattering center location peaks that is meaningful.

Although it requires more intensive processing to execute, the high-resolution spectral estimation processing technique offers two key advantages:

1. The resolution of the estimate is considerably better for closely spaced scattering centers because the side-lobe reduction weighting required for conventional DFT processing is not used, and the spectral estimation processing techniques inherently provide better resolution. In fact, the newly developed techniques

closely approach the Cramer-Rao bounds for location estimation for a given signal-to-noise ratio.

2. More importantly, the complex amplitudes (D_n), $n = 1, \ldots, N$ determined using the high-resolution estimator are phase matched to the data block, in contrast to the relative phase extracted from a conventional range, range-rate image. This property is extremely important when constructing a model of the data: for example, for use in the data extrapolation techniques developed in chapter 12 using the (D_n) along with the signal poles extracted from the data block, or in the measurements-based signature modeling techniques introduced in chapters 15 and 16. From an electromagnetic scattering theory viewpoint, the extracted complex amplitudes (D_n) are directly related to the geometrical theory of diffraction (GTD) diffraction coefficients for the isolated scattering component evaluated at the look angle to the target and averaged over the frequency band.

Application of either spectral estimation technique requires that an appropriate value of the number of scattering centers, N, contained in the data block be used. The choice governs the size of the observable set $(R_n, \dot{R}_n, D_n)$, $n = 1, \ldots, N$, that is extracted. Because the objective in this text is target modeling, an analysis-based objective, estimating an appropriate value for N is straightforward: examine a set of sparsely spaced (in time) range, range-rate images and examine the number of peaks present in the images.

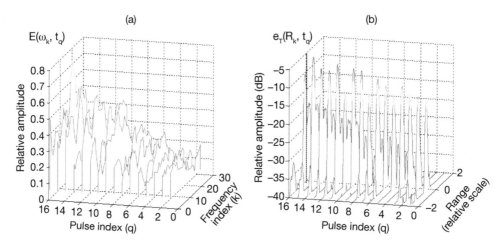

Figure 1.13 Data blocks of 16 pulses. (a) Uncompressed, frequency domain pulses given by $|E(\omega_k, t_q)|$. (b) Compressed range profile pulses given by $|e_T(R_k, t_q)|$.

Choosing a value of N to be slightly greater than the maximum of the number of peaks present in the image sequence typically offers a reasonable choice. One can then examine the change in the resulting spectral estimation sequence as N is increased to determine the appropriate value of N to use. For example, compare two sequences using values of N and $N + 1$, respectively, and observe any notable differences in the sequence. An alternative criterion for choosing N is discussed in [40].

1.4   Ultrawideband Scattering Phenomenology

1.4.1   RCS Scattering Responses versus Operational Frequency Band

For applications for which the target is detected and tracked at very long ranges, because of transmit power limitations it may be impractical to place a sufficient amount of energy on the target over a broad enough bandwidth to characterize the target adequately. Thus, it is logical to ask the following question: Is it possible to segment a very wide bandwidth, so that high power can be placed in each of the subbands, possibly using separate sensors, and combine the received waveforms coherently or noncoherently? Factors that influence the potential frequency subband choices are examined in this section. Fusing signature data acquired over different sparse subbands is discussed in section 1.4.2 and sensor fusion techniques are developed in chapter 13.

A detailed discussion of the frequency regime decomposition characterized by all-pole models is developed in section 2.1. Using the decomposition developed there, it is possible to summarize various modes of scattering phenomenology that occur as a function of frequency regions. These regions are typically referred to as the low-frequency region, the transition region, and the high-frequency region. Such a qualitative characterization is illustrated in figure 1.14, where contributions to various regions of the ultrawideband response can be described as a function of frequency regimes. Note that the example frequency bands shown in the figure for each regime are a function of target size expressed in wavelengths, and the boundaries depicted in the figure are for targets approximately 1–10 m in length. However, the general categorization of low-frequency, transition, and high-frequency bands is valid, and scaled appropriately with target size.

For excitation in the low-frequency regime, the wavelength is on the order of the body size, and the currents induced on the target radiate collectively from the entire target. This type of scattering is termed volume scattering, and is essentially a resonance phenomena. This frequency regime is also sensitive to material types that make up the target. Excitation in this regime provides a good measure of the stored energy, Q, of the target, as defined in [41]. In the high-frequency regime the target is many

• Scattering phenomenology

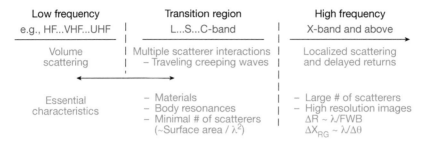

Figure 1.14   Ultrawideband scattering phenomenology. Nominal frequency bands for targets 1–10 m in length.

wavelengths in size, and for wideband excitation (e.g., 10 percent bandwidth) the scattering emanates from specific discontinuities and localized components (e.g., antenna ports, joins) on the target. The scattering resolution in range is given by λ/FBW, where λ denotes the wavelength, and FBW is the fractional bandwidth. If a number of pulses are processed as the target rotates over an angle $\Delta\theta$ a 2D image of the target can be formed having cross-range resolution given by $\lambda/(\Delta\theta)$. Thus, scattering centers can be isolated in both range and cross range with an accuracy that increases as the wavelength decreases, that is, the operational frequency increases. Finally, the transition regime contains both types of responses, and is particularly useful in characterizing multiple interactions between scattering centers. Here one can also identify other scattering mechanisms that occur, such as traveling and creeping waves that transit different paths along the target surface.

These various trade-offs, and the characteristics of the various scattering mechanisms are discussed in later chapters. It is important to understand the interaction of the radar waveform with the target, because this leads to critical choices in the various mathematical signal models used for technique development. In chapter 6, a detailed characterization of some of these scattering mechanisms is developed. Chapters 15 and 16 are dedicated to understanding the various electromagnetic scattering models from a prediction and measurements viewpoint, as well as the connection between scattering phenomenology and signal modeling.

1.4.2   Synthetic Ultrawideband Processing

Although conventional radars provide a high degree of range resolution, important target features are often smaller than conventionally processed resolution cells. To improve the range resolution of the sensor, the bandwidth of the radar can be increased. Alternately, the received signal can be processed with enhanced resolution techniques using the single block HRSE approach developed in this text to provide an increase in resolution. Using the $(R_n, \dot{R}_n, D_n)$ observables extracted for a given data block, signal models can be constructed to extrapolate the data beyond the observation region. When this extrapolation is carried out over frequency and/or angle space, the resultant data are referred to in this text as *synthetic data*. Another term commonly used for this type of processing carried out in frequency space is *bandwidth extrapolation* [22], shown as BWE in figure 1.1. When data extrapolation is applied to the output of the HRSE technique, the results lead to increased sharpness in the appearance of the features on the target, suggesting that the estimated location of a single isolated scattering center is improved. Depending on the signal-to-noise ratio, and the relative proximity of the scattering centers, the resolution between adjacent scattering centers is not guaranteed by application of the HRSE processing technique. Sharpness improves the appearance of the result, while resolution improves the ability to detect closely spaced individual scattering centers.

Enhanced resolution processing techniques using synthetic data are developed and characterized in later sections of the text, and are considered in detail in chapter 12. However, even using enhanced resolution processing only yields on the order of 30 percent fractional bandwidth extrapolation, whereas bandwidths much larger than this are required for true ultrawideband scattering center characterization. However, appropriate processing of data obtained from multiple sensors operating over different frequency bands can yield effective fractional bandwidths approaching 50 percent or greater (referred to as ultrawide bandwidths in this text), and the amplitude of an individual scattering center can exhibit significant variation with frequency over the larger bandwidth. Spheres, edges, and surface joins are examples of scattering centers that exhibit such behavior. Achieving such high fractional bandwidths is typically impractical for a single fielded radar system. Yet, exploiting target's ultrawideband radar signature can have significant payoff from a target modeling viewpoint. Not only would extremely fine range resolution be obtained, but the amplitude behavior of isolated scattering centers can help in typing the scattering center. Many canonical scattering centers are known to exhibit $f^{\tilde{\alpha}}$-type scattering behavior, where the parameter $\tilde{\alpha}$ is characteristic of specific scattering component types. For example, the RCS of flat plates, singly curved surfaces (cone sections), and doubly curved surfaces (sphere) vary as f^2, f^1, f^0, respectively. The RCS of a curved edge varies as f^{-1}, whereas a cone vertex may be characterized with an f^{-2} RCS frequency dependence. One of the advantages of ultrawideband processing is the ability to estimate the frequency-dependent terms using

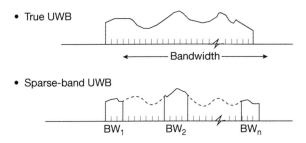

Figure 1.15   Synthetic ultrawideband processing.

the measured data, and use the estimates to identify the scattering center type for incorporation into the radar signature model of the target. Because it is difficult to obtain such ultrawide bandwidths from a single sensor, a concept that fuses the data from a collection of sensors that together cover the ultrawide bandwidth of interest has been developed. The resulting processed waveform is referred to as having synthetic ultrawide bandwidth.

Specific processing techniques for achieving synthetic ultrawide bandwidth are considered in chapter 13. Figure 1.15 illustrates the basic concept.

Through the application of such sparse-band fusion techniques, the true ultrawideband target response can be approximated by fusing the data from sparsely separated frequency bands.

2

Target Signature Modeling and the Role of Sequential Estimation Processing

In this and the following two chapters, a set of processing tools are developed that are critical to the new framework. This chapter introduces an all-pole model to represent the measurement data, and discusses its relationship to the electromagnetic geometrical theory of diffraction (GTD) scattering theory. Because signature modeling—particularly wideband signature modeling that covers the operational frequency regions illustrated in figure 1.14—is the focus of the framework, the chapter begins by establishing the role of the target signature model using fundamental concepts of ultrawideband scattering theory. The all-pole model is used to fit the measured radar data [organized in a two-dimensional (2D) frequency-time data block as depicted in figure 1.10], and provides estimates for the range, range-rate, and complex diffraction coefficient observables associated with each scattering center characterizing the data block, as viewed along the radar line of sight (LoS) to the target. The all-pole model parameters characterizing the data matrix are estimated from the data using the 2D HRSE spectral estimation techniques developed in appendixes B and C. By considering a sequence of frequency-time data blocks, a time sequence of range, range-rate and amplitude observables is obtained that contain information about the motion of the target as well as the diffraction coefficient variation of each scattering center as a function of look angle. The observable sequence is critical to obtaining solutions to the joint target-motion problem considered in subsequent chapters.

The remainder of chapter 2 compares the performance of the 2D HRSE techniques developed in appendix B and C to that of conventional Fourier-based processing using simulation data generated from a GTD-based scattering model. Techniques for estimating the acceleration of each scattering center are developed in chapter 3, and then used to refine the quality of the range, range-rate estimates. In chapter 4, a set of tools are developed that provide a quantitative measure for the degree of coherency of the data contained within a given data block. The measures are employed to examine the fundamental limits on the type of processing (e.g., coherent versus noncoherent) that can be applied to the data block.

2.1   Target Representation and All-Pole Modeling

In the early 1960s, Kennaugh and Moffett [43] introduced a novel approach to characterize the time response of the electromagnetic field scattered from a given target. These authors, as well as others, treated the scattering response from a target at a specified look angle as a *linear system* that could be analyzed using electrical network theory concepts. Using linear systems network theory, a linear time-invariant system can be characterized by its impulse response. Denote the response of the system to a unit impulse, $\delta(t)$, by $h(t)$, and $H(\omega)$ its Fourier transform. Because the Fourier transform of $\delta(t)$ contains all frequencies equally weighted, knowledge of $H(\omega)$ determines the response of the system over all frequencies. Thus, the analytic nature of $H(\omega)$ provides insight into the signal models appropriate for modeling the received waveforms over the operational frequency bands of interest. In particular, the realization of an all-pole signal model as it relates to the structure of $H(\omega)$ is discussed in detail in [44]. Assume that the target is interrogated by a unit impulse, and the scattered field measured at a notional incidence angle, ϕ, is denoted by $E(\omega, \phi)$. Based on much research in this area, general statements can be made with respect to the information content of $E(\omega, \phi)$ as ω ranges from the low- to high-frequency regimes indicated in figure 1.14. References [44, 45, 46] present a detailed overview of the ultrawideband characterization of $E(\omega, \phi)$. It is shown there that for each polarization of the scattered waveform at a specified incidence angle, the frequency variation of $E(\omega, \phi)$ decomposes into the general form (dropping the angle dependence for simplicity)

$$E(\omega) = H_1(\omega) + H_2(\omega) \tag{2.1}$$

where $H_1(\omega)$ is an all-pole frequency domain model of the form

$$H_1(\omega) = \sum_k \frac{A_k}{(s - s_k)} \tag{2.2}$$

and $s = \sigma + j\omega$ is the traditional complex frequency used to characterize the Laplace transform s-plane frequency response. Thus, $H_2(\omega)$ can be shown [44] to be an entire function of complex frequency, characterized by exponential functions of frequency, and thus has no poles in the complex frequency plane. Consequently, $H_2(\omega)$ can be expressed in the form

$$H_2(\omega) = \sum_n D_n e^{-(\alpha_n + j\tau_n)\omega} \tag{2.3}$$

where τ_n represents the delay associated with direct and mutual coupling between various scattering centers located on the target, and α_n is the localized approximation to each scattering center amplitude frequency response. Because $H_2(\omega)$ is an entire function of frequency, its inverse, $H_2^{-1}(\omega)$, is characterized by an all-pole function. Because of the duality between H_1 and H_2, both the s-plane pole functional model and the exponential function of frequency model are referred to in this text as all-pole target models.

Figure 2.1 illustrates a representation of the s-plane frequency domain poles of $H_1(\omega)$ compared to the exponential frequency domain poles defined by $H_2(\omega)$ relative to the unit circle in exponential space.

The $H_2(\omega)$ characterizes the high-frequency steady-state response of the target, and $H_1(\omega)$ characterizes the low-frequency resonances of the target. Applying the Fourier transform to each response function, $h_1(t)$ shows time domain responses of the form $e^{-s_k t}$ that generally damp out exponentially in time, and $h_2(t)$ has time domain responses of the form of $\delta(t - \tau_n)$, assuming $\alpha_n = 0$. The delayed impulse response characterizes

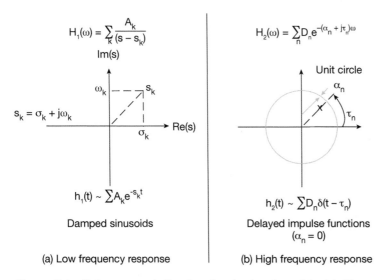

Figure 2.1   Pole representation for all-pole signal models. (a): The s-plane pole representation. (b) Exponential poles relative to the unit circle.

the delayed returns associated with the locations of the various scattering centers on the target relative to the radar LoS. For finite but very large bandwidths, these responses are modulated by an appropriate envelope function having width inversely proportional to bandwidth. The impulse response function $h_1(t)$ is often termed the *long-time response* of the target, due to the very long delay returns typically associated with target resonances, and $h_2(t)$ is called the *short-time response* of the target. In the transition between the two regions illustrated in figure 1.14, either signal model might be used to characterize the target. Although the poles of $H_2(\omega)$ can be associated with corresponding scattering centers on the target, the s-plane poles of $H_1(\omega)$ are more complex in nature, and are characteristic of the size of the target and its material composition. The poles associated with $H_1(\omega)$ have been studied extensively in the literature in relation to potential low-frequency target identification schemes, and numerous techniques have been developed for extracting these poles from measurement data [47, 48, 49, 50, 51]. The frequency region generally associated with the lower frequency applications lies in the range $0.2 < k_0 a < 4$, where $k_0 = 2\pi/\lambda$, and a denotes the radius of the maximum dimension of a sphere enclosing the target. For applications considered in this text, the operational frequencies of interest are much higher, and a signal model of the type $H_2(\omega)$ is used. Typically, $h_2(t)$ decomposes into two contributions: one corresponding to direct scattering from components present on the target (the short-time response); the second being the delayed mutual coupling between these centers, most predominant in the midfrequency band region (the midband or transition region response). A more general discussion of the delayed return scattering mechanism is presented in chapter 6. All-pole signal models associated with $H_2(\omega)$ are well suited for ultrawideband processing, and accurately characterize the target by a superposition of discrete scattering centers, each with its own frequency-dependent term. While all-pole models are best matched to signals that grow or decay exponentially with frequency, they can also be used to accurately characterize $f^{\tilde{\alpha}}$ scattering behavior characteristic of a class of canonical scattering components such as ridges, joins, and spheres over limited bandwidths. Examples that illustrate these types of scattering centers using ultrawideband processing are developed in [52] and again in chapter 13.

The decomposition of $H(\omega)$ discussed previously can be illustrated using the ultrawideband scattering response for two simple targets: a perfectly conducting sphere and a thin linear wire. Consider first the frequency response for the radar cross section (RCS) of a perfectly conducting sphere of radius a. This is illustrated in figure 2.2 for frequencies covering the very low to very high-frequency regime.

Note that in addition to $H(\omega)$ as discussed previously, in the very low-frequency region where the sphere radius is very small relative to wavelength, the scattering is

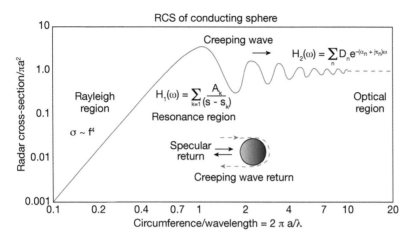

Figure 2.2   The ultrawideband frequency response of a perfectly conducting sphere having radius a.

characterized by what is referred to as the Rayleigh region for which $\sigma \sim f^4$, and is essentially independent of target geometry. Transitioning from the resonance region characterized by $H_1(\omega)$ centered about $2\pi a/\lambda \approx 1$, the scattering response changes to a multiple pole frequency response of the form of $H_2(\omega)$ characterized by the interaction between the direct return from the sphere with the delayed return referred to as the *creeping wave* return. The creeping wave transits the circumference of the sphere (perhaps multiple times depending on the sphere radius) and coherently interferes with the direct return radiated back to the sensor. Finally, in the very high-frequency region, the effect of the creeping wave damps out and only the constant amplitude direct return remains, characterized by an RCS given by $\sigma = \pi a^2$. Observe that for this simple target, in the resonance region, the frequency response can be approximately modeled using a single complex pole associated with $H_1(\omega)$. Similarly, the transition region can be modeled either by introducing multiple poles associated with either $H_1(\omega)$ or $H_2(\omega)$. Finally, in the high-frequency region, the scattering is modeled by a single constant amplitude pole associated with $H_2(\omega)$.

Now consider the second example target. Figure 2.3 illustrates the ultrawideband frequency response for the backscattered field from a thin wire of length $L = 0.5$ m, having radius 0.001 m, viewed at a scattering angle of 45° relative to the wire axis (courtesy of A. Dumanian). Both $H(\omega)$, and $h(t)$ are plotted. Because of computational time required using a method of moments computational code, the signature samples

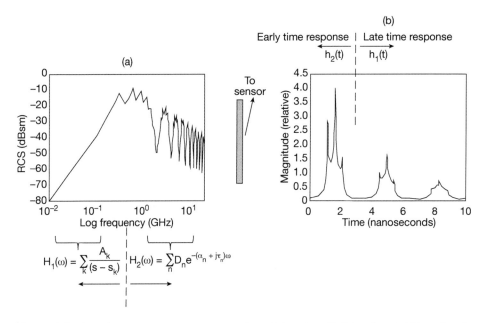

Figure 2.3   Impulse response components for a 0.5-m length thin wire at 45° angle of incidence relative to wire axis. (a) Frequency response, indicating the complementary behavior of $H_1(\omega)$ and $H_2(\omega)$. (b) Impulse response $h(t) = h_1(t) + h_2(t)$.

were computed at a sparse sample rate, but indicative of the nature of the frequency response.

The Rayleigh region characterizes the frequency response for $H(\omega)$ at the lowest end of the band below target resonance, over which the wire length is much smaller than a wavelength. In this region, only the length of the wire significantly affects the response, and the scattered field dependence versus frequency is quadratic in nature. When the wire length approaches a wavelength, a transition into the resonance region characterized by $H_1(\omega)$ occurs. For this example, transitioning out of the very high frequency (VHF) region (300 MHz), the ends of the wire are strongly coupled, resulting in numerous multiple bounces between the end points. The oscillations in the frequency response are caused by constructive and destructive interference between these scattering terms. The corresponding sequence of delayed returns exhibited in $h(t)$ is evident in the temporal evolution of the time response indicted in the figure. In the transition region, either $H_1(\omega)$ or $H_2(\omega)$ can be used to model the data, and the pole model order required using $H_2(\omega)$ would be large. At the high end of the band, the wire endpoints become uncoupled and the frequency response is characterized by isolated two point-like scattering centers, and a two-pole target model is adequate to characterize the scattering phenomenology. Other examples for the ultrawideband response of various targets are presented in [55].

In the next section the foundations of the GTD are used as the high-frequency RCS scattering formulation to develop an all-pole model characterizing the received waveform. Using this formulation a block of uncompressed received pulses can be represented by a 2D all-pole signal model in frequency-time space, consistent with the target model represented by $H_2(\omega)$. The representation of this all-pole signal model is particularly suited for using the advanced spectral estimation techniques developed in appendixes B and C to obtain high-resolution estimates of the signal poles exhibited in the data[1].

2.2   The 2D All-Pole Model

As discussed in section 1.3.1, the received radar signal contains information about both the motion and signature of the target. Corrections must be applied to the received signal so that it will accurately represent the target RCS. This process, commonly referred to as sensor calibration, was discussed in detail in section 1.3.1. In applying the techniques developed in this text, it is assumed that the measurement data are fully calibrated, recorded, and processed collectively in order to develop the signature model of the target.

There are two common techniques for measuring the target's radar signature: RCS static range measurements and field measurements. The processing techniques differ for each of these venues, but ultimately the data must be registered into a single common framework. Static range measurements are taken on a standard RCS range having controlled, known motion, and where the center of rotation (CoR) of the target is determined by the target's position on the measurement pylon. Conversely, field measurements are obtained during an actual flight test for which the motion and CoR of the target are dependent on the mass distribution of the target and must be inferred from measurements. The objective of the measurements-based model (MBM) signature modeling technique is to integrate both the static range and field measurement data into a common signature model, referenced to a specific origin and coordinate

---

1. Throughout the text, the phrases *signal model* and *signature model* are both used to represent a model representation of the data. Signal typically refers to the progression of data through the receiver chain; signature refers specifically to a representation of the calibrated signature data.

system fixed to the target. The location of the CoR on the target for each of these measurement and computational situations, as well as the orientation of their respective coordinate axes, are generally different. As a consequence, throughout the text it is important to keep track of the particular measurement conditions in order to properly transform the data into the target-centered system.

In the ideal case, the time history of the field data collection can be used to remove the effects of the translational motion from the recorded data, and the adjusted data set further processed to characterize the radar viewing angles to the target resulting from the rotational motion. The rotational motion is governed by torques acting on the target referenced to a coordinate system origin located at the center of mass and defines the CoR characterizing the angular change in the target signature [38, 39]. The notional incidence angle ϕ for the measured RCS is defined relative to the CoR of the target. Techniques for removing the translational motion from the field data are developed in appendix F. Removing the translational motion, in effect, registers the field measurement data as if it were obtained from the target positioned on a measurement pylon analogous to a static range measurement system.

In order to exploit the received data set effectively, the rotational motion must be defined in terms of a coordinate system fixed to the target having origin located at the target CoR. In this frame of reference, the sensor can be viewed as rotating about the target, and viewing angles to the sensor can be characterized using standard spherical coordinates (θ, φ) defined relative to the origin of the system as illustrated in figure 2.4. Contrast this coordinate system that is fixed to the target to the radar fixed coordinate system where the target motion is defined relative to an observer located at the radar and the target moves relative to this radar centered system. When integrating measurement data into the computational MBM signature model, it is necessary to recognize that two separate target-fixed coordinate systems are needed: one in which the measurement data are taken and processed, and the other for the computational model that is used to model the target signature. For each of the measurement and computational situations the location of the CoR and orientation of the coordinate axes can be different, and must be taken into account. Each system uses spherical coordinates; however, the origin and axis orientation of the measurement system is generally displaced and rotated relative to the origin and axes orientation of the target model system. When contrasting these two coordinate systems, the two coordinate systems are distinguished by notation, where the measurement system is denoted by the primed notation (x', y', z') having angles (θ', φ'), and the target-fixed computational system is characterized by the unprimed notation (x, y, z) and (θ, φ). When considered separately, the (x, y, z) notation is used generically to characterize the coordinate system

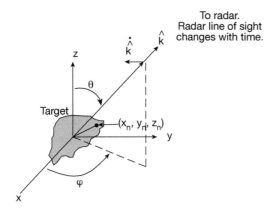

To radar.
Radar line of sight
changes with time.

Figure 2.4   Target space scattering center coordinate system. Radar LoS changes with viewing angle to sensor. Target remains fixed.

under consideration. Chapter 9 and appendix F address this difference and develop the techniques required to integrate the field and static range measurements on the target into the computational signature model.

In this section, a linear all-pole HRSE model representative of the measurement data is developed. The scattered field and sensor measurement are defined in an (x, y, z) coordinate system. Because the field scattered from the target is referenced to the location of the CoR on the target, the all-pole model fit to the received and sampled radar data of the scattered field is referenced to the same CoR location. The parameters of the all-pole model are estimated using the HRSE processing introduced in this chapter, and by minimizing the mean square error (MSE) between the model generated data and the measurement data block.

Consider figure 2.4 that depicts a coordinate system fixed to the target with its origin at the target's CoR. In this system the radar can be viewed as if it were moving around the target relative to an observer fixed at the origin of the target-fixed reference system.

In this frame of reference, $\hat{k}$ is a unit vector pointing toward the radar, in the direction of the scattered field, and represents the viewing angle in the target frame of reference. In this case, $\dot{\hat{k}}$ is a vector perpendicular to the radar LoS, and indicates the direction of the *cross range* apparent motion of the sensor due to target motion. Standard spherical coordinates (θ, φ) characterize the look angle to the radar. The rotational motion characterized by the angles (θ, φ) provides the sensor with a unique observation history of the target over a range of viewing aspect angles. It is important to recognize that the broader the viewing angle coverage, the more information contained in the data set available to characterize the target. The advantages of using the target-fixed coordinate system are two-fold. First, the location of the scattering centers on the target remains fixed with time, independent of the target motion (assuming a rigid body having fixed-point scattering center locations and noting that scatterer occlusion can obstruct the radar's view of some scatterers over limited time spans). The second advantage is that it provides consistency with the computational coordinate system used in various electromagnetic signature modeling codes (perhaps having a different CoR location on the target), allowing a straightforward association of the all-pole HRSE model parameters with the computational model. Expressions for $\hat{k}$ and its time derivative $\dot{\hat{k}}$, are given by

$$\hat{k} = \sin\theta\cos\varphi\hat{x} + \sin\theta\sin\varphi\hat{y} + \cos\theta\hat{z} \tag{2.4}$$

$$\dot{\hat{k}} = [\dot{\theta}\hat{\theta} + \dot{\varphi}\sin\theta\hat{\varphi}] \tag{2.5a}$$

where the notation "∧" denotes a unit vector, and the super "." denotes a time derivative. The expressions for $\hat{\theta}$ and $\hat{\varphi}$ are given by

$$\hat{\theta} = \cos\theta\cos\varphi\hat{x} + \cos\theta\sin\varphi\hat{y} - \sin\theta\hat{z}$$
$$\hat{\varphi} = -\sin\varphi\hat{x} + \cos\varphi\hat{y} \tag{2.5b}$$

The target coordinates are defined relative to the origin of this system, and an arbitrary scattering center on the target is denoted as (x_n, y_n, z_n). The mapping from time to motion is parametric in nature, so that in the target-centered reference frame, the look angle of the radar changes parametrically with time, that is, $\theta = \theta(t)$, $\varphi = \varphi(t)$. The mapping from time to angles is referred to as the *motion solution*. If $\theta(t)$ and $\varphi(t)$ are known, the case is considered to have a known motion solution. Otherwise, $\theta(t)$ and $\varphi(t)$ must be inferred from the radar measurements. Techniques for estimating the motion solution from radar measurement data using the processing tools developed in the framework are discussed in detail in chapters 8–10.

The all-pole model used in this text is based on the GTD scattering model introduced by Keller [56, 58, 59, 60, 61]. The all-pole model is defined in the frequency domain and used to model a 2D measurement data block consisting of Q_0 uncompressed radar pulses. Because the measurement data are labeled as a function of time, denote the signature model as $E(f, t)$. In the context used here, t is sampled at the pulse repetition interval (PRI) of the radar and represents slow time, t_q, as defined in section 1.2. The all-pole model formulation is adapted from [57]. Using the formulation of the GTD, the monostatic field, scattered from a collection of fixed-point[2] scattering centers is given by

$$E(f,t) = \sum_n E_n(f,t) = \sum_n D_n(f, \theta(t), \varphi(t)) e^{-j\frac{4\pi f}{c}\hat{k}(t)\cdot \underline{r}_n} \tag{2.6}$$

where $c = 3 \times 10^8$ m/s, f denotes the RF frequency,[3] $r_n = (x_n, y_n, z_n)$ denotes the vector location of each scattering center, and $D_n(f, \theta, \varphi)$ is the GTD scattering diffraction coefficient introduced by Keller [56]. Equation (2.6) represents the field in the frequency domain for a single pulse at a given look angle to the target at a specified time t. As noted in chapter 1, targets demonstrate polarimetric behavior [42]. In general, the RCS is a function of the antenna transmit and receiver polarization. The scattered field, $E(f, t)$, is defined relative to this polarization match. The GTD scattering model defined by equation (2.6) can be used to characterize a wide variety of scattering center types, as discussed in later chapters.

Assuming that the radar is fully calibrated, $E(f, t)$ defines a signature model representation of the scattered field characterizing the complex RCS σ_c defined in equation (1.1). For a plane wave incident on the target (i.e., the target is in the far field of the radar), the signature model characterizing the scattered field depends on the projection of each scattering center's location onto the LoS vector between the origin of the target-centered coordinate system and the sensor, and the amplitude of each scattering component. Because the field scattered from the target is inherently referenced to the location of the CoR on the target, the signature model representation of this scattered field is also referenced to the same CoR. The signature model used to represent the measurement data over a 2D data block centered about $f=f_0$, $t=t_0$ is represented by equation (2.6) and evaluated over the frequency-time data block limits. The GTD scattering diffraction coefficient $D_n(f, \theta, \varphi)$ encompasses a broad class of scattering components. In chapters 15 and 16 it is directly related to the amplitude and phase of each scattering component extracted from measurement data using the 2D all-pole HRSE techniques developed in appendixes B and C.

---

2. The location of certain types of scattering centers, referenced to target space, can vary with time. One of these is referred to as a *slipping scattering center* and requires special treatment. This class of scattering center is discussed later in this chapter as well as in chapter 6.

3. Note, using the notation introduced in equation (1.2), section 1.3.1, f denotes the baseband frequency and ideally one should carry the notation (RF_frequency) $=f+f_1$ throughout the text to be consistent with this notation. However, for the remainder of the text, with limited exceptions, the notation f is used to represent the RF frequency in order to conform to standard electromagnetic scattering theory notational conventions.

In general, the diffraction coefficient $D_n(f, \theta, \varphi)$ is a complex function having a frequency and angle variation that is considerably slower than the total phase variation in equation (2.6). Thus a single block of data provides information only on a small local region of $D_n(f, \theta, \varphi)$, which in its entirety is a wide-angle, broadband function. Wide-angle characterization of $D_n(f, \theta, \varphi)$ requires processing a sequence of data blocks that cover a time window over which changes in $\theta(t)$, $\varphi(t)$ are adequate to observe the angular changes in $D_n(f, \theta, \varphi)$.

Because the radar return pulses are processed as a function of time, the changes in (θ, φ) must be inferred from the received signal. In frequency-time space, the signature model component $E_n(f, t)$ used to represent the measurement data corresponding to each individual scattering center, expressed in the form of equation (2.6), has an amplitude D_n and phase ψ_n, where $\psi_n(f, t)$ and $D_n(f, t)$ are implicit functions of $\hat{k}(\theta, \varphi)$ and $\underline{r}_n$ according to the following:

$$\psi_n(f, t) = \frac{4\pi f}{c}\,\hat{k}(t) \cdot \underline{r}_n \tag{2.7}$$

$$D_n(f, t) = D_n(f, \hat{k}(t)). \tag{2.8}$$

In order to apply the signature model defined in equation (2.6) to accommodate variations in frequency and viewing angles, it is necessary to consider local approximations for ψ_n and D_n over both frequency and time. Equations (2.7) and (2.8) can be used to develop an all-pole model representation for the wideband, wide-angle field scattered from each scattering center. First consider a local approximation to the diffraction coefficient. Over limited frequency-time intervals, centered about the frequency-time reference point (f_0, t_0), $D_n(f, t)$ can be locally approximated as having an exponential behavior, that is,

$$D_n(f, t) \sim D_n e^{-\alpha_n(f - f_0)} e^{-\beta_n(t - t_0)}, \tag{2.9}$$

where the real constants α_n and β_n are used to approximate the local frequency-time variation of $D_n(f, t)$, and D_n is a complex constant. This particular local approximation is convenient for applying 2D all-pole modeling spectral estimation techniques to targets that exhibit complex poles in the measurement data. Now consider the phase variation ψ_n approximated over the same frequency-time interval. Local approximations to equation (2.6) are obtained using a 2D Taylor series expansion of the form

$$\psi(f, t) = \psi(f_0, t_0) + \frac{\partial \psi}{\partial f}(f - f_0) + \frac{\partial \psi}{\partial t}(t - t_0). \tag{2.10}$$

Using this expansion, the phase behavior $\psi_n(f, t)$ is first approximated over frequency at fixed t_0 and then over time at fixed f_0:

$$\text{Fixed } t_0. \text{ Vary } f = f_0 + k\Delta f, \quad k = 1, \dots, n_f$$

where n_f denotes the number of discrete frequencies covering the radar bandwidth BW. The $\psi_n(f, t)$ can be expanded in the form

$$\psi_n(f_k, t_0) = \psi_n(f_0, t_0) + \left(\frac{\partial \psi}{\partial f} \Delta f \right) k . \tag{2.11}$$

and t_0 corresponds to $\theta_0 = \theta(t_0)$, $\varphi_0 = \varphi(t_0)$.

$$\text{Fixed } f_0. \text{ Vary } t = t_0 + q\Delta t, \quad q = 1, \dots, Q_0$$

For this case $\psi_n(f_0, t)$ is expanded in the form

$$\psi_n(f_0, t_q) = \psi(f_0, t_0) + \left(\frac{\partial \psi}{\partial t} \Delta t \right) q \tag{2.12}$$

Using the local approximations in equations (2.9), (2.11), and (2.12), the result can be combined into the form of a linear 2D all-pole model. In this case, the total scattered field $E(f, t)$ can be locally expressed as a sampled representation of $E(f, t)$, denoted by $E(k, q)$, written in the form

$$E(k, q) = \sum_n E_n(k, q) = \sum_n \left\{ D_n e^{-j\psi(f_0, t_0)} \right\} s_n^k p_n^q \equiv \sum_n D'_n s_n^k p_n^q , \tag{2.13}$$

where the complex poles s_n and p_n are given by

$$s_n = \exp\left(-\Delta f \left[\alpha_n + j\left(\frac{\partial \psi_n}{\partial f} \right) \right] \right)$$

$$p_n = \exp\left(-\Delta t \left[\beta_n + j\left(\frac{\partial \psi_n}{\partial t} \right) \right] \right) . \tag{2.14}$$

Equation (2.13) defines the linear all-pole HRSE model representation of the measurement data. The GTD amplitude D_n and phase $\psi_n(f_0, t_0)$ are incorporated into the HRSE model amplitudes D'_n. It is important to distinguish the GTD signature model amplitudes D_n from HRSE model amplitudes D'_n. The HRSE model amplitudes D'_n are extracted directly from the measurement frequency-time data block; the

GTD model amplitudes, D_n, are integrated into the MBM GTD signature model introduced in chapter 15. Once the HRSE model amplitudes are determined, the GTD signature model amplitudes are obtained from the phase relationship

$$D_n = D'_n e^{+j\psi(f_0, t_0)}.$$

This transformation is discussed in chapter 3, equation (3.25) and again in chapters 15 and 16 for the development of the MBM signature model.

The form of equation (2.13) is now consistent with the all-pole model $H_2(\omega)$ discussed in section 2.1. The pole pairs can be extracted from the data block using either the Fourier-based or HRSE technique introduced later in this chapter. Once the N pole pairs, (s_n, p_n), $n = 1, \ldots, N$, are determined, the HRSE model amplitudes D'_n, $n = 1$, $\ldots, N$, can be estimated using a mean square error (MSE) fit of the HRSE model as per equation (2.13) to the measurement data set. The direct fit of the all-pole model to the measurement data assures that the resulting amplitudes are amplitude and phase matched to the data. For convenience, throughout the text, because $|D_n| = |D'_n|$, the generic notation D_n is used to indicate either D_n or D'_n unless the distinction is specifically required.

The N pole pairs extracted from the 2D frequency-time data block can be used to estimate the range, range-rate, and amplitude observables $(R_n, \dot{R}_n, D_n)$, $n = 1, \ldots, N$, characterizing the instantaneous motion of each scattering center located on the target relative to the radar LoS to the target. These estimates are referred to in the text as the extracted observables. The 2D pole mapping to accomplish the range, range-rate estimate is developed in the next section.

2.3   2D All-Pole Mapping to the Range, Range-Rate Observable Space

Conversion of the poles s_n and p_n to range, range-rate space is accomplished by applying the natural logarithm operator to equation (2.14):

$$\ln s_n = -\left(\alpha_n + j\frac{\partial \psi_n}{\partial f} \right) \Delta f$$

$$\ln p_n = -\left(\beta_n + j\frac{\partial \psi_n}{\partial t} \right) \Delta t .$$

$$(2.15)$$

Using equation (2.7)

$$\frac{\partial \psi_n}{\partial f} = \frac{4\pi}{c} \hat{k} \cdot \underline{r}_n$$

$$\frac{\partial \psi_n}{\partial t} = \frac{4\pi f_0}{c} \dot{\hat{k}} \cdot \underline{r}_n, \qquad (2.16)$$

where $\dot{\hat{k}} \equiv d\hat{k}/dt$ denotes the change in look angle versus time. Define the range and range-rate observables $(R_n, \dot{R}_n)$ according to

$$R_n \equiv \hat{k} \cdot \underline{r}_n$$

$$\dot{R}_n \equiv \dot{\hat{k}} \cdot \underline{r}_n. \qquad (2.17)$$

Noting that $\lambda = c/f_0$, combining equations (2.15), (2.16), and (2.17) achieves the desired pole pair mapping to $(R_n, \dot{R}_n)$ observable space:

$$R_n = -\frac{c}{4\pi \Delta f} \mathrm{Im}(\ln s_n)$$

$$\dot{R}_n = -\frac{\lambda}{4\pi \Delta t} \mathrm{Im}(\ln p_n), \qquad (2.18)$$

where $(R_n, \dot{R}_n)$ are the range, range-rate observables as viewed along the radar LoS to the target that correspond to the nth scattering center located on the target at (x_n, y_n, z_n), time referenced to the center of the data block. The notation $\mathrm{Im}(\cdot)$ denotes the imaginary part of a complex number. Note that relative to figure 2.4, R_n is defined in the target coordinate system as the relative range from target to radar.

Figure 2.5(a) illustrates graphically the decomposition of a single data block of Q_0 uncompressed pulses into the all-pole model, while figure 2.5(b) illustrates the all-pole extraction methodology and subsequent conversion to range, range-rate space. The pole pair (s_n, p_n) is extracted from the data matrix, and converted to range, range-rate observables using equation (2.18). The reference time associated with the observables extracted from a given data block is at the center of the block, and this center block time represents *block time*.

Once the observables $(R_n, \dot{R}_n)$, $n = 1, \ldots, N$, are extracted from the data block, equation (2.17) can be used to estimate the scattering locations. Equation (2.17) defines the range, range-rate mapping to the target space scattering center locations, $\underline{r}_n$, $n = 1, \ldots, N$,

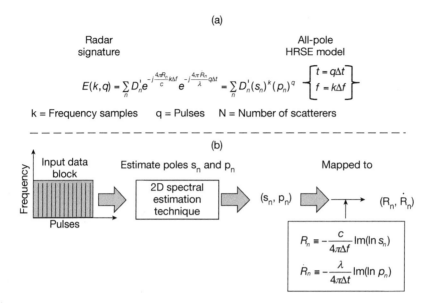

(a)

Radar
signature

All-pole
HRSE model

$$E(k,q) = \sum_n D'_n e^{-j\frac{4\pi R_n}{c}k\Delta f}\, e^{-j\frac{4\pi \dot{R}_n}{\lambda}q\Delta t} = \sum_n D'_n (s_n)^k (p_n)^q \quad \begin{bmatrix} t = q\Delta t \\ f = k\Delta f \end{bmatrix}$$

k = Frequency samples   q = Pulses   N = Number of scatterers

(b)

Frequency

Input data
block

Estimate poles s_n and p_n

Mapped to

Pulses

2D spectral
estimation
technique

(s_n, p_n)

$(R_n, \dot{R}_n)$

$$R_n \equiv -\frac{c}{4\pi\Delta f}\,\mathrm{Im}(\ln s_n)$$

$$\dot{R}_n \equiv -\frac{\lambda}{4\pi\Delta t}\,\mathrm{Im}(\ln p_n)$$

Figure 2.5 High-resolution 2D single block spectral estimation methodology.
(a) Development of the linear all-pole model. (b) Single block of uncompressed
pulses showing pole mapping extraction to obtain the range, range-rate observables.

for each of the N scattering centers on the target. They set the framework used throughout the text to obtain three-dimensional (3D) estimates of the scattering center locations in the fixed target coordinate system. Special cases of this mapping are considered
in chapters 5–10. It is important to recognize that the mapping is a function of $\hat{k}$ and
$\dot{\hat{k}}$, which in turn are a functions of $\theta, \varphi, \dot{\theta}, \dot{\varphi}$, and define a nonlinear set of equations
coupling the motion and target scattering center coordinates. A case of particular
interest occurs if the motion is known, for which the following comments apply:

For a given scattering center, $\underline{r}_n$ is defined by three (x_n, y_n, z_n) coordinates, whereas
equation (2.17) provides only two equations. Thus, in general, solutions for
(x_n, y_n, z_n) cannot be obtained using the observables extracted from a single data
block.

The $(R_n, \dot{R}_n)$ range, range-rate pair in equation (2.17) are unique to a specific scattering center. When associating the $(R_n, \dot{R}_n)$ estimates from one data block to
those extracted from a subsequent data block, it is imperative that the labeling

represented by the index n in the subsequent data block corresponds to the same scattering center. This association is referred to as *scattering center correlation*. By applying an $(R_n, \dot{R}_n)$ correlation technique to the $(R_n, \dot{R}_n)$ observables extracted from a sequence of data blocks, equation (2.17) can be applied to generate additional equations that unambiguously estimate the 3D locations (x_n, y_n, z_n) for each scattering center. In general, a minimum of two independent data blocks separated in time are required. The framework for 3D target characterization developed in section 5.5 is based on such sequential processing and correlation.

Observe that the pole to observable mapping defined by equation (2.18) uses only the imaginary part of the poles s_n and p_n. Characterization of the diffraction coefficient $D_n(f, t)$ approximated by equation (2.9) is obtained from the real part of the poles according to

$$\alpha_n = -\frac{1}{\Delta f} \mathrm{Re}(\ln s_n)$$

$$\beta_n = -\frac{1}{\Delta t} \mathrm{Re}(\ln p_n).$$

(2.19)

In this manner, the general characterization of $D_n(f, t)$ can be determined using piecewise segments of data.

Due to the mathematical complexity of the one-dimensional (1D) and 2D HRSE estimation techniques, their complete development is contained in appendixes B and C. Two estimation techniques are considered, each based on different mathematical structures, one developed by Piou [26, 122], the other by Burrows [54]. For data analysis purposes, each technique is treated as a *library routine* (i.e., a MATLAB subroutine), where the input is the 2D data block $E(k, q)$ or 1D for narrowband sensors, $E(k_0, q)$. The output is the range, range-rate observable estimate $(R_n, \dot{R}_n)$, $n = 1, \ldots, N$, of each scattering center extracted from the 2D data block, or $(\dot{R}_n, D_n)$, $n = 1, \ldots, N$ for the narrowband case. The 2D HRSE amplitudes are determined applying a least squares fit of the HRSE model defined by equation (2.13) to the data block $E(k, q)$.

The scattered field $E(f, t)$ observed by the radar is defined in terms of a 2D frequency-time space, and the next section addresses the extraction of the observables $(R_n, \dot{R}_n)$ from a sequence of 2D blocks of radar data. Chapters 5–10 address the mapping between the observable sequence and target space to characterize the individual scattering centers in 3D target space.

2.4   Sequential Estimation Processing

Equations (2.18) and (2.19) provide estimates of the range, range-rate and complex amplitude observables $(R_n, \dot{R}_n, D_n)$, $n = 1, \ldots, N$ using the N pole pairs extracted from a single data block. However, it is the time evolution of these observables over a sequence of data blocks that provides the most complete target characterization. The processing of this time evolving sequence versus the block time t_b, $\{(R_n, \dot{R}_n, D_n)_b, n = 1, \ldots, N\}$, $b = 1, \ldots, B$ for B data blocks is referred to in the text as *sequential spectral estimation processing*. The sequential processing concept is illustrated conceptually in figure 2.6.

To illustrate the sequential processing methodology consider each block as consisting of high pulse repetition frequency (PRF) data collected on an RCS range. The RCS range is the simplest example of the implementation of a radar system used solely for

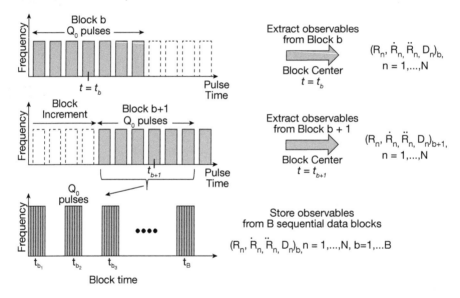

Figure 2.6   Sequential estimation processing methodology. Each data block consists of Q_0 uncompressed pulses. The $R_n, \dot{R}_n, \ddot{R}_n$ observables are extracted from each data block, and correlated versus increasing block time, so that the scattering center index n in data block b corresponds to the same scattering center index n in data block $b + 1$ as the block time increases. The time evolution of the observables $R_n, \dot{R}_n, \ddot{R}_n$ over multiple data blocks centered at times $t = t_b$, $b = 1, \ldots, B$, provides the highest quality target characterization. Note that the figure has been generalized to include the extraction of the additional observable $\ddot{R}_n$ as developed in chapter 3.

data collection, and eliminates the complexity introduced by the tracking and other operational radar functions. The target is mounted on an RCS pylon (a holding structure) and is rotated at a slow rate to cover 360° in viewing angle. The transmitter sends out a sequence of frequency-stepped pulses (short time), where the time duration of each pulse is short enough to freeze the target motion. The PRF chosen for the transmitter is governed by the requirement to sample the rotational motion of the target so that pulse-to-pulse phase changes in the received data are small, preventing phase ambiguities in the processed data. The CoR of the target is determined by the placement of the target on the pylon, independent of the target mass distribution. Fusion of the static range data with data collected on a flight test using the same target for which the received data are registered to a CoR location determined by the center of mass of the target is discussed in appendix F. The static range received pulses are recorded in a frequency versus time (angle) format over a complete rotation period and used for further processing.

Because the motion is known, this example falls into the category of the *known motion solution*. Assuming a target-centered coordinate system, a viewer fixed in the target-centered system would perceive the radar orbiting uniformly about the target. The received pulses can be organized so that a block of Q_0 pulses is arrayed into a single uncompressed data block, and B sliding blocks are indexed versus time, each time increment in the sequence referenced to the center time of each block. The organization of the block sequence is determined by the end goal of the processing. For example, one case would be to increment each block by a single PRI time interval. This results in a smoothly varying extracted range, range-rate sequence of observables because in this case $Q_0 - 1$ of the pulses are repeated in the subsequent block. This block organization is typically used for developing the target signature model. Another option would be to organize the block sequence sparsely, with much larger time spacing between blocks. This option might be used in a radar target simulation scenario to represent a radar system periodically transmitting high PRF bursts of pulses.

In the remainder of this chapter, a performance comparison between the two signal processing schemes indicated by paths A and B in figure 1.10 is presented. The resolution obtained using the Fourier-based spectral estimator applied to the data block versus the HRSE technique applied to the same data block is examined using the extracted range, range-rate observable sequence as the performance metric. The simulation data are generated using the GTD-based scattering model introduced in the next section. The example target introduced in the next section is used throughout the text to illustrate applications of the various processing techniques discussed in later chapters.

2.5   Example Target and Simulation Motion Parameters

In this section a simulation example is introduced to illustrate the sequential processing concept. The simulation is designed to illustrate five key points: (1) to show the utility of using the GTD scattering formulation defined by equation (2.6) to generate simulation data; (2) to illustrate the nature of the observable sequence $\{(R_n, \dot{R}_n, D_n)_b, n = 1, \ldots, N\}, b = 1, \ldots, B$ as the block sequence evolves in time; (3) to contrast the precision obtained using the Fourier-based spectral estimator to that of the newly developed HRSE estimation techniques; (4) to introduce two unique scattering phenomena commonly observed in field data, that is, slipping scattering centers and shadowing; and (5) to compare the characterization of the target scattering center locations as defined in the target-centered coordinate system versus radar observable space. The methods used to incorporate these scattering mechanisms into a GTD-based scattering model are introduced, and greater detail is provided in chapter 6.

A cylindrical target is used in the simulation example and consists of 10 scattering centers: eight fixed point and two slipping scattering centers, each located at the ends of the cylinder. The target, 2 m in length and 0.8 m in diameter, has four fixed scattering centers located at each end, equally spaced around the circumference, as well as two slipping scattering centers modeling each of the cylinder edges, located at each end of the cylinder. The four fixed-point scattering centers located on the top edge of the cylinder are displaced 20° in roll angle from those on the bottom edge of the cylinder. The target geometry is illustrated in figure 2.7.

The motion for this example consists of pure spin, having no precession; thus the viewing aspect (θ) to the target z-axis is constant. Because the aspect angle is constant, the spin motion maps directly into a rotation of the roll angle relative to the target z-axis. The slipping scattering center has the unique property that, as viewed on the target in radar observable space, for pure roll angle motion, the scattering from the slipping scattering center appears to the radar to emanate from a fixed point on the target independent of roll angle change. As a consequence, the measured range from the radar to the slipping cylinder edges does not change with time. However, when the slipping scattering center location is referenced to the target-centered coordinate system, the location of each slipping scattering center must change with roll angle in order to follow the LoS to the radar. Thus, when developing the GTD scattering model defined in the target-centered coordinate system, the locations of the point-like scattering centers are fixed and the location of the slipping scattering centers is time varying. In target space, the location, $r_s(\varphi)$, of the slipping scattering centers is given by the expression

$$r_s(\varphi) = \rho_0 \hat{\rho} + z_0 \hat{z}, \tag{2.20}$$

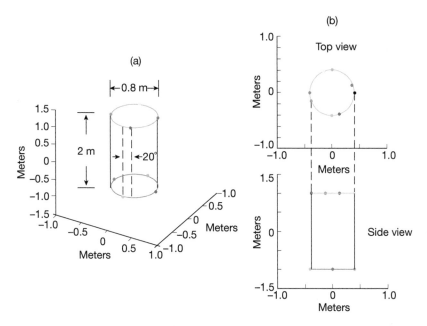

Figure 2.7   Cylinder with fixed-point scattering centers on each cylinder end. (a) 3D view of target. (b) Top and side views of target. Scattering centers on bottom cylinder edge rotated 20° relative to point scattering centers on top edge.

where $\hat{\rho} = \cos\varphi\hat{x} + \sin\varphi\hat{y}$ is the radial unit vector, and $\rho_0 = 0.4$ and $z_0 = \pm 1$ for each respective cylinder end. Each slipping scattering center appears to rotate in exactly the same manner as the radar rotates about the cylinder, presenting the same range measurement to the radar versus roll angle. For the more general motion case where the target exhibits both spin and precession, the target space representation of a slipping scattering center defined by equation (2.20) remains valid, but the range as well as the range-rate will vary as characterized by equation (2.17).

Scattering centers are *shadowed* when their LoS to the radar is obstructed. This occurs for a physical target when the scattering centers become invisible to the sensor due to the target motion. For this simulation, a simplistic approach to illustrate the impact on the temporal evolution of the observable sequence due to the appearance and disappearance of scattering centers caused by shadowing is implemented by discarding those scattering centers that lie on the back of the cylinder as viewed perpendicular to the cylinder axis as the target rotates about the z-axis. Though not representative for the scattering centers at the top portion of the cylinder that are not shadowed for the 30° radar viewing aspect, it does properly represent the shadowing effects at the lower end of the cylinder. This constraint is implemented by applying the mathematical test condition

$$[\hat{k}_q]_\perp \cdot r_n < 0 \tag{2.21}$$

where $[\hat{k}_q]_\perp$ corresponds to the look-angle vector $\hat{k}_q$ at time $t = t_q$ projected to the x, y-plane: that is,

$$[\hat{k}_q]_\perp = \sin\theta_0 \cos\varphi_q \hat{x} + \sin\theta_0 \sin\varphi_q \hat{y}.$$

This constraint is applied to each viewing angle as the simulation data are generated.

The GTD scattering model is now applied to this example. Using equation (2.6) the monostatic scattered field is expressed in the form

$$E(f, t) = \sum_{n=1}^{10} D_n e^{-j\frac{4\pi f}{c}\hat{k}(t)\cdot r_n}$$

where the look angle to the sensor, $\hat{k}(t)$, is defined by equation (2.4). For convenience, choose the D_n to have unit amplitude and constant phase, chosen randomly for each scattering center. When the viewing angle for a specific scattering center satisfies the test condition defined by equation (2.21) its value of D_n is set to zero. Note that in generating simulation data, $E(f, t)$ is represented directly in the frequency domain, as would be the case for the uncompressed received waveform. This is consistent with the fact that many RCS scattering codes use scattering concepts formulated in the frequency domain. When using the Fourier-based spectral estimator, the windowed pulse compressed data, $e_T(R_k)$ defined in section 1.5.2, is used to form each data block as illustrated in figure 1.13(b). For the high-resolution spectral estimation technique, the uncompressed, frequency domain samples as illustrated in figure 1.13(a) are used directly by the HRSE estimator.

The rotational motion considered in this example can be cast as a special case of Euler motion [30, 31] as described in appendix A. For Euler motion, the motion $[\theta(t), \varphi(t)]$, is fully determined by the functional mapping.

$$F\{\kappa, \theta_p, \psi, \phi_p, t\} \rightarrow [\theta(t), \varphi(t)], \tag{2.22}$$

where κ is the localized aspect angle between the radar LoS and the target angular momentum vector, θ_p is the precession cone angle relative to target-motion angular momentum vector, ϕ_p is the precession rotation angle, and ψ is the spin-angle.

For targets that are nearly rotationally symmetric about the target's z-axis, that is, having nearly symmetric moments of inertia $I_x \sim I_y$ the precession angle θ_p is nearly constant and the motion can be further simplified [30] by introducing the precession frequency and phase, f_p, α_p, and the spin frequency and phase f_s, α_s. In this case the Euler angles ϕ_p and ψ become linear in time $\phi_p = 2\pi f_p t + \alpha_p$ and $\psi = 2\pi f_s t + \alpha_s$, and the rotational motion is characterized by the Euler parameters θ_p, f_p, α_p, f_s, α_s, and κ, the look angle to the target's angular momentum vector.

The functional mapping to spherical coordinate angles developed in appendix A is characterized by equations (A.8) and (A.16). The aspect and roll angles $[\theta(t), \varphi(t)]$ are given by

$$\cos\theta = \cos\kappa \cos\theta_p + \sin\kappa \sin\theta_p \sin\phi_p \tag{2.23}$$

$$\varphi = -\psi + \tan^{-1}\left(\frac{\cos\kappa \sin\theta_p - \sin\kappa \sin\phi_p \cos\theta_p}{\sin\kappa \cos\phi_p} \right). \tag{2.24}$$

Note the utility of using equations (2.23) and (2.24) for simulating Euler motion. There is no need to resort to a dynamic trajectory simulation that includes both the translational as well as the rotational motion of the target. A simple specification of the Euler parameters and κ described previously is sufficient to characterize the rotational motion. Note also the inverse is true for studies that attempt to estimate the motion from measurement data. In chapter 8 it is shown that coupling equations (2.23) and (2.24) with the extracted range, range-rate observable sequence, once the translational motion is removed, provides a powerful technique for estimating the target rotational motion.

2.6   Performance Comparison

For the simulation, choose a radar viewing angle of $\theta_0 = 30°$ relative to the z-axis of the cylinder as illustrated in figure 2.8(a). In applying the mapping defined in equations (2.23) and (2.24), there is no precession so that $\theta_p = 0$ and $f_p = 0$. Choose a spin frequency $f_s = 1$ Hz. The spinning motion is then characterized by $\theta_0 = 30°$ and $\varphi(t) = 2\pi t$. The remaining sensor and viewing parameters used in the simulation are illustrated in table 2.1.

A data block consisting of 16 pulses is used, consistent with the illustration in figure 1.13, so that the uncompressed and compressed data blocks appear similar in nature to figures 1.13(a) and 1.13(b). The 2 GHz bandwidth waveform is sampled at 48 frequencies so that each 2D uncompressed data block size is $[48 \times 16]$. For the pulse compressed data blocks, a discrete Fourier transform (DFT) size $N_{fft} = 256$ is used, resulting in a Fourier-based block size of $[256 \times 16]$. A small number of 16 pulses are chosen for

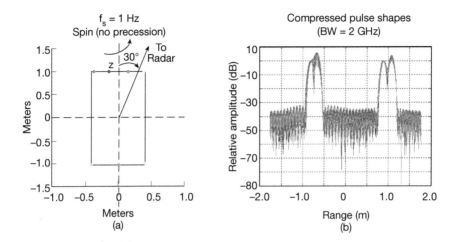

Figure 2.8 Superimposed pulse shapes (30° incidence angle) for backscatter from spinning cylinder. (a) Side view of target with radar at 30° incidence angle to target axis. (b) Superposition of 16 time-arrayed compressed pulses.

Table 2.1 Simulation parameters for example 1

BW	2 GHz
PRF	1,000 Hz
Q_0	16 pulses per block
Kappa (κ)	30°
f_0	10 GHz

each data block to test the robustness of the HRSE estimator when compared to the Fourier-based image peak extraction estimator described in figure 1.10. For this example, the Fourier-based Doppler resolution using this small number of pulses is poor, emphasizing the enhanced Doppler resolution achieved using the HRSE estimator.

A typical set of 16 compressed pulses generated using the simulation data applied to a typical Fourier-based data block are superimposed and illustrated in figure 2.8. Note that the 16 pulse shapes depict only a slight change in range over the time frame of the data block: for the Fourier-based spectral estimator, the peaks of these pulse shapes determine the range poles, and the phase change in the data across the pulses in the range gates located about the peaks sets the range-rate poles, paired to the range poles.

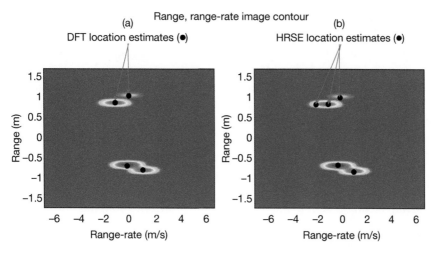

Figure 2.9 Range, range-rate observable estimates using (a) DFT peak extraction, and (b) 2D HRSE extraction overlayed on a typical range, range-rate image.

The comparison in range, range-rate performance for the two estimators for a typical data block can be illustrated graphically by overlaying the HRSE range, range-rate location estimates onto a Fourier-based range, range-rate image. Figure 2.9 illustrates a range, range-rate image for a single data block in which the five HRSE estimated range, range-rate pairs, $(R_n, \dot{R}_n), n = 1, \ldots, 5$ are overlaid onto the range, range-rate image generated by the Fourier-based estimator. The Fourier-based peak extraction range, range-rate pairs are illustrated in figure 2.9(a) and the HRSE extracted range, range-rate pairs are shown in figure 2.9(b). Note that the range, range-rate image characterized by the Fourier-based technique smears the peaks for closely spaced adjacent scattering centers. The HRSE estimator provides significantly better location estimates in this region of the image. Of course, eventually a resolution limit is reached where neither technique can resolve the scattering centers.

The enhanced estimation performance obtained using the HRSE estimator can be illustrated more clearly using the sequential spectral estimation processing scheme described previously. Consider the case for which blocks of 16 pulses are sequentially incremented by a single pulse, that is, the block time is incremented sequentially by Δt according to $t_{b+1} = t_b + \Delta t$. The increments are carried out covering two complete 360° roll cycles of the target motion. The extracted observable sequence, $(R_n, \dot{R}_n)_b, n = 1, \ldots, 6, b = 1, \ldots, B$, for the Fourier-based estimator is shown in figure 2.10 and for the HRSE estimator in figure 2.11. The results are plotted as a function of sliding block time, where the block time is referenced to the center of each block. (Note: frequency-time data from the same block time is used by each estimator as the simulation evolves in time.) A maximum of six poles are extracted from each data block. The six pole pairs extracted from each data block as the block time is incremented sequentially are not necessarily associated with the same scattering centers due to shadowing and the occasional visibility of more than six scattering centers relative to the radar LoS.

The observable sequence $\{(R_n, \dot{R}_n)_b, n = 1, \ldots, B\}$ as illustrated in figures 2.10 and 2.11 are directly extracted from the sliding data blocks using each respective spectral estimation technique applied to each data block and no correlation is applied to successive block estimates. Before discussing the performance of the HRSE and Fourier-based estimation techniques, note the overall temporal behavior coupling the motion and target scattering center locations illustrated by the time evolution of the observable sequence. Examination of figures 2.10 and 2.11 shows an unambiguous visual

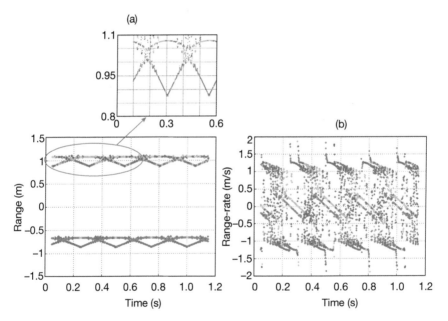

Figure 2.10 Sequential block processing using DFT peak extraction (extracted range, range-rate sequence vs. time): $N = 6$, $Q_0 = 16$ pulses in block, $n_f = 48$ frequency samples, 2 GHz bandwidth at X-band. (a) Range sequence versus time. (b) Range-rate sequence versus time. No correlation of the sequence from block-to-block pole extraction.

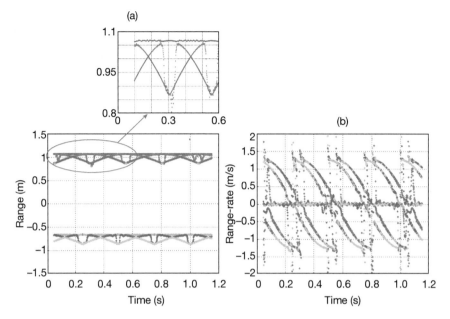

Figure 2.11   Sequential block processing using HRSE spectral estimation (extracted range, range-rate sequence vs. time): $N = 6$, $Q_0 = 16$ pulses in block, $n_f = 48$ frequency samples, 2 GHz bandwidth at X-band. (a) Range sequence versus time. (b) Range-rate sequence versus time. No correlation of the sequence from block-to-block pole extraction.

association of the time evolution of each specific scattering center observable with a unique trajectory over a limited time interval. Observe how the paired range and range-rate sequences, as they evolve in time, provide an overall picture of the coupling between the target's motion and its dimensional characteristics to present a unique view of the motion-target interaction as viewed by the radar.

Although the time evolution of the observables associated with a specific scattering center can be visually associated over specific time intervals, the output of each spectral estimation technique does not consistently associate the $(R_n, \dot{R}_n)_b$ observables from data block to data block. This is indicated by the changes in color along a visually associated scattering center trajectory. The color identification on the graphs corresponds to the labeling for which the extracted observables are input to the MATLAB analysis plot routine. No correlation of the extracted observables from successive data blocks was applied. Although the range, range-rate observables extracted from a given data block are properly paired, the correlation to the labeling associated with same scattering center observable extracted from a subsequent data block, as indicated by the change in color along the sequence plotted on the graph, is not necessarily preserved.

In order to explain the color changes occurring along a specific trajectory, let the scattering center range extractions from a data block, assuming three scattering centers at time t_{b_1}, be represented by $(R_1, R_2, R_3)_{b_1}$. The first plot value for R_1 uses the color blue, the color green for R_2, and so forth. For the extractions in a subsequent block $(R_1, R_2, R_3)_{b_2}$ corresponding to block time t_{b_2}, if the first value R_1 corresponds to a different scattering center (because of noise, shadowing, or perhaps because the value of N is chosen as too small or large) it will still be labeled as blue, but will appear at a different value on the range scale. Thus the paths traced by the range and range-rate observables often change color as a function of time because of an inconsistency in the labeling associated with the outputs of the spectral estimation techniques. The observable pairs as plotted are correct; only the color labeling and block-to-block association might change. Hence the application of a tracking and association technique is often applied to the sequence of observables before processing. Such an association technique is discussed in section 5.5.

Now compare the detailed differences in the range and range-rate sequences extracted using both the Fourier-based and the HRSE estimators: figure 2.10(a) versus figure 2.11(a) for the range observable sequences, and figure 2.10(b) versus figure 2.11(b) for the range-rate sequences. First, observe that the effect of shadowing as implemented in the simulation using equation (2.21) destroys the perfect sinusoidal behavior of the range and range-rate sequences associated with each fixed-point scattering center as the cylinder rotates over a complete 360° rotation period. Scattering centers on the rear side of the cylinder relative to the radar LoS satisfying equation (2.21) are not visible to the radar. Also note that, although the resolution of the Fourier-based estimator is adequate to resolve scattering centers located at one end of the cylinder relative to the other end, the HRSE estimator performs significantly better in resolving the set of four scattering centers positioned around each edge of the cylinder. As the cylinder rotates, as viewed relative to the radar LoS, the location of adjacent ring scattering centers merge into a common location and then diverge as the rotation continues. Comparing the magnification of the range sequence for the upper portion of the cylinder, figure 2.10(a) relative to figure 2.11(a), it is clear that the HRSE estimator provides considerably better performance relative to isolating both the number of ring scattering centers present, as well as resolving their relative locations. Referring to the inset of figure 2.11(a), the presence of the slipping scattering center at the upper edge of the cylinder is clearly evident, as indicated by the constant range variation versus time for this scattering center. It is also well resolved from the rotating fixed-point scattering centers. The dispersion in the estimator depicted in the magnified inset in figure 2.10(a) using the Fourier-based estimator is far worse than that obtained by the HRSE technique.

The HRSE estimator produces dramatically better results for the range-rate sequence [figure 2.10(b) vs. figure 2.11(b)]. This occurs because, for $Q_0 = 16$ pulses, the Fourier-based resolution in Doppler is not generally adequate to resolve all the scattering centers in the Doppler domain, whereas the high-resolution estimator does significantly better in this regard. The Doppler resolution of the Fourier-based spectral estimator can be improved by increasing the number of pulses in the data block. A small value for Q_0 has been deliberately selected for this example. However, if the radar PRF is limited relative to the target motion, one might have only a limited number of pulses to process for fast-rotating targets. In this situation, Q_0 cannot be chosen as arbitrarily large because the scattering centers can migrate into adjacent resolution cells. This effect is known as *Doppler smearing*. The upper limits on Q_0 required for linear imaging are discussed in detail in chapter 4. Because of the optimal nature of the HRSE techniques, the resolution achievable for a given block size will always be better when compared to Fourier-based processing.

The sequential nature of the extracted observables characterized by the sequence $(R_n, \dot{R}_n)_b, n = 1, \ldots, 6, b = 1, \ldots, B$ can be exploited to provide a corporate memory of

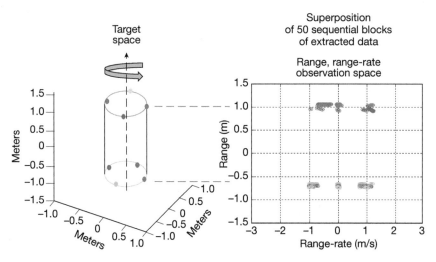

Figure 2.12   Time evolution of sequential range, range-rate pairs superimposed onto a range, range-rate space coordinate system. Each block incremented by one pulse incremental time.

the extracted observables versus sliding block time. This information can be accumulated by providing a graphical display of the time evolution of the extracted observables $(R_n, \dot{R}_n)_b$ on a 2D $R, \dot{R}$ grid space, and thus presents a short-time picture of the motion of the target as viewed in radar observable space. It would be equivalent to watching a movie of the range, range-rate image as it progresses in time for a sequence of sliding blocks, with only the essential contents of the image displayed. A simple depiction of this time evolution is illustrated in figure 2.12 for 50 sets of the range, range-rate observables extracted from the simulation values in figure 2.10, where the $50 \times 6 = 300$ range, range-rate pairs are superimposed on the $R, \dot{R}$ grid space. The figure also suggests that by following an individual trace in this $R, \dot{R}$ space, one can implement a simple nearest neighbor technique to correlate and track individual scattering centers as they evolve in time. The nearest neighbor correlation technique is exploited in some detail in section 5.5.

3

Acceleration Estimation: Extending the All-Pole Model

3.1 Introduction

In chapter 2 a linear all-pole model was introduced and used to develop a HRSE technique for extracting the range, range-rate, and amplitude observables for each scattering center on the target. In that development it was assumed the phase behavior of each scattering center was adequately modeled as a point undergoing constant velocity during the observation interval. In this chapter a quadratic term is added to the linear model in order to model the effect of scattering center acceleration. This extra term models the phase change due to scattering center acceleration and the resulting phase, $\chi_n(t)$, is given by

$$\chi_n(t) = \frac{4\pi}{\lambda}\left(R_n + \dot{R}_n t + \frac{1}{2}\ddot{R}_n t^2\right) \tag{3.1}$$

The inclusion of the nonlinear acceleration term allows development of techniques that yield improved estimation accuracy for the scattering center observables.

The constant velocity assumption is satisfied if the time extent of the data block used for the Fourier-based or the HRSE technique is sufficiently small such that the quadratic phase term $\frac{4\pi}{\lambda}\left(\frac{1}{2}\ddot{R}_n T^2\right)$ associated with each scattering center is negligible. This size constraint on the image data block is standard for the generation of Fourier-based range, range-rate (range, Doppler) images, and the image processing is often referred to as linear imaging. Limits on the processing time T for specific linear imaging examples are discussed in chapter 4.

In order to develop techniques that yield improved estimation accuracy, it is useful to consider the case where the scattering center motion is assumed to have constant acceleration $A = \ddot{R}$ over the data block processing interval. By extending the block processing time T beyond the linear imaging limits, techniques for estimating the scattering center acceleration can be developed for both the Fourier-based and HRSE techniques.

There are other approaches that might be used to estimate the acceleration associated with each scattering center. For example, a tracking filter might be implemented and applied to the time sequence of range, range-rate estimates. Such an approach is examined in [62, 63, 64, 65] and is used in appendix G to develop a suboptimal estimate for $(\ddot{R}_n)$. However, this approach becomes problematic for targets with many closely spaced scattering centers, or for scattering centers that have very low signal-to-noise levels, because the sequence of range, range-rate observables from each and every scattering center must be associated correctly over the entire observation interval. A number of scattering center acceleration estimation approaches were evaluated, and the best results were obtained using the technique presented in this chapter. It employs a sequence of symmetrically expanding data blocks to iteratively solve for the nonlinear model that achieves the best fit to the observation data block. The extra degree of freedom possessed by the nonlinear model results in a better fit to the data when the N observables $(R_n, \dot{R}_n, \ddot{R}_n, D_n)$ are applied to the larger measurement data block. The enhanced technique also provides an improved estimate for the $(R_n, \dot{R}_n)$ pair associated with $(\ddot{R}_n)$ compared to the estimates obtained using the baseline estimators that assume a constant scattering center velocity.

Figure 3.1 illustrates the method used for increasing the block size when estimating scattering center acceleration. A set of increasingly larger data blocks are constructed in such a manner that the data contained in the outer sectors of the expanded data block contribute to the acceleration component of the phase of the nonlinear model for each scattering center. The procedure begins by selecting an initial frequency-time data block centered at $t_b = t_0$ with a time span $2T_0$ sufficiently short such that the data contribution to the nonlinear model acceleration phase term over this time interval is negligible. The range, range-rate observables $(R_n, \dot{R}_n)$ are estimated from the initial data block using either the Fourier-based or the HRSE estimators. The block size is then expanded symmetrically to cover an increasingly larger time span out to a limiting value $2T$ with each block increment remaining centered about the initial data block. For the Fourier-based technique, the block size is increased by two pulses for each increment, and the acceleration associated with each scattering center is determined by exploiting the changes induced in the phase of the peaks in the updated range, range-rate image as the data block size is increased. For the HRSE technique the block size is increased by a larger number of pulses, and the acceleration is estimated by exploiting the changes induced in the nonlinear model applied over the larger block size.

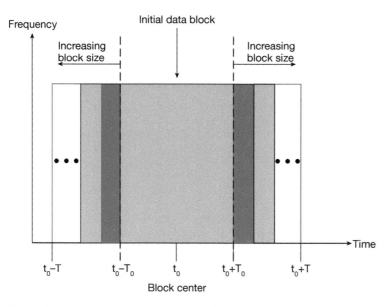

Figure 3.1   Incrementally expanding 2D data block size used for scattering center acceleration estimation.

It is desirable to choose the initial data block to cover the longest time interval over which the linear phase behavior of $\chi_n(t)$ is valid. This results in the greatest amount of integration gain that could be realized in the presence of noise without introducing undesirable smearing artifacts. Analytical bounds characterizing the size of the initial data block are developed in chapter 4.

The sequential processing concept introduced in section 2.4 can be applied to the enhanced HRSE model containing the added acceleration term. The estimate of $(\ddot{R}_n)$ associated with each scattering center results in the expanded observable sequence $(R_n, \dot{R}_n, \ddot{R}_n, D_n)_b$, $n = 1, \ldots, N, b = 1, \ldots, B$ used to characterize the N scattering centers over B sequential data blocks. Throughout the text, when the sequential observable sequence includes $\ddot{R}_n$, it is understood that the block time duration T used to extract the observables has been extended beyond the linear imaging limit and consequently uses more pulses. It is shown in section 5.6 that including the acceleration estimate $\ddot{R}_n$ in the observable sequence leads to a more robust solution for the scattering center location estimates.

It is instructive to develop the Fourier-based version of the technique first because it most clearly illustrates how the phase associated with each Fourier-based range-Doppler peak in an image is affected by the acceleration contribution as the blocks are incrementally extended.

3.2   Estimating $\ddot{R}$ Using Fourier-Based Processing

Consider the geometrical theory of diffraction (GTD) scattering model introduced in chapter 2. Using equation (2.6), the wideband scattered field can be expressed in the form

$$E(f,t) = \sum_n D_n(t) e^{-j\frac{4\pi f}{c} R_n(t)} \tag{3.2}$$

where $R_n(t) \equiv \hat{k}(t) \cdot \underline{r}_n$ is the range to the nth scattering center along the radar line of sight (LoS) to the target defined relative to the target center of rotation. The range $R_n(t)$ for each scattering center will change in concert with the aspect and roll angles to the target as governed by the target motion. Consider an arbitrary data block centered about the block time $t_b = t_0$ as illustrated in figure 3.1. The variation of $D_n(t)$ is slow relative to the phase changes caused by $R_n(t)$, and is assumed constant over the data block. Expand $R_n(t)$ in a Taylor series over the data block (including the acceleration term)

$$R_n(t) = R_n + \dot{R}_n(t - t_0) + \frac{1}{2}\ddot{R}_n(t - t_0)^2. \tag{3.3}$$

Both acceleration estimation techniques are dependent on exploiting the quadratic nature of the acceleration phase contribution to $E(f, t)$.

In developing the Fourier-based technique, it is instructive to first consider a one-dimensional (1D) (i.e., narrowband) example because the effect of a finite integration time on the time-limited data block can be clearly illustrated. It is easily extended to the more general two-dimensional (2D) data block by simple deduction.

A narrowband example (1D data block)

The 1D example is also useful for several other reasons. First, it illustrates how incrementing the data block size in small steps allows a direct estimate of acceleration. Second, because the example assumes a sequence of narrowband waveforms typical of what might be used in the sensor track mode, it illustrates that the general technique is applicable to estimating $\ddot{R}_n$ for targets in the narrowband track, and is not restricted to characterizing wideband scattering center behavior. Finally, it provides an analytical formulation for estimating $\ddot{R}_n$ as a function of increasing block size.

In order to develop an intuitive appreciation for the estimation procedure, consider a noise-free example that has a target moving with speed $\dot{R}_n = V$, acceleration $\ddot{R}_n = A$, and is tracked by a narrowband sensor operating at frequency $f = f_0$. The objective is to estimate V and A. For the narrowband case, the continuous time function $e(t)$ is introduced to characterize the narrowband scattered field

$$e(t) \equiv E(f_0, t) = D_0 e^{-j\frac{4\pi}{\lambda}Vt} e^{-j\frac{4\pi}{\lambda}\left(\frac{1}{2}At^2\right)} \tag{3.4}$$

where $\lambda = c/f_0$, and D_0 is a complex constant. Assume the radar transmits a sequence (pulse train) of narrowband radar waveforms, and each received pulse is sampled at the range gate containing the target at times determined by the pulse repetition frequency (PRF) of the radar. The data samples $e(t_q)$ correspond to the samples of a specific row of the data matrix, $E(k_0, q)$, from which the 2D wideband all-pole model is derived, where $k = k_0$ corresponds to $f = f_0$ as indicated in figure 3.2, that is, the narrowband data block is 1D. For the following development, it is useful to consider the sample time as a continuous variable.

Define the variables ω_d and ω_a

$$\omega_d = \frac{4\pi V}{\lambda}; \omega_a = \frac{4\pi A}{\lambda} \tag{3.5}$$

The expression for $e(t)$ can be written as

$$e(t) = D_0 e^{-j\omega_d t} e^{-j\frac{\omega_a}{2}t^2} \tag{3.6}$$

For a general time function having quadratic phase, defined over the interval $-\infty < t < \infty$, the Fourier transform of $e(t)$ is governed by the relationship [66]

$$e^{-j\omega_d t} e^{-j\frac{\omega_a}{2}t^2} \leftrightarrow \sqrt{\frac{2\pi}{\omega_a}} e^{-j\pi/4} e^{j(\omega - \omega_d)^2/2\omega_a}, \tag{3.7}$$

where the notation "$\leftrightarrow$" denotes a Fourier transform pair. This transform pair relationship indicates that the Fourier transform of a quadratic time function exhibits a quadratic phase behavior. The phase variation of the spectrum in the region about $\omega = \omega_d$ is characterized by the acceleration contribution ω_a. Because $e(t)$ is time limited over an interval determined by the number of pulses in each data block defining $e(t_q)$, simple analytic functions for the Fourier transform over this finite time interval are not available. However, based on the phase dependence on ω_a in the region $\omega = \omega_d$ indicated by equation (3.7), one would anticipate that, as the block processing time interval indicated in figure 3.2 transitions incrementally from the linear region to the nonlinear region, that is, from $T_0 \to T$ as shown in figure 3.2, the phase at the peak of the Doppler spectrum associated with each data set would incrementally show a strong dependence on ω_a. The following development characterizes this dependence and is used to develop the Fourier technique for estimating the scattering center acceleration.

Assume V and A are to be estimated using samples from a data interval defined about $t_b = t_0$ over a time span $2T$. Then equation (3.6) can be written in the form

$$e_T(t) = D_0 e^{-j\omega_d(t-t_0)} e^{-j\frac{\omega_a}{2}(t-t_0)^2} \tag{3.8}$$

where $t_0 - T < t < t_0 + T$ and for simplicity assume $D_0 = 1$. To characterize the phase behavior of the finite time Fourier transform for increasing data lengths, consider a set of data intervals that increase in duration, for $T > T_0$, symmetric about $t = t_0$, as illustrated using the block window function in figure 3.2

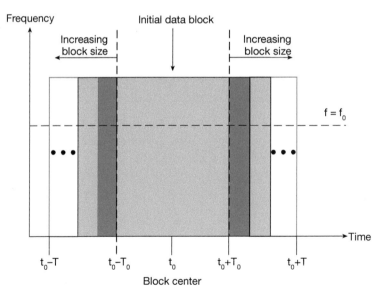

Figure 3.2   Expanding data blocks for acceleration estimation (single frequency, $f = f_0$).

Let $E_T(\omega)$ be the symmetric Fourier transform of $e_T(t)$ over the time interval $(t_0 - T, t_0 + T)$

$$E_T(\omega) = \int_{t_0-T}^{t_0+T} e^{-j\omega_d(t-t_0)} e^{-j\frac{1}{2}\omega_a(t-t_0)^2} e^{+j\omega t} dt,$$
(3.9)

Using a change of variable $t' = t - t_0$ equation (3.9) can be written as

$$E_T(\omega) = e^{+j\omega t_0} \int_{-T}^{T} e^{+j(\omega-\omega_d)t'} e^{-j\frac{1}{2}\omega_a t'^2} dt'.$$
(3.10)

Comparing equation (3.10) to the Fourier transform pair in equation (3.7), $E_T(\omega)$ is the Fourier transform of the time-limited representation of $e(t)$, denoted as $e_T(t)$. Now examine the phase behavior of $E_T(\omega)$ as T increases for $T > T_0$ where T_0 is limited such that the contribution from acceleration over the initial data block is negligible. The maximum of $E_T(\omega)$ occurs at $\omega = \omega_d$, and provides the estimate of V. (Note the analogy to the 2D Fourier-based range, range-rate spectral estimation technique, where the estimate of $(R_n, \dot{R}_n)$ is extracted from the range, range-rate image peaks.) Consider the phase behavior at the peak of $E_T(\omega)$, $\omega = \omega_d$, as a function of increasing T. Denote the phase of the Doppler peak as

$$\xi(T) = phase[E_T(\omega = \omega_d)].$$
(3.11)

From equation (3.10),

$$E_T(\omega = \omega_d) = e^{+j\omega_d t_0} \int_{-T}^{T} e^{-j\frac{1}{2}\omega_a t'^2} dt'.$$
(3.12)

The integral in equation (3.12) is related to the Fresnel integral, $Fr(x)$ [67], defined by

$$Fr(x) \equiv \sqrt{\frac{2}{\pi}} \int_0^x e^{+jy^2} dy.$$
(3.13)

After some manipulation, equation (3.12) can be expressed in the form

$$E_T(\omega = \omega_d) = e^{+j\omega_d t_0} \left[Fr^*\left(\sqrt{\frac{\omega_a}{2}}T\right) - Fr^*\left(-\sqrt{\frac{\omega_a}{2}}T\right) \right]$$
(3.14)

where the superscript "*" indicates complex conjugate. For ω_a sufficiently small, $Fr(x)$ can be approximated [67] by the series

$$Fr(x) \sim \sqrt{\frac{2}{\pi}}(x + jx^3/3) \tag{3.15}$$

Substituting equation (3.15) in equation (3.14), after some manipulation, one obtains

$$\xi(T) = \omega_d t_0 + \omega_a T^2/6. \tag{3.16}$$

Referencing the phase $\xi(T)$ to $\xi(T_0)$ as T increases results in the phase change:

$$\Delta\xi(T) = \xi(T) - \xi(T_0) = \frac{\omega_a}{6}(T^2 - T_0^2). \tag{3.17a}$$

Define $T = T_0 + \Delta T$, where ΔT denotes the block time increment. Equation (3.17) can be expressed in the form

$$\Delta\xi(\Delta T) = \frac{\omega_a}{6} 2T_0\Delta T + \frac{\omega_a}{6}(\Delta T)^2. \tag{3.17b}$$

Thus the change in phase behavior at the peak of $F(\omega = \omega_d)$ as the block size increases exhibits a quadratic dependence versus increasing block size, ΔT, and provides a direct measure of target acceleration.

It is important to recognize that each time sample in the data block represents the peak complex amplitude of an individual compressed narrowband radar waveform return from the target. Thus the technique is coherently processing a pulse train of increasing duration. Assume the length of the data set grows symmetrically by time steps $\Delta T = q\Delta t$, $1 = 1, \ldots, Q$ over a series of Q block size increments. Using equation (3.17b), a simple technique for estimating ω_a follows:

1. For each data block over the interval $(t_0 - T_0 - q\Delta t, t_0 + T_0 + q\Delta t)$, perform the Fourier transform of the data samples.
2. For each Fourier transform in step 1, isolate the peak $E_T(\omega = \omega_d)$. Extract the phase as a function of $\Delta T = q\Delta t$, and reference the phase, $\Delta\xi(\Delta T)$ to the phase extracted from the initial data block.
3. Fit a second-order polynomial to the variation of $\Delta\xi(\Delta T)$ versus $q\Delta t$, $q = 1, \ldots, Q$. Denote the coefficient of the ΔT^2 term as p_1.

4. Using equation (3.17b) and equation (3.5), the acceleration estimate is given by

$$\hat{A} = \frac{3\lambda}{2\pi} p_1.$$ (3.18)

Effect of Doppler weighting

In order to keep the Doppler sidelobes low, the data samples over each Fourier trans-form interval $(t_0 - T_0 - q\Delta t,\ t_0 + T_0 + q\Delta t)$ are typically multiplied by a weighting function, $w_T(t)$, before taking the Fourier transform. The weighting results in a modi-fication of the phase behavior $\Delta\xi(\Delta T)$. It can be shown empirically that for Taylor weighting, equation (3.18) must be modified according to

$$\hat{A} = \beta_w \frac{3\lambda}{2\pi} p_1$$ (3.19)

where $\beta_w \approx 1.9$ for Taylor weighting.

A narrowband illustration (1D compressed waveform data block)

To illustrate the application of the technique, consider a noise-free example having parameters $V = 2.5$ m/s and $A = 3$ m/s² at X-band, $f_0 = 10$ GHz. Choose a PRF of 1,000 Hz ($\Delta t = .001$ s) and an initial block consisting of 16 samples of narrowband data. Expand the block length symmetrically about the center, in increments $\Delta T = \Delta t$ up to a maximum of $Q = 80$ data samples. Figure 3.3(a) illustrates the compressed pulse train, using Taylor weighting, for the initial train of 16 pulses (red), and final train of 80 pulses (blue). The horizontal axis is scaled to Doppler frequency $f_d = 2\hat{V}/\lambda$, with the estimate $\hat{V} = V$ extracted from the maximum of the compressed waveforms. Note the increase in gain and enhanced Doppler resolution as the number of pulses processed is increased from 16 to 80. In figure 3.3(b) the progression of compressed waveforms for each incre-ment Δt in block size is illustrated. As the integration time increases, the compressed waveforms demonstrate corresponding increase in gain and resolution on the target.

Of prime interest for estimating the acceleration A is the phase behavior at the peak of each compressed pulse. This phase dependence, plotted versus increasing block size, is illustrated in figure 3.3(c). The quadratic nature of the phase behavior of $\Delta\xi(\Delta T)$ defined by equation (3.17a) is illustrated versus increasing block size, $q\Delta t$, where $q = 1, \ldots, 32$, is evident. The quadratic fit to the extracted phase (indicated by the "o" data points) is shown by the smooth curve. After extracting p_1 from the quadratic data fit, and using equation (3.19), the estimate of A is given by

$$\hat{A} = 2.88 \,\mathrm{m/s^2} \tag{3.20}$$

which corresponds well to the exact value $A = 3.0$.

Now consider the accuracy of the estimate of the acceleration A as A ranges from 0 to increasing larger values. Define the parameter

$$Arg \equiv \frac{2\pi}{\lambda} \ddot{R}_n \Delta t^2 q_{MAX}^2 \tag{3.21}$$

which represents the phase contribution of the acceleration term $e(t)$ in equation (3.3), evaluated at the edge of the largest data block, denoted as q_{MAX}. The value of Arg characterizes the limits of the small angle approximation to the Fresnel integral function. Figure 3.4 compares the estimate of A to the true value versus the acceleration phase limit, Arg, as Arg increases. Clearly, in the region $Arg > 1$, the acceleration estimate begins to degrade.

Extension to wideband processing (2D compressed waveform data block)

Extrapolating the 1D example to the 2D wideband, N point scattering center case is straightforward. An initial block of compressed pulses defined over the interval $(t_0 - T_0,$

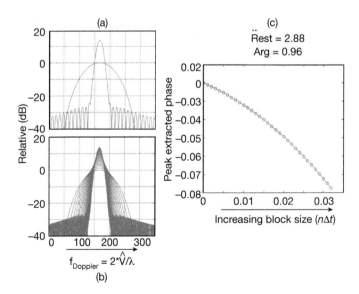

Figure 3.3 Doppler spectrum of the compressed pulse train and phase change versus increasing block size. (a) Red = 16 pulses; blue = 80 pulses. (b) Overlay of sequence of compressed pulse train for block size increasing from 16 to 80 pulses. (c) Differential phase extracted versus block size increase scaled to $n\Delta t$.

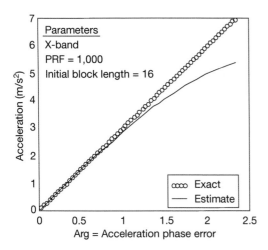

Figure 3.4 Acceleration estimate versus phase limit.

$t_0 + T_0$) is processed using Fourier-based processing to develop a range, range-rate image. The N range, range-rate image peaks are isolated and the $(R_n, \dot{R}_n), n = 1, \ldots, N$ estimates are obtained. The phase behavior at each peak, as the data block size is increased will exhibit the same quadratic behavior as observed in the 1D example. The 2D acceleration estimation technique for the Fourier-based approach can then be summarized as follows:

Using the initial data block, form the 2D range, range-rate image, isolate the peaks of the image and extract the estimates for $\{R_n, \dot{R}_n, |D_n|\}$ for each scattering center in the image.

To obtain the estimate of $\ddot{R}_n$ corresponding to $(R_n, \dot{R}_n)$, apply steps 1–4 that lead up to the narrowband acceleration estimate defined in equation (3.19), except now the 1D peak defined by $E_T(\omega = \omega_d)$ in step 1 is replaced by the 2D peak in the range, range-rate image. As long as the scattering centers remain isolated, each peak will exhibit the differential quadratic phase behavior as a function of increasing data block size. The resultant $\ddot{R}_n$ estimate is then appended to the range and range-rate estimate $(R_n, \dot{R}_n)$ forming a data set $\{R_n, \dot{R}_n, \ddot{R}_n, |D_n|\}$ to characterize each scattering center present in the image.

However, note that the Fourier-based technique performs poorly when two scattering centers are closely spaced in Doppler and occur in the same range gate. This is a consequence of the interference between the two scattering centers that destroys the quadratic phase behavior at the peaks of the Fourier transform. The high-resolution technique developed in the next section circumvents this problem by fitting the non-linear model directly to the uncompressed frequency-time data. Furthermore, the Fourier-based estimation technique is subject to the condition $Arg \ll 1$, and limits the size of the expanding data block. Each of these limitations is relaxed using the HRSE technique developed in the next sections.

3.3   Estimating $\ddot{R}_n$ Using the HRSE Technique

To develop the scattering center acceleration estimate using the HRSE approach, the linear all-pole model discussed in chapter 2 is modified by incorporating the nonlinear acceleration term, $(\ddot{R}_n)$, into the model. The development proceeds as follows: first, the all-pole model for $E(k, q)$ defined in equation (2.13) is generalized to include the acceleration contribution; this is followed by the development of the $(\ddot{R}_n)$ estimation technique assuming a small acceleration approximation analogous to that used in section 3.2. Subsequently, the small acceleration constraint is removed in section 3.5 by introducing an iterative approach that yields significantly better performance.

3.3.1   The Nonlinear Model

Using the expression for $R_n(t)$ defined in equation (3.3), for a data block centered about the block time $t_b = t_0$, the expression for the wideband scattered field defined in equation (3.2) can be expressed in the form

$$E(f,t) = \sum_n D_n(t) e^{-j\frac{4\pi f}{c}\left[R_n + \dot{R}_n(t - t_0) + \frac{1}{2}\ddot{R}_n(t - t_0)^2\right]} \qquad (3.22)$$

It is useful to write the frequency and time variables in equation (3.22) in discrete form:

$$\begin{aligned} f &= f_0 + k\Delta f \\ t &= t_0 + q\Delta t \end{aligned} \qquad (3.23)$$

Substituting equation (3.23) into equation (3.22), and combining terms, the general form for $E(k, q)$ can be written as:

$$E(k,q) = \sum_n D'_n e^{-j\frac{4\pi R_n}{c}k\Delta f} \, e^{-j\frac{4\pi}{\lambda}\dot{R}_n q\Delta t} \, e^{-j\frac{4\pi}{\lambda}(kq)\dot{R}_n \Delta t \Delta f / f_0} \, e^{-j\frac{4\pi}{\lambda}\frac{1}{2}\ddot{R}_n(q\Delta t)^2}, \qquad (3.24)$$

where, for convenience, the complex amplitude D_n in equation (3.22) is rescaled to include the reference phase independent of the (k, q) indices:

$$D_n e^{-j\frac{4\pi R_n}{\lambda}} \rightarrow D'_n \tag{3.25}$$

Equation (3.25) defines the phase relationship between the HRSE model pole amplitudes D'_n and GTD-based scattering center diffraction coefficients D_n as discussed in section 2.2. This transformation will be used later in chapters 15 and 16 to transform the HRSE model pole amplitudes D'_n to the GTD model amplitudes D_n. For now, as discussed in chapter 2, it is most convenient to drop the primed notation and defer the primed notation to chapters 15 and 16. Note in equation (3.24) the factor containing $\dot{R}_n \Delta f \Delta t$ in the exponent might be considered negligible, but it is retained here to improve the accuracy of the acceleration estimate.

To relate the nonlinear expression for $E(k, q)$ in equation (3.24) to the linear all-pole model defined by equation (2.13), the variables s_n, p_n and p'_n are introduced:

$$s_n = e^{-j\frac{4\pi R_n}{c}\Delta f} \tag{3.26}$$

$$p_n = e^{-j\frac{4\pi \dot{R}_n \Delta t}{\lambda}} \tag{3.27}$$

$$p'_n = e^{-j\frac{2\pi}{\lambda}\ddot{R}_n \Delta t^2} \tag{3.28}$$

Then equation (3.24) can be written in compact form

$$E(k, q) = \sum_n D_n s_n^k p_n^{q(1 + k\Delta f/f_0)} (p'_n)^{q^2}. \tag{3.29}$$

Equation (3.29) represents the nonlinear HRSE model representation of the measurement data. Estimating $\ddot{R}_n$ is equivalent to estimating p'_n in equation (3.29). The nonlinear model is now used to develop a technique for estimating $\ddot{R}_n$ using a small acceleration approximation.

3.3.2 Estimating $\ddot{R}_n$

The estimate of $\ddot{R}_n$ used in the HRSE approach is derived using the nonlinear model for the uncompressed data block, $E(k, q)$, expressed in equation (3.29), following the methodology briefly described in section 3.1. Before examining the more general case, this section describes a basic approach using only two data blocks: an initial data block over which the acceleration contribution is negligible; and a second, larger data block over which the acceleration is small but not negligible so that the nonlinear model (i.e., constant acceleration) is appropriate. The restriction on small acceleration is removed in section 3.5 using an iterative scheme to generalize the two-block example described in this section.

The process is initialized using a data block of Q_0 pulses over which the acceleration contribution is negligible. Because the contribution to the nonlinear model due to $\ddot{R}_n$ over the initial data block is negligible, valid estimates for the poles (s_n, p_n, D_n), $n = 1, \ldots, N$ are obtained using the linear all-pole model. To estimate p'_n, a second larger data block, expanded symmetrically as indicated in figure 3.1 is used, over which the contribution from $\ddot{R}_n$ to the nonlinear model is no longer negligible. However, as contrasted to the expanding data blocks used in the Fourier-based technique developed of section 3.2, the incremental expansion of the data block used for the HRSE technique is typically larger than the $2\Delta t$ time increments used there. Smaller increments for the Fourier-based technique are used to aide in the accuracy of the polynomial fit required to estimate the quadratic polynomial coefficient p_1. For the HRSE technique, no curve fit is required and larger block size increments can be utilized.

Choose the larger data block incremented in size symmetrically by N_0 pulses, resulting in $Q_0 + 2N_0$ pulses. Assuming $\ddot{R}_n$ is sufficiently small over the larger data block, $(p'_n)^{q^2}$ can be expanded in the form

$$(p'_n)^{q^2} = e^{-j\frac{4\pi}{\lambda}\frac{1}{2}\ddot{R}_n(q\Delta t)^2} \approx 1 - j\frac{2\pi}{\lambda}\ddot{R}_n q^2 \Delta t^2. \tag{3.30}$$

Denote the data over the larger block as $E_d(k, q)$. Denote by $E_0(k, q)$ the nonlinear model extrapolated over the larger data block size using the set (s_n, p_n, D_n), $n = 1, \ldots, N$ estimated from the initial data block, assuming $\ddot{R} = 0$. Substituting equation (3.30) into (3.29), the difference between the data and model extrapolated over the expanded data block, denoted by $\Delta(k, q)$, can be expressed as

$$\Delta(k, q) = E_d(k, q) - E_0(k, q) = \sum_n A_n \psi_n(k, q), \tag{3.31}$$

where

$$A_n \equiv -j\frac{2\pi}{\lambda}\Delta t^2 D_n \ddot{R}_n. \tag{3.32}$$

and the basis set $\psi_n(k, q)$ is given by

$$\psi_n(k, q) \equiv \left\{ q^2 s_n^k p_n^{q(1 + k\Delta f/f_0)} \right\}. \tag{3.33}$$

The amplitudes A_n in equation (3.32) are obtained by determining the set (A_n), $n = 1, \ldots, N$ that minimizes the mean square error $| \Delta(k, q) - \sum_n A_n \psi_n(k, q) |^2$ over the larger data block, using the functional expansion defined in equation (3.31)

$$\underset{(A_n)}{Min} | \Delta(k, q) - \sum_n A_n \psi_n(k, q) |^2, k = 1, \ldots n_f, q = 1, \ldots, Q_0 + 2N_0 \tag{3.34}$$

In equation (3.34), n_f denotes the number of frequencies, so that there are $n_f(Q_0 + 2N_0)$ total complex samples in the expanded data block. The estimate of $\ddot{R}_n$ is determined from the real part of the solutions to equation (3.32). After some manipulation, it can be shown that solutions for $\ddot{R}_n$ are given by

$$\ddot{R}_n = - \mathrm{Im} \left\{ \frac{\lambda}{2\pi\Delta t^2} (A_n / D_n) \right\}, \tag{3.35}$$

where D_n was estimated from the initial block[1]. Note that the $\ddot{R}_n$ estimate from equation (3.35) is correctly associated with $\{R_n, \dot{R}_n, \text{and } D_n\}$ because the pole pairs (s_n, p_n), $n = 1, \ldots, N$ are used as the expansion basis set in equation (3.31).

The estimate for $\ddot{R}_n$ using equation (3.35) is dependent on the condition $Arg \ll 1$, analogous to the Fourier-based small acceleration approximation limit defined in equation (3.21). The limits of applicability defined by $Arg \ll 1$ are expanded using an iterative scheme introduced in section 3.5. However, beforehand it is instructive to consider a simulation comparing the Fourier-based and HRSE estimation techniques, both assuming the small acceleration approximation.

3.4   An Example Using Simulated Data

In this section simulated data are used to compare the extracted observables $(R_n, \dot{R}_n, \ddot{R}_n)_b$ obtained using the Fourier-based and HRSE acceleration estimation techniques developed in sections 3.2 and 3.3 to each other as well as to truth. A simulated target having three fixed-point scattering centers undergoing complex spin-precession motion is

---

1. Note that because the formulation operates on the instantaneous range time variation it correctly predicts the triplet for fixed point as well as slipping (time-varying) scatterers.

used. The target and motion parameters are presented in table 3.1, where the scattering center coordinates are specified in cylindrical coordinates (ρ_n, φ_n, z_n). The simulated data are generated using a GTD-based scattering model similar to the technique introduced in section 2.5. The simulation data assumes radar operating parameters at X-band ($f_0 = 10$ GHz) using a PRF = 2,000 Hz and a 1 GHz bandwidth waveform.

Once the Euler parameters and Kappa are specified, the motion of the target defined by $\theta(t)$, $\varphi(t)$ is determined using equations (2.23) and (2.24).

Figure 3.5(a) illustrates a three-dimensional (3D) view of the three fixed-point scattering center locations. Figure 3.5(b) illustrates a typical range, range-rate image at a particular instant of block time. Note that the three scattering centers for this example are widely spaced relative to each other, so the Fourier-based spectral estimation technique can be used to extract the range, range-rate observables from the initial data block.

The simulation used for this example assumes the radar is transmitting at a PRF of 2,000 Hz. The range, range-rate observable sequence $(R_n, \dot{R}_n, \ddot{R}_n, D_n)$ $n = 1, \ldots, N$, for $b = 1, \ldots, B$ is estimated from each data block assuming $Q_0 = 32$ pulses. The block center, t_b, is shifted by 0.005 s (10 pulses) for each successive block in the sequence for a total of $B = 150$ processed data blocks. Figures 3.5(c) and 3.5(d) illustrate the estimated range and range-rate observables extracted from the simulation data for each of the three scattering centers using the Fourier-based spectral estimation technique. The estimated values (dots) are compared to the true observables $(R_n, \dot{R}_n)_b$ (solid lines), obtained using the simulation motion and the true scattering center locations by evaluating the expressions

$$R_n(t_b) \equiv \hat{k}_b \cdot \underline{r}_n \tag{3.36}$$

$$\dot{R}_n(t_b) \equiv \dot{\hat{k}}_b \cdot \underline{r}_n \tag{3.37}$$

In equations (3.36) and (3.37), $\hat{k}$ is defined by equation (2.4), $\dot{\hat{k}}$ is defined by equation (2.5), and the values of $\hat{k}$ and $\dot{\hat{k}}$ are evaluated at the block time $t = t_b$ using the known motion $\theta(t_b)$, $\varphi(t_b)$. Note on examination of figure 3.5(c), the motion of the scattering center fixed on the z-axis is independent of roll and the range, range-rate observable sequence follows that of the 0.5 s precession period; the motion of the other two fixed scattering centers follows a complex spin-precession type variation. As expected, the range and range-rate observables estimated from the data compare well with the simulated values.

Now consider the comparison of the HRSE and Fourier-based scattering center acceleration estimates for this example. For each initial data block of size Q_0 pulses

Table 3.1   Simulation parameters: target and motion

	ρ (m)	ϕ (°)	z (m)
Scatterer #1	0	0	−1.0
Scatterer #2	0.25	320	−0.2
Scatterer #3	0.25	140	+0.5
Kappa (°)		30	
θ_p (°)		20	
T_p (s)		0.5	
T_s (s)		0.2	
α_p (°)		0	
α_s (°)		90	

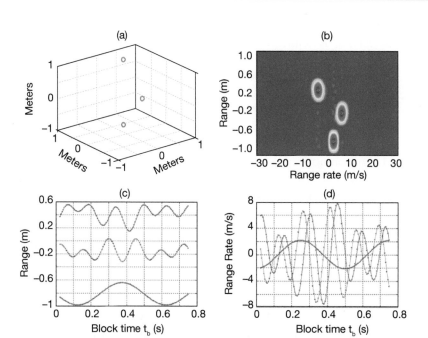

Figure 3.5   Three scattering center simulations illustrating the range, range-rate observable sequence extracted using the initial data block consisting of $Q_0 = 32$ pulses. (a) Scattering center 3D locations. (b) A typical range, range-rate image. (c) Extracted range sequence versus sliding block time (solid = simulation, dotted = extracted). (d) Extracted range-rate sequence versus sliding block time (solid = simulation, dotted = extracted).

centered at $t = t_b$ used in obtaining the range, range-rate observable sequence, a larger data block of symmetrically spaced pulses appropriate to each of the Fourier-based and HRSE estimation techniques as described in sections 3.2 and 3.3 is used to estimate $(\ddot{R}_n)_b$. Figure 3.6(b) compares the acceleration observable sequence estimates using the Fourier-based and HRSE techniques to truth (simulation), indicated by the solid line. The truth behavior of the range-acceleration sequence $(\ddot{R}_n)_b$ for fixed-point scattering centers is obtained by evaluating the expression

$$\ddot{R}_n(t_b) \equiv (\ddot{\underline{k}}_b)_{FP} \cdot \underline{r}_n \tag{3.38}$$

where for fixed-point (FP) scattering centers it can be shown that $\ddot{\underline{k}}$ is given by

$$\ddot{\underline{k}}_{FP} = (\ddot{\theta} - \dot{\varphi}^2 \sin\theta\cos\theta)\hat{\theta} + [\ddot{\varphi}\sin\theta + 2\dot{\theta}\dot{\varphi}\cos\theta]\hat{\varphi} - [(\dot{\theta})^2 + \dot{\varphi}^2\sin^2\theta]\hat{k} \tag{3.39}$$

Note that equation (3.39) is valid for FP scattering centers, for which $\underline{r}_n$ is fixed in target space and independent of time. As discussed in section 2.6, slipping scattering centers appear time varying as defined in the target-centered system and their target space location is defined by equation (2.20). Thus, for a slipping scattering center, equation (3.39) requires modification. It can be shown that for a slipping scattering center having radius ρ_{0n}, the expression for $\ddot{\underline{k}}_{FP} \cdot \underline{r}_n$ must be modified according to

$$\ddot{\underline{k}} \cdot \underline{r}_n = \ddot{\underline{k}}_{FP} \cdot \underline{r}_n + \rho_{0n}\dot{\varphi}^2\sin\theta \tag{3.40}$$

where $\underline{r}_n$ is defined by equation (2.20). The different expressions characterizing $\ddot{\underline{k}} \cdot \underline{r}_n$ for fixed versus slipping scattering centers illustrates the need to treat each type of scattering center separately in later chapters, addressing joint solutions for estimating both motion and target scattering center locations. An example of the impact of this difference is discussed in appendix D.3.1.

The points marked with a circle in figure 3.6(b) are obtained using the Fourier-based estimator, and those marked with a dot are the HRSE estimates. Figure 3.6(a) illustrates the value of *Arg* that characterizes the small acceleration approximation as the sequence progresses in time. The comparisons of the Fourier-based estimates and the HRSE estimates are essentially identical except in the regions where the value of *Arg* approaches 0.4. In this region the Fourier-based estimate provides slightly greater accuracy as illustrated in the expanded inset of figure 3.6(b). This occurs because the validity of the small argument approximation imposed using equation (3.30) for the HRSE estimator is more constrained relative to the small argument approximation using the Fresnel integral small argument expansion in equation (3.15) for the Fourier-based

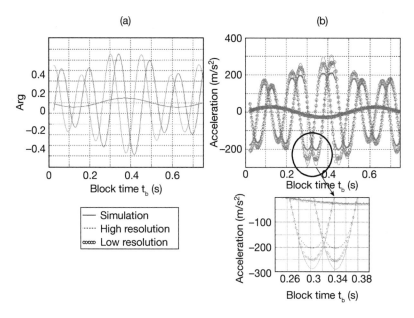

Figure 3.6 Comparison of the scattering center acceleration estimation sequences using the Fourier-based and HRSE techniques. (a) Value of *Arg* defined by equation (3.21) versus sliding block time. (b) Sequential acceleration estimate versus sliding block time (solid = simulation; circle = estimated, Fourier-based; dotted = estimated, HRSE).

estimator. However, for either case the comparison to truth is very good. In the next section an iterative scheme is developed that relaxes the limitation on the HRSE technique.

3.5 Extensions for Large Acceleration

In this section an iterative scheme is developed that significantly relaxes the limits on acceleration values for which the HRSE technique described in section 3.3.2 is applicable. The technique is initialized using the single block iterative scheme described in section 3.3.2: an initial frequency-time data block centered at $t_b = t_0$ with a time span $2T_0$ is selected such that the acceleration phase term contribution is small. The range, range-rate and complex amplitude observables $(R_n, \dot{R}_n, D_n)$, $n = 1, \ldots, N$ are estimated from the initial data block using the linear all-pole model. The block size is then increased symmetrically to cover a larger time interval and an initial estimate of each $\ddot{R}_n$ is obtained as described in section 3.3.2. The iterative technique continues this initial estimation scheme by processing a set of increasing larger data blocks, constructed in such a manner that the collective set of range, range-rate and range-acceleration observables are used in the nonlinear model to optimize the fit of the nonlinear model to the measurement data, defined by equation (3.34), over the expanded data blocks. The eventual data fit of the nonlinear model to the measurement data using the complete triplet of observables over the time interval provides the highest quality estimate.

The iterative approach is based on the nonlinear model defined by equation (3.29) and repeated here to illustrate those factors contributing to the nonlinear model used in the iteration scheme.

$$E(k, q) = \sum_n D_n \underbrace{e^{-j\frac{4\pi}{c}R_n(k\Delta f)}}_{E_n} \underbrace{e^{-j\frac{4\pi}{\lambda}\dot{R}_n q\left(1 + k\frac{\Delta f}{f_0}\right)\Delta t}}_{F_n} \underbrace{e^{-j\frac{2\pi}{\lambda}\ddot{R}_n q^2 \Delta t^2}}_{G_n} \tag{3.41}$$

Consider the three factors inside the summation, defined as E_n (a function of R_n) F_n (a function of $\dot{R}_n$), and G_n (a function of $\ddot{R}_n$), indicated within the brackets in equation (3.41). Each factor will be treated iteratively, holding the other two fixed, for iteration on each of the observables, with the goal of optimizing the nonlinear model fit to the data as a function of incremental block size. For each of these three observables, the iterative change can be written as

$$R_n = R_{n_0} + \Delta R_n$$
$$\dot{R}_n = \dot{R}_{n_0} + \Delta \dot{R}_n \tag{3.42}$$
$$\ddot{R}_n = \ddot{R}_{n_0} + \Delta \ddot{R}_n.$$

The iteration scheme used for each observable is identical in method to equations (3.30) through (3.35) with only a change in functional basis set $\psi_n(k, q)$ used in equation (3.31) to optimize the data minus model difference. To determine the iterative change in each observable, an expansion analogous to equation (3.30) is used, except the correction term is now defined relative to the current (iterative) value of each respective R_n, $\dot{R}_n$, and $\ddot{R}_n$ observable. In order to emphasize the similarity of the single block iteration scheme defined in equations (3.30) to (3.35) as it applies to each of the observables, the same notation for the set of coefficients, A_n, and the functional expansion set $\psi_n(k, q)$ is used. For example, the iteration on the R_n estimate results in the small argument expansion

$$e^{-j\frac{4\pi}{c}R_n(k\Delta f)} \approx e^{-j\frac{4\pi}{c}R_{n_0}(k\Delta f)}\left[1 - j\frac{4\pi}{c}\Delta R_n(k\Delta f)\right]. \tag{3.43}$$

When iterating on R_n, the expression for the nonlinear model $E_0(k, q)$ defined in equation (3.31), is given by

$$E_0(k, q) = \sum_n D_n E_n F_n G_n \Big|_{\substack{Evaluated\ at \\ \{R_{n_0}, \dot{R}_{n_0}, \ddot{R}_{n_0}\}}}, \tag{3.44}$$

and the expansion basis set used to match the mean square difference between data and nonlinear model, $| \Delta(k, q) - \sum_n A_n \psi_n(k, q) |^2$ is given by

$$\psi_n(k, q) = \{k s_n^k p_n^{q(1 + k\Delta f/f_0)} (p_n')^2\}_n. \tag{3.45}$$

The set (A_n) obtained for the range iteration is then determined from the optimization defined in equation (3.34). Once the A_n are determined, the estimate for R_n is updated analogous to equation (3.35)

$$\Delta R_n = -\frac{c}{4\pi\Delta f} \text{Im}\{A_n / D_n\}. \tag{3.46}$$

The iteration on $\dot{R}_n$ is similarly determined using the expansion basis at

$$\psi_n(k, q) = \{q s_n^k p_n^{q(1 + k\Delta f/f_0)} (p_n')^2\}_n. \tag{3.47}$$

from which the $\dot{R}_n$ estimate is updated using

$$\Delta \dot{R}_n = -\frac{\lambda}{4\pi\Delta t} \text{Im}\{A_n / D_n\} \tag{3.48}$$

The iteration on $(\ddot{R}_n)$ is obtained using the basis set defined by equation (3.33) as described in section 3.3.2, appropriately modified to include the factor $(p_n')^{q^2}$. Note also that the weight D_n used in the model fit to the data over the initial data block is updated as the values for R_n and $\dot{R}_n$ are updated.

Figure 3.7 illustrates the overall methodology for estimating $(\ddot{R}_n)$ used in the iteration scheme. Using the figure as a reference, the following methodology is employed:

1. Assume an initial block size consisting of Q_0 pulses. Using this data block extract the initial range, range-rate and amplitude estimates $(R_n, \dot{R}_n, D_n), n = 1, \ldots, N$.
2. Determine the initial estimate for $\ddot{R}_n$: Choose a larger data block incremented in size symmetrically by N_0 pulses, having size $Q_0 + 2N_0$ pulses. Apply the iteration scheme defined by equations (3.43) through (3.48), for each of the range, range-rate and range-acceleration observables, where the nonlinear model is

extrapolated over the larger data block using the current values for $(R_n, \dot{R}_n, D_n)$ but constraining $\ddot{R}_n = 0$.

3. Expand the data block size symmetrically by another set of N_0 pulses, now consisting of $Q_0 + 4N_0$ pulses. Repeat the iteration scheme defined in step 2, except now, for each iteration, use the most current values $(R_n, \dot{R}_n, \ddot{R}_n, D_n)$ in the nonlinear model which is extrapolated over the larger data block.

4. Keeping the same block size defined in step 3, carry out the iterations on $(R_n, \dot{R}_n, \ddot{R}_n, D_n)$ a fixed number of times, designated in figure 3.7 as "iteration within blocks." Typically, two iterations are adequate for each of the observables $R_n, \dot{R}_n,$ and $\ddot{R}_n$.

5. Increase the block size a fixed number of times, denoted by the parameter Nit, and repeat steps 3 and 4 for each block size increase.

The initial extraction of the linear HRSE model poles can either be obtained using a Fourier-based or HRSE estimator. Typically, if the bandwidth is sufficiently large and the scattering centers of interest are resolved within a range, range-rate image, then the Fourier-based estimator will be adequate. If the scattering centers are not resolvable using Fourier-based processing, then the HRSE technique should be used.

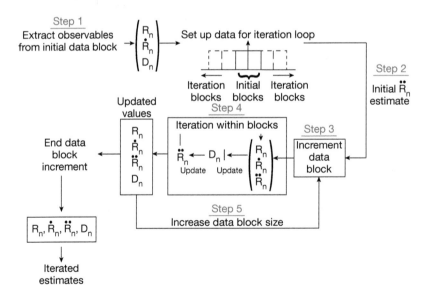

Figure 3.7 Iteration methodology used for large acceleration estimation scheme.

As an example of the iteration scheme, the simulation data for the spinning cylinder example illustrated in chapter 2, figure 2.7, is used to extract the triplet sequence $(R_n, \dot{R}_n, \ddot{R}_n)_b$, $n = 1, \ldots, N$, $b = 1, \ldots, B$. The target geometry is illustrated in figure 2.7(a). The target motion $[\theta(t), \varphi(t)]$, is characterized by equations (2.23) and (2.24), consisting of pure spin, where the sensor has a constant 30° LoS to the z-axis of the cylinder as illustrated in figure 2.8(a). Assume an initial block size of $Q_0 = 32$ pulses and increase each data block symmetrically by $N_0 = 10$ pulses. Use the Fourier-based spectral estimator for the initial $(R_n, \dot{R}_n)$ observable estimates. Assume four iterations at each block time, t_b, so that $Nit = 4$. Due to the shadowing constraint imposed on the simulation defined by equation (2.21) as the target rotates, only six of the 10 possible scattering centers are visible to the radar at any given time, except for the special case where two opposing diagonal scatterering centers enter and exit the field of view at the same time. Thus, an order of six is selected to obtain the initial spectral estimate.

The results of the simulation are illustrated in figure 3.8 for data blocks centered at $t = t_b$, where t_b covers the time interval (0.1, 1.1) seconds. To obtain the sequential estimates, the observable sequence block time t_b is incremented by Δt, and the label t_{b+1} corresponds to block time $t_{b+1} = t_b + \Delta t$. Note that when increasing the observable reference time by one pulse, all of the pulses minus one of the pulses in the initial and expanded data blocks associated with the previous reference time are repeated in the new set of data blocks. This results in a highly correlated set of sequential observable estimates making it easier to keep the individual scattering centers properly associated. Figure 3.8(a) illustrates the estimated values of the range $(R_n)_b$ as a function of block reference time $b\Delta t$, $b = 1, \ldots, B$; figure 3.8(b) illustrates the time evolution of the iterated values of the $\{\dot{R}_n\}_b$; and figure 3.8(c) the time evolution of the $\{\ddot{R}_n\}_b$.

The truth values for $(R, \dot{R}_n, \ddot{R}_n)_q$ obtained using equations (3.36), (3.37), and (3.38) are illustrated in figure 3.8 as solid lines, compared to the estimated values indicated by dots. The estimated results compare quite well to the true values, and generally degrade only in the regions where two scattering centers approach each other in the range, range-rate image plane.

As an illustration of the utility of the iteration process in improving the quality of the nonlinear model by the incorporation of $(\ddot{R}_n)$, figure 3.9 illustrates the absolute value of the data minus model difference, $|\Delta(k, q)|^2$, versus number of pulses in the iteration data block under consideration for two cases: figure 3.9(a) applies the linear all-pole HRSE model characterized by $\ddot{R}_n = 0$ over the initial expanded data block used for the first iteration, and figure 3.9(b) applies the current observables to the nonlinear HRSE model over the expanded data block after three iterations. The graph values shown in figure 3.9(a) for $|\Delta(k, q)|^2$, which use the data block for the initial iteration,

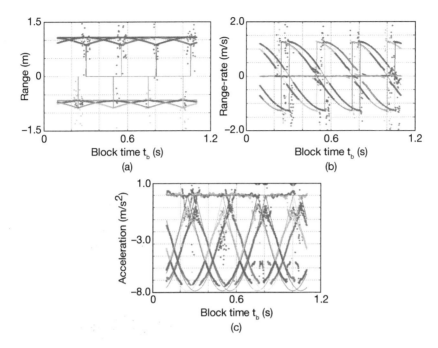

Figure 3.8   Estimated range, range-rate and range-acceleration sequence versus truth for the spinning cylinder example illustrated in chapter 2. Values used for the simulation are as follows: $Nit = 4$, $Q_0 = 32$, $N_0 = 10$, and $N = 6$. (a) Sequential evolution of $(R_n)_b$ versus block time. (b) Sequential evolution of $(\dot{R}_n)_b$ versus block time. (c) Sequential evolution of $(\ddot{R}_n)_b$ versus block time. Solid curve = simulation. Dotted curve = extracted.

are superimposed graphs for each of the 48 frequencies over the data block—that is, the cancellation (data minus model), $|\Delta(k_0, q)|^2$, is plotted versus initial block pulse index q for $k_0 = 1, \ldots, 48$. Figure 3.9(b) illustrates the same results for the expanded data block. Note the block size used for the initial iteration is of length $Q_0 + 2N_0 = 52$ pulses, and the data block size used for the third iteration is of length $Q_0 + 6N_0 = 92$ pulses, as indicated on the horizontal axis of the figures. In figure 3.9(a), for the linear all-pole model applied over this larger data block, the contribution of the acceleration term at the edges of the data block is clearly evident. The linear all-pole model provides good cancellation only over the region defined by the initial data block. Applying the estimated observables $\{R_n, \dot{R}_n, \ddot{R}_n\}_b$, $n = 1, \ldots, N$ to the nonlinear model, indicated in figure 3.9(b), results in good cancellation over the entire expanded data block, indicating confidence in the estimated values.

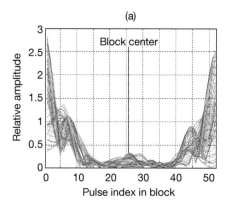
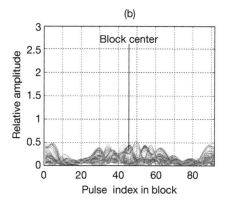

Figure 3.9 Cancellation (data minus model) for initial and final iteration data blocks.
(a) Data minus model for the linear all-pole HRSE model characterized by $\ddot{R}_n = 0$ over the
initial expanded data block used for the first iteration (b) Data minus model for the
nonlinear model final iteration data block after three iterations.

Finally, it is useful to compare the final estimates for the $(R_n, \dot{R}_n)_b$ sequence after
iteration to those obtained using HRSE for the initial pole estimates. Consider, for
example, a value of $Q_0 = 16$ pulses used for the initial data block size corresponding to
the value of Q_0 used in chapter 2 for the simulation results illustrated in figures 2.10
and 2.11. The comparative results using the iterative acceleration estimation technique
over a short duration time interval are indicated in figure 3.10. Figure 3.10(a) illustrates the initial linear model pole estimates for $(\dot{R}_n)_b$ over the expanded time scale
0.1 s to 0.22 s; figure 3.10(b) shows the resultant estimates for $(\dot{R}_n)_b$ over the same
time interval after four iterations using an incremental block size of $N_0 = 10$ pulses.
Clearly, the adjustments to $(\dot{R}_n)_b$ required to fit the more robust nonlinear model result
in more accurate, smoother sequential values when compared to the initial estimates.
The same conclusion also holds for the range values.

Although the values for the sequence $(R_n, \dot{R}_n)_b$ are generally more precise when compared to those corresponding to the initial extractions, observe from figure 3.10(b)
that occasionally spurious values occur in the sequence. These spurious values result
when pairs of range, range-rate observables occur in close proximity, leading to an
ambiguous nonlinear model. Techniques for compensating for these spurious values
when applied to estimating the scattering center locations are presented in chapter 5.
Additionally, because estimating $(\ddot{R}_n)$ requires a much larger block size compared to
that for the linear model, it is more sensitive to scattering centers that are not

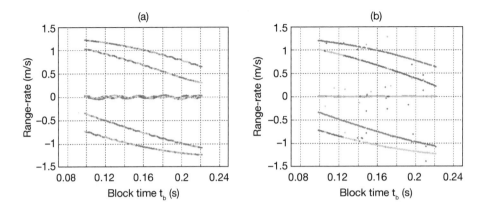

Figure 3.10   The $(\dot{R}_n)$ sequential spectral estimates versus time expanded over the time interval 0.1 s to 0.22 s. (a) Initial $(\dot{R}_n)$ estimate using the linear all-pole model for data block having $Q_0 = 16$ pulses. (b) The $(\dot{R}_n)$ estimate using the nonlinear model for three iterations of $N_0 = 10$ pulses and $Nit = 3$.

persistent over the extent of the larger block. MATLAB code for the implementation of the HRSE iteration scheme illustrated in figure 3.7 is presented in appendix E.

The following chapter introduces a set of tools that provide a quantitative measure of coherency for the received radar data contained within a given data block. These coherency tools are applied to a canonical target data set in appendix E to illustrate how they can be used to set the initial and final block size input parameters for the iteration portion of the acceleration estimate.

4

Autocorrelation Measures and Signal Coherence

4.1   Introduction

The new framework developed in the first three chapters exploits the information contained in a sequence of radar pulses reflected from a target. The processing techniques introduced in chapters 1–3 operate on two-dimensional (2D) blocks of radar data ordered by pulse transmit time as outlined in figure 1.10. Each pulse of the data array may be represented by either uncompressed waveform data sampled in the frequency domain, or Fourier compressed waveform data organized and sampled by range resolution cells. In both cases, the performance of the respective techniques depends on the coherency of the data received over the sequence of pulses.

This chapter introduces a set of tools that provide a quantitative measure of coherency for the received radar data contained within the data block. The techniques operate on the one-dimensional (1D) or 2D blocks of compressed radar data ordered by pulse transmit time. Two different scenarios for target and motion are considered: multiple targets in the radar transmit/receive beam, each having independent motion; and a single rigid body target in the radar transmit/receive beam consisting of multiple scattering centers having correlated rotational motion. For a narrowband waveform, for either scenario, the multiple targets or individual scattering centers are not resolved and reside in a single range gate, thus the data block is 1D. The time samples represent the peak of the compressed waveform for each pulse in the data block. For a wideband waveform the bandwidth is set so that the multiple targets or individual scattering centers are resolved in separate range resolution cells, and the time samples for each row in the matrix corresponds to the samples from a given range gate.

The fundamental limits on the type of processing (e.g., coherent vs. noncoherent) that can be applied to the data block are also examined. Three metrics characterizing the data block coherency are developed and used to identify the parameters applicable to either coherent or noncoherent processing.

Given that the transmitted waveform sequence is coherent, the coherency of the received data sequence is inherently related to the target motion and radar pulse repetition frequency (PRF). In section 1.3.1 the concept of determining the motion solution for a target observed by the radar was introduced. A complete motion solution describes the time history of the radar to target viewing angle. However, it is not necessary to develop a complete motion solution to exploit data coherency. Some degree of insight regarding the data coherency can be gained by using the statistical correlation properties embedded in the time history of the radar cross section (RCS) fluctuations, a concept familiar to radar system analysts.

Historically, this concept has been applied to the study of target detection probabilities, leading to the familiar target fluctuation models introduced by [68], and described in detail in [69]. These models have been developed for amplitude fluctuations only. When examining the coherency of data contained within a data block, it is the phase of the data that dominates data coherence. The measures of the *degree of coherency* for a time sequence of data samples developed in this chapter use both amplitude and phase information. Discussion relating the degree of coherency to estimating some general characteristics of target motion is also provided.

The coherency metrics are applied to time sequences of compressed radar data organized in the block form described in section 1.3.3 and illustrated in figure 1.10b. For a typical data block of Q_0 compressed pulses, where each pulse consists of N_{ffiW} range gates, the data matrix $e_T(R_k, t_q)$ is of size $[N_{ffiW} \cdot Q_0]$. For each row of the data matrix there are Q_0 complex time samples. Denote these Q_0 time samples as

$$[x_1, x_2, \ldots, x_{Q_0}]_{k_0} \equiv e_T(R_{k_0}, t_q) \tag{4.1a}$$

where each row defined at $k = k_0$ corresponds to a specific range gate R_{k_0}. In the radar context there are two target-motion processing scenarios for which the concept of coherency provides useful insight.

The first case addresses targets that reside in a single range resolution cell. The situation typically arises with narrowband waveforms used during a search function. The time history of radar returns in that cell sampled over a sequence of radar pulses yields a 1D data block for processing. Adjacent resolution cells contained in the received narrowband waveform could contain other targets, but the data would be statistically independent. Hence, each range gate can be processed independently with no loss of information.

The second case involves a single rigid body target possessing multiple scattering centers for which the body motion imposes correlation between scattering centers located on the target. In the wideband application, these centers reside in different resolution cells, and the correlation behavior within each range gate is collectively exploited. In such a case, it is important to constrain the 2D data block to cover the range extent of the target as described earlier in the text. The following analysis applies to either case.

For each row of the data matrix, the Q_0 time samples can be processed to yield a statistical measure of the coherency of the data in the specified range gate. In range gates that contain the target, the time samples will exhibit some degree of coherency depending on target type and motion; for range gates containing only noise, the time samples will be uncorrelated.

In traditional signal processing the autocorrelation function, $R_a(\tau)$, of the received signal characterizes its coherency and is governed by a combination of the target's signature variation with observation angle and its associated motion. It provides a measure of coherence between data samples in the time series that are separated by a specified lag time τ. Whether or not the data samples exhibit a strong degree of coherence will be a function of the changes in RCS with motion as determined by the sampling interval Δt, determined by the radar PRF. For any target and motion combination, there exists a high enough sampling rate for which the data samples will appear coherent—that is, the processed returns will appear well correlated over a given number of pulses, corresponding to a correlation time, T_c. Because the motion of each target as well as its RCS changes versus observation angle are different, T_c can be expected to be different for each target-motion combination.

It is interesting to note that there exists only one completely coherent object, the sphere, because its RCS is constant over all possible aspect angles. Figure 4.1 depicts the correlation function of two hypothetical targets and how their correlation properties vary with the sampling interval. If pure white noise (an infinite bandwidth process) were depicted in the figure it would be represented by an impulse function at the origin because all noise samples are completely uncorrelated with each other.

For very short sampling lag times, the RCS time series of both targets show a high degree of coherency and it is very difficult to differentiate between them. However, there exists an interval of lag time for which the correlation properties of the two target classes differ significantly. Finally, for very long lag times, data samples spaced greater than T_{c_2} appear uncorrelated for both targets. It is possible to exploit the correlation lag transition region from coherence to noncoherence to establish a measure of coherence for a given data block.

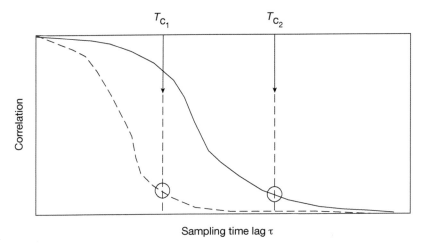

Figure 4.1   Autocorrelation versus sample time lag for two different targets.

The degree of coherency of the time series as characterized by the autocorrelation function strongly affects the ability to coherently sum the data samples to improve the signal-to-noise ratio. In the next section a generalized complex correlation matrix, R, is introduced [equation (4.2)] and the connection between coherence as defined by the generalized autocorrelation function, and the potential signal-to-noise ratio gain obtained using coherent or noncoherent integration is discussed. In the case where the compressed waveform for a complex target is resolved into a number of individual range resolution cells, the concept of a wideband *autocorrelation contour map*, is introduced (section 4.4) and compared to the range, range-rate image contour obtained from a shorter duration subset of the data block. It is then used to quantify the coherency of the data contained in the data block, and to illustrate the relationship between data coherency and target motion.

4.2   The Generalized Autocorrelation Matrix

It is instructive to consider a generalized form of the row vector $[x_1, x_2, \ldots, x_{Q_0}]$ defined in equation (4.1a), modified according to

$$\underline{x} = [x_1, x_2 e^{j\Delta}, x_3 e^{j2\Delta}, \ldots, x_{Q_0} e^{j(Q_0 - 1)\Delta}] \tag{4.1b}$$

where Δ represents a fixed observable that is to be estimated from the data: for example, choose Δ to be proportional to Doppler frequency $\Delta = (2\pi f_d)\Delta t$, or range-rate $\Delta = (4\pi \dot{R}/\lambda)\Delta t$, where $\dot{R} = \lambda f_d / 2$.

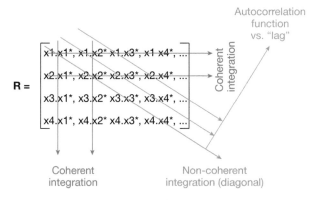

Figure 4.2 Illustration of the form of the generalized autocorrelation matrix **R** for $\Delta = 0$. Three possible processing schemes using cuts through the **R** matrix are indicated by the following: green = noncoherent integration; red = coherent integration; and blue = autocorrelation and signal spectral density.

Various signal processing techniques associated with $\underline{x}$ can be illustrated by introducing the generalized autocorrelation matrix $\boldsymbol{R} = (\underline{x}\dagger\underline{x})$. The elements of $\boldsymbol{R}$ are given by

$$R(q, q') = (\underline{x}\dagger\underline{x})_{q,q'} = [x_q^* x_{q'} \, e^{+j(q'-1)\Delta} e^{-j(q-1)\Delta}] \tag{4.2}$$

where "$\dagger$" indicates complex conjugate transpose and x^* is the complex conjugate of x. An illustration of $\boldsymbol{R}$ is provided in figure 4.2 for $\Delta = 0$, and provides a useful way of interpreting the autocorrelation matrix elements in terms of a number of standard radar signal processing concepts. Of particular interest is the relationship between various methods of conventional processing to the generalized autocorrelation function using cuts through the $\boldsymbol{R}$ matrix as indicated in the figure. In particular, three processing schemes can be identified as the summation of specific elements $\boldsymbol{R}(q, q')$:

1. Noncoherent integration: Matrix elements are summed along the diagonal of $\boldsymbol{R}$.

$$I_{nc} = \sum_{q=1}^{Q_0} R(q, q) = \sum_{q=1}^{Q_0} |x_q|^2 \tag{4.3}$$

2. Coherent integration (Fourier transform of the data samples): Matrix elements of $\boldsymbol{R}$ are summed over each column (or row).

$$I_c(\Delta, q) = \sum_{q'=1}^{Q_0} R(q, q') = x_q^* e^{-j(q-1)\Delta} \sum_{q'=1}^{Q_0} x_{q'} e^{+j(q'-1)\Delta} \tag{4.4}$$

3. Signal autocorrelation: Matrix elements of $\boldsymbol{R}$ are processed by summing the off-diagonal lags of $\boldsymbol{R}(q', q'+q-1)$, denoted by $\hat{R}_a(q)$, for a given lag q

$$\hat{R}_a(q) = \sum_{q'=1}^{Q_0/2} R(q', q'+q-1) \tag{4.5}$$

where for a matrix of size $Q_0 \times Q_0$ the sum necessarily ends at a maximum lag $Q_0/2$ for a time series Q_0 samples long.

4.2.1   Coherent Integration

Let $F(\omega)$ represent the Fourier transform of $x(t)$:

$$F(\omega) = \int x(t)e^{+j\omega t}dt \tag{4.6}$$

The integral can be approximated as a discrete sum, where $t = (q'-1)\,\Delta t$, $q' = 1,\ldots, Q_0$ and $x_{q'} = x(t_{q'})$ to yield

$$F(\omega) \approx \Delta t \sum_{q'=1}^{Q_0} x_{q'}e^{+j(q'-1)(\omega\Delta t)} \tag{4.7}$$

Comparing equations (4.7) and (4.4), it follows that the magnitude of $F(\omega)$ is proportional to $I_c(\Delta, q)$ in the form

$$|F(\omega)| \propto |I_c(\Delta, q)| \tag{4.8}$$

where $\Delta = \omega\Delta t$. Note that equation (4.8) is valid independent of the column (or row) of $\boldsymbol{R}$ that is processed, and only the constant of proportionality $|x_q^*|$ changes. Because equation (4.4) is equivalent to Doppler (range-rate) processing of the time series expressed as function of $\Delta = 2\pi f_d \Delta t$, the peaks of $|I_c(\Delta, q)|$ viewed as a function of Δ identify the Doppler-bins for which the data summation is maximized. Thus, processing down the column or across the row of R is equivalent to the Doppler processing discussed in earlier chapters.

4.2.2   Autocorrelation and Signal Spectral Density

The noncoherent and coherent processing schemes represented by equations (4.3) and (4.4) are the conventional methods used to enhance the detection of targets (or scattering centers) in the presence of noise. The choice of method employed depends on the coherence of the target over the data processing interval. A precise measure of coherence can be obtained by summing over the off-diagonal elements of R, as indicated in equation (4.5) and illustrated in figure 4.2, depicting the result as a function of the off-diagonal lag.

The general measure of the temporal cross-correlation [70] between two complex time functions $x_1(t)$ and $x_2(t)$ is defined by

$$R_{12}(\tau) = \frac{1}{2T} \int_0^{2T} x_1(t) x_2^*(t + \tau)\, dt \tag{4.9}$$

where τ denotes the time offset (lag) relative to t, and $2T$ is the averaging time. The autocorrelation of $x(t)$ is defined using equation (4.9) with $x_1(t) = x_2(t)$. Thus, the complex autocorrelation function is defined as

$$R_a(\tau) = \frac{1}{2T} \int_0^{2T} x(t) x^*(t + \tau)\, dt \tag{4.10}$$

where the subscript denotes autocorrelation. In discrete form $t = (q' - 1)\Delta t$, and $\tau = (q - 1)\,\Delta t$, such that for Q_0 data samples $R_a(\tau)$ takes the discrete form

$$R_a(q) = \frac{2}{Q_0} \sum_{q'=1}^{Q_0/2} x_{q'} x_{q'+q-1}^* \tag{4.11}$$

where q indicates the time offset $(q - 1)\Delta t$. Using equation (4.2) and equation (4.5), the expression for $\hat{R}_a(q)$ reduces to the form

$$\hat{R}_a(q) = e^{jq\Delta} \sum_{q'=1}^{Q_0/2} x_{q'} x_{q'+q-1}^* \tag{4.12}$$

Thus off-diagonal processing of the generalized correlation matrix is equivalent to conventional autocorrelation processing:

$$|R_a(q)| = \frac{2}{Q_0} |\hat{R}_a(q)| \tag{4.13}$$

Furthermore, the discrete Fourier transform of $\hat{R}_a(q)$ provides a direct estimate of the signal spectral density $S_a(\Delta)$ of $x(t)$ [70] according to

$$S_a(\Delta) = \sum_q e^{jq\Delta} R_a(q) = \frac{2}{Q_0} \sum_q \hat{R}_a(q) \tag{4.14}$$

Thus there are three metrics that can be used to characterize the degree of coherency of the data block represented by $\boldsymbol{R}$:

The coherently integrated Doppler spectrum $I_c(\Delta)$

The autocorrelation time scale T_c determined from $\boldsymbol{R}_a(q)$

The autocorrelation spectral density estimate $S_a(\Delta)$

It is useful to compare these three metrics relative to the degree of coherency exhibited by the generalized autocorrelation matrix $\boldsymbol{R}$ illustrated in figure 4.2. If one views processing along the diagonal of $\boldsymbol{R}$ as completely noncoherent processing, and along each row or column of $\boldsymbol{R}$ as completely coherent processing, it follows that processing along the off-diagonal corresponds to *partially coherent processing*, a concept that will be discussed and quantified in the following sections.

4.3   Data Block Size Limits for Linear and Nonlinear Signature Models

In chapter 3, two signature models were introduced to model subblocks of measurement data contained within the 2D data block illustrated in figure 3.1. The linear signature model was applied to the initial data block to estimate the scattering center location in range and its Doppler velocity. The nonlinear signature model was then iteratively applied to increasingly larger data blocks to obtain estimates of the scattering center's acceleration. In this section the three coherency metrics discussed are used to examine the data block size appropriate for each of these signature models. Once the maximum block size limit is established for the linear model, increasing the data block size beyond this time extent assures that the expanded data block will provide an appropriate data set for estimating the acceleration term in the nonlinear model. Examples of using this coherency measure to set the processing block parameters when estimating $\ddot{R}$ are presented in appendix E.

For simplicity, the behavior of the coherence metrics are first examined for a single target residing in a range resolution cell. In section 4.3.3 the more general two-target scenario is considered. The two-target scenario is illustrated in figure 4.9. Different uncorrelated motions are assigned to each target to develop an understanding of the relationship between coherency and the correlation metric. Introducing the second target examines the case where, using a narrowband waveform, more than a single target occupies the range resolution cell to show how interference between targets can affect the narrowband coherency metric. The two-target example for the wideband waveform is considered in section 4.4.

The geometrical theory of diffraction (GTD) scattering model for each target used to simulate the measurement data consists of a single point scattering center undergoing orbital motion about a center point, each having independent motion parameters. For the single target example depicted in figure 4.3 it is shown that the coherence time T_c of the associated data sequence $[x_1, x_2, \ldots, x_{Q_0}]$ is identical to the time, T_i, over which the linear signature model is valid. For example, when applying the acceleration estimation technique developed in chapter 3, the coherence time sets the maximum number of pulses that should be used in the initial data block.

As a corollary, it is shown that the measurement of T_c leads to a novel concept in which T_c can be used to characterize the dynamics of the target. This concept will be explored further with a wideband coherence measure introduced in section 4.4.

4.3.1 The Single Target Example

In this section a target consisting of a single scattering center undergoing orbital motion about a center point (depicted in figure 4.3) is used to examine the behavior of the different coherence metrics.

Assume the motion M_1 of target S_1 is defined by $\varphi_1(t)$. The motion is described by $\varphi_1(t) = \dfrac{2\pi}{T_{0_1}} t$, where T_{0_1} denotes the rotation period of the single point scattering center. Denote the viewing angle of the sensor to the target as $\hat{k} = \hat{x}$ along the x-axis. The position of the scattering center relative to its center of rotation (CoR) is given by

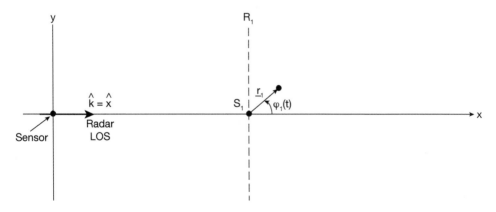

Figure 4.3 An orbiting point target, S_1, centered on the x-axis, having motion M_1.

$$\underline{r}_1(t) = a_1 \cos(\varphi_1(t))\hat{x} + a_1 \sin(\varphi_1(t))\hat{y},$$

where a_1 denotes the radius of the rotating scattering center. Using the GTD-based scattering formulation defined in section 2.2, it follows that, using equation (2.6), because $\hat{k} \cdot \hat{y} = 0$, the scattered signal $E(f_0, t)$ can be written in the following form:

$$E(f_0, t) = D_1 e^{-j\frac{4\pi f_0}{c} a_1 \cos\left(\frac{2\pi}{T_{0_1}}t\right)}. \tag{4.15}$$

Assume the radar transmits a sequence (pulse train) of narrowband waveforms, and each received pulse is sampled at the peak of the range gate containing the target. The sampling time $\Delta t = 1/\mathrm{PRF}$, such that $t = (q-1)\Delta t$, $q = 1, \ldots Q_0$. Evaluating $E(f_0, t)$ at $t = t_q$ leads to the Q_0 time samples $(x_1, \ldots x_{Q_0})$ defined by

$$x_q = D_1 e^{-j\frac{4\pi f_0}{c} a_1 \cos\left(\frac{2\pi}{T_{0_1}}(q-1)\Delta t\right)} \tag{4.16}$$

Examine the metrics defined in section 4.2 to assess the coherency of the data samples as Q_0 increases. Assume the scattering center rotates about its local origin at a radius $a_1 = 0.1\,\mathrm{m}$ with a rotation period $T_{0_1} = 0.1\mathrm{s}$. Assume a $\mathrm{PRF} = 2,000$ Hz ($\Delta t = .0005\mathrm{s}$), such that over 200 time samples the target will undergo a complete rotation. Assume an operating frequency of 10 GHz and a waveform of bandwidth 10 MHz—that is, a 0.1 percent fractional bandwidth. Divide this band into 32 frequency bins, so that $n_f = 33$. Each single-pulse compressed waveform on reception covers a range extent of $c/2\Delta f = 480$ m, typical of a narrowband waveform. The single-pulse compressed waveform using a Taylor weighting function to reduce the range sidelobes is illustrated in figure 4.4.

First, examine the behavior of $R_a(q)$ for a data sequence of 400 time samples (400 narrowband pulses) covering two complete rotation periods, so that the correlation value for lag times up to a single rotation period may be calculated. Figure 4.5(a) illustrates the autocorrelation function $R_a(q)$ for this data sequence plotted in terms of the lag time $(q-1)\Delta t$. Note that $R_a(q)$ is periodic with period T_{0_1} and the rapid decay of $R_a(q)$ versus lag time indicates that the data are correlated (coherent) over only a very short time period. Figure 4.5(b) illustrates the corresponding spectral density $S_a(\Delta)$ plotted versus $\dot{R}$, where the scale for $S_a(\Delta) \Rightarrow S_a(\dot{R})$ is chosen to be the instantaneous range-rate defined with the mapping $\Delta = (4\pi\dot{R}/\lambda)\Delta t$. The spectral density of the autocorrelation function is quite broad and uniformly distributed over the range-rate

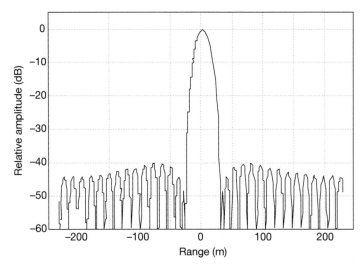

Figure 4.4 Narrowband compressed waveform using a Taylor weighting function. $BW = 10$ MHz, range window = 480 m.

interval bounded by $\pm \dot{R}_{max}$, where the maximum range-rate for the orbiting scattering center, $\dot{R}_{max}$, is given by $\dot{R}_{max} = 2\pi a_1 / T_{0_1} = 2\pi$ m/s. The large spread of $S_a(\dot{R})$ in range-rate indicates that the data sequence is mostly incoherent when data are processed over a complete rotation cycle. For comparison, consider coherently processing a shorter interval of 20 pulses using coherent integration to obtain the range-rate estimate as described by equation (4.4). For the 20 pulses selected, the resultant peak in the range-rate spectrum would characterize the instantaneous velocity [relative to the radar line of sight (LoS)] of the scattering center in its orbit around the CoR at the time referenced to the center of the coherent 20 pulse data block. If the center time of the 20-pulse data block is incremented along the total 200 pulse data set, the range-rate estimated from each short coherent data block would increment along the noncoherent spectrum illustrated in figure 4.5(b), eventually covering the entire spectrum. It follows that the spectrum determined from the autocorrelation function provides a measure of the *long-duration* behavior of the motion, including all velocities, and the *short-duration* coherent processing characterizes the instantaneous measure of the scattering centers range-rate at a particular point on its orbit.

It is interesting to compare the coherent integration metric defined by $I_c(\Delta)$ to the spectral density function $S_a(\Delta)$ by coherently processing the data set using $I_c(\Delta)$ over

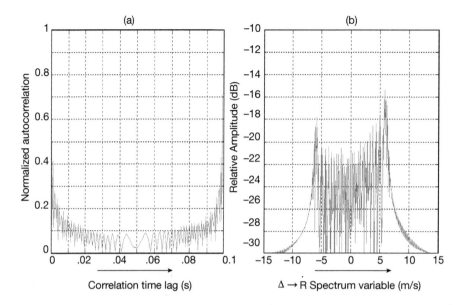

Figure 4.5 (a) Autocorrelation function $R_a(\tau)$, and (b) spectral density $S_a(\dot{R})$ for a single rotating target having rotation period $T = 0.1$ s for a data sequence of 400 time samples.

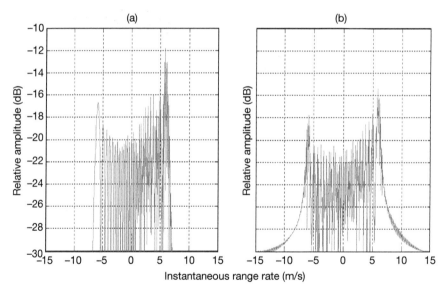

Figure 4.6 (a) Coherent integration spectrum $I_c(\Delta)$, for the single rotating target having rotation period $T = 0.1$ s by coherently processing the data over the complete rotation period and (b) spectral density $S_a(\Delta)$ applied to same data set.

the large 200 pulse processing interval. This is illustrated in figure 4.6 where $I_c(\Delta) \Rightarrow I_c(\dot{R})$ and $S_a(\dot{R})$ are plotted versus $\dot{R}$. Apart from an amplitude scaling factor, both spectra provide essentially the same information. Coherent processing of the entire 200 pulse data set is essentially equivalent to superimposing the range-rate estimates for the sliding 20 pulse data sets as they traverse the 200-pulse data block.

4.3.2   The Connection between T_i and T_c

Thus, the question arises: How does the correlation time T_c determined from the auto-correlation function compare to the to the maximum integration time interval, T_i, used for coherent integration of the data without loss in processing gain? Intuitively, one expects that adding data to the sequence used for coherent processing that is uncorrelated to data over the initial portion of the sequence would degrade the signal processing gain. The answer to the previous question is equivalent to determining the length of the data sequence for which the appropriate signal model used to reconstruct the data transitions from the linear model to the nonlinear model. In applying the nonlinear signal model to obtain the acceleration estimate using the technique developed in chapter 3, one anticipates incrementing the data block size consistent with the number of pulses contained within the interval characterizing T_i, assuring the transition to the nonlinear signal model. A simple analysis demonstrates this connection between T_c and T_i.

As discussed in chapter 3, for linear model to be valid, the time span associated with the data block is constrained so that the acceleration phase term in equation (3.3) is negligible. Evaluating equation (3.3) at the limit of the time interval $(0, T_i)$, where T_i denotes the coherent processing time interval, equation (3.3) takes the form

$$R(T_i) = R_0 + \dot{R}T_i + \frac{1}{2}\ddot{R}T_i^2. \tag{4.17}$$

For the linear signal model to remain valid, the time T_i must be small enough that the quadratic phase term, $\frac{4\pi}{\lambda}\left(\frac{1}{2}\ddot{R}T_i^2\right)$ corresponding to the acceleration phase contribution to $E(f, t)$ is negligible. Thus for coherent processing to provide an accurate estimate of $\dot{R}$, the integration time T_i is constrained by the criteria $\frac{2\pi}{\lambda}\ddot{R}_{max}(T_i)^2 \ll 1$, leading to the limiting value for the integration time:

$$T_i \ll \sqrt{\frac{\lambda}{2\pi\ddot{R}_{max}}}. \tag{4.18}$$

For the rotating scattering center, it can be shown that

$$\ddot{R}_{max} = \left(\frac{2\pi}{T_{0_1}}\right)^2 a_1.$$ (4.19)

Thus the integration limit T_i for linear imaging is constrained by

$$(T_i)_{max} << \frac{T_{0_1}}{2\pi}\sqrt{\frac{\lambda}{2a_1}}.$$ (4.20)

For this example at X-band ($\lambda = 0.03$ m), using $a_1 = 0.1$ m and $T_{0_1} = 0.1$ s, the maximum data block time span for linear imaging is

$$(T_i)_{max} \leq 0.0062 \, \text{sec}.$$ (4.21)

For $\Delta t = 0.0005$ s, equation (4.21) indicates the data block should consist of no more than 12 pulses. The effect caused by expanding the data block beyond this limit are considered later in the section.

Now compare the integration time limit $(T_i)_{max}$ to values of T_c characterizing the decay of the autocorrelation function versus time lag. Figure 4.7 shows an expanded graph of $R_a(\tau)$ versus correlation time lag τ, illustrating the decay behavior of $R_a(\tau)$. Define the correlation time as that value for which the behavior of the autocorrelation function $R_a(\tau)$ has just reached its near steady-state level. This corresponds to lag $\tau = T_c \approx 0.006$ s corresponding to nearly the same value for $(T_i)_{max}$. Because the value of T_c is essentially the same as $(T_i)_{max}$, the connection between $R_a(\tau)$ and $I_c(\Delta)$ is established: the correlation time T_c that provides a good measure of the data coherence time is essentially the same time $(T_i)_{max}$ that characterizes the limiting time for coherent integration of the data samples.

$$T_c \approx (T_i)_{max} = \frac{T_{0_1}}{2\pi}\sqrt{\frac{\lambda}{2a_1}}$$ (4.22)

Consequently, adding additional samples in the data sequence used for coherent integration beyond the limit expressed in equation (4.22) can only reduce the quality of the Doppler spectrum estimate because the target velocity begins to migrate into adjacent Doppler resolution cells. The phase of the additional samples is uncorrelated to the initial samples in the sequence, and reduces the processing gain.

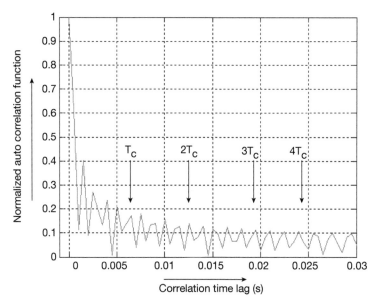

Figure 4.7 Expanded view of $R_a(\tau)$ illustrated in figure 4.5(a) illustrating the decay behavior of $R_a(\tau)$.

The connection relating the coherence time T_c to the integration time limit $(T_i)_{max}$ characterized by equation (4.22) leads to a novel, and useful concept: the measurement of T_c can be used to discern differences in the rotational motion of one target relative to another. Observe from equation (4.22) that T_c is directly proportional to the target rotational period T_{0_1}. For fixed PRF, if T_{0_1} decreases (a faster rotation rate) T_c will decrease, and the corresponding length of the data sequence that consists of only coherent samples will decrease. Measurement of T_c is thus directly related to the dynamics of the target. This concept will be explored further with the wideband coherence measure introduced in section 4.4.

It is instructive to examine the range-rate spectrum for the single target example using the coherent processing metric for data sequences of increasing length Q_0, where Q_0 covers integration times T_c, $2T_c$, $3T_c$, and $4T_c$, respectively. One should expect that the addition of uncorrelated data samples into the data sequence would adversely affect the range-rate spectrum. Figures 4.8(a–d) illustrate the coherent spectrum, $I_c(\dot{R})$, evaluated over a typical time interval along the orbit, for integration times progressing through these four integration limits. For integration time $T_i = T_c$ the spectrum is well

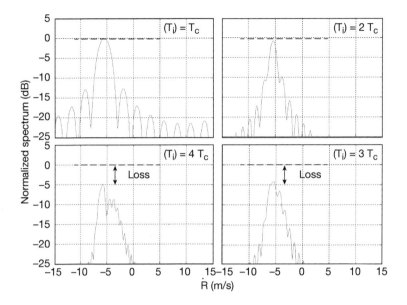

Figure 4.8   Coherent spectrum $I_c(\dot{R})$ evaluated over a typical time interval along the orbit for varying integration times. The $T_i = T_c$ corresponds to the bound on T_i for linear imaging to be valid.

focused and the velocity of the orbiting scattering center relative to the radar LoS is readily determined; for integration time $T_i = 2T_c$, although the velocity resolution is enhanced, the spectrum begins to defocus and exhibit some spectral spreading with increased side-lobe levels; for $T_i > 2T_c$ the spectrum becomes considerably distorted and exhibits reduced integration gain, as well as spectral spreading.

These results can be quantified by tabulating the values of three key metrics usually extracted using coherent integration: integration gain (loss), range-rate estimation, and spectral spreading. These are tabulated in table 4.1. The values indicated are approximate, but indicate the general trends as the spectrum begins to lose coherence over the expanding data blocks. For $T_i = T_c$ and $2T_c$, the range-rate estimation for the target is well-defined. The table values for integration loss (dB) and normalized spectral width are tabulated relative to the coherent case $T_i = T_c$.

It is also interesting to contrast the information content contained in $R_a(\tau)$ versus $I_c(\dot{R})$ relative to the discussion characterizing figure 4.5(b) earlier in this section: the correlation time extracted from $R_a(\tau)$ is proportional to T_{0_1}, the target rotation period (i.e., the long-duration behavior of the motion), and $I_c(\dot{R})$ provides a measure of the

Table 4.1   Degradation of key observables versus integration time (normalized to T_i = multiples of T_c)

Integration Time, T_i	Integration Gain (Loss) (dB)	$\dot{R}$ Estimate (m/s)	Normalized Spectral Spread
$T_c/6$	0	−5.5	1
T_c	0	−5.5	1
$2T_c$	−0.82	−5.4	1.2
$3T_c$	−4.2	−5.4	1.3
$4T_c$	−5.0	−5.8	2.3
$5T_c$	−5.6	−6	4
$11T_c$	−10.5	−5.9	6
$21T_c$	−12	−5.9	7

instantaneous motion of the scattering center (i.e., the short duration behavior) at a particular location on its orbit about the CoR.

4.3.3   Effects of Multitarget Interference

The previous section examined the behavior of the coherency metrics defined in section 4.2 for a single target. In practice, it is not uncommon to encounter multiple targets that under conditions of limited bandwidth will interfere with each other within a single resolution range gate. To appreciate the effects of such interference a slightly more complex simulation model is used to discuss such effects. Consider an augmented simulation model that consists of two targets, each consisting of a scattering center orbiting about a center point, as illustrated in figure 4.9. The two point targets, denoted by S_1 and S_2, are aligned to the radar LoS and have motions M_1 [defined by $\varphi_1(t)$] and M_2 [defined by $\varphi_2(t)$], and are assumed to be in the far field of the sensor. Each point target is assumed to orbit about the center of its own localized coordinate system, separated by range $\Delta R = R_2 - R_1$, and the motion is described by $\varphi_1(t) = \dfrac{2\pi}{T_{0_1}}t$, and $\varphi_2(t) = \dfrac{2\pi}{T_{0_2}}t$, where T_{0_1} and T_{0_2} denote the rotation periods respectively. Because the viewing angle to each of the localized coordinate systems is the same, the backscattered field can be expressed as the superposition of two wavefronts displaced by range delay $\Delta R/c$. Denote the viewing angle as $\hat{k} = \hat{x}$ along the x-axis. The position of each scattering center relative to each respective CoR is given by

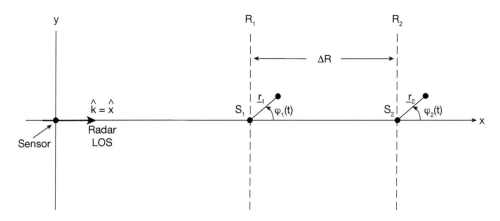

Figure 4.9 Two orbiting point targets, S_1 and S_2, centered on the x-axis, having motions M_1 and M_2 separated by range ΔR.

$$\underline{r}_1(t) = a_1 \cos(\varphi_1(t))\hat{x} + a_1 \sin(\varphi_1(t))\hat{y}, \quad \underline{r}_2(t) = a_2 \cos(\varphi_2(t))\hat{x} + a_2 \sin(\varphi_2(t))\hat{y},$$

where a_1 and a_2 denote the radius for each of the rotating scattering centers.

$$\Delta R = R_2 - R_1.$$

Using the GTD-based scattering formulation defined in section 2.2, the scattered signal for these two targets can be written in the form

$$E(f,t) = D_1 e^{-j\frac{4\pi f}{c}\hat{k}\cdot\underline{r}_1} + D_2 e^{-j\frac{4\pi f}{c}\Delta R}\, e^{-j\frac{4\pi f}{c}\hat{k}\cdot\underline{r}_2}. \tag{4.23}$$

Referring to figure 4.9, it follows that because $\hat{k}\cdot\hat{y} = 0$,

$$\begin{aligned}\hat{k}\cdot\underline{r}_1(t) &= a_1 \cos\phi_1(t) \\ \hat{k}\cdot\underline{r}_2(t) &= a_2 \cos\phi_2(t).\end{aligned} \tag{4.24}$$

Equation (4.23) reduces to the simple form

$$E(f,t) = D_1 e^{-j\frac{4\pi f}{c}a_1\cos\varphi_1} + D_2 e^{-j\frac{4\pi f}{c}\Delta R}\, e^{-j\frac{4\pi f}{c}a_2\cos\varphi_2} \tag{4.25}$$

Assume the two targets are separated by $\Delta R = 4\,\mathrm{m}$. Each scattering center orbits at a different rate in separate target-centered coordinates as depicted in figure 4.9. For this example, the same parameters are used as the single target example in section 4.3.1, except now the parameters $a_2 = 0.3$ m and $T_{0_2} = 0.7$ s are included for the second target. Target 1 is rotating seven times faster than target 2. Scattering from the orbiting point for each target is of unity amplitude, but a random phase is added to the second— that is, $D_2 = e^{j\psi_2}$, where ψ_2 denotes the phase assigned to the second orbiting scattering center. The scene observed by the sensor then consists of two targets, one exhibiting a fast rotation and the second a slow rotation, respectively, and it is of interest to extract the relative motion characteristics of each target.

First consider the behavior of the narrowband autocorrelation metric applied to the two-target scenario (i.e., the waveform bandwidth is inadequate to resolve the 4-m range difference between the two targets). Consistent with the previous example, assume a sampling time of $\Delta t = 1/\text{PRF}$, so that $t = (q-1)\Delta t$, $q = 1, \ldots, Q_0$. Evaluating $E(f_0, t)$ at $t = t_q$ leads to the Q_0 time samples $(x_1, \ldots, x_{Q_0})$ defined by

$$x_q = D_1 e^{-j\frac{4\pi f_0}{c}a_1 \cos\left(\frac{2\pi}{T_{0_1}}(q-1)\Delta t\right)} + D_2\, e^{-j\frac{4\pi f_0}{c}\Delta R}\, e^{-j\frac{4\pi f_0}{c}a_2 \cos\left(\frac{2\pi}{T_{0_2}}(q-1)\Delta t\right)}. \qquad (4.26)$$

Assume the same radar parameters as for the single target example, except in order to accommodate the slower rotational motion, choose a data sequence length of 4,000 pulses corresponding to 2 s of high PRF data. Figure 4.10 illustrates the autocorrelation function of the entire data sequence, where the lag time has been extended over the interval $(0, 1)$ seconds. The rotation periods $T_{0_1} = 0.1$ s and $T_{0_2} = 0.7$ s are now clearly evident. However, the time decay associated with $R_a(\tau)$ for the slower target is overshadowed by the periodic repetition resulting from the fast-rotating target, and one must wait the full rotation cycle of the slower target to ascertain its rotation rate. Thus the main drawback of using the narrowband autocorrelation metric applied to two or more interfering targets is that one must process the data over a time span that encompasses the period of the slower target. The correlation time of the slower target is masked by the correlation properties of the faster one. The autocorrelation function of the faster target effectively masks the early lag time information content contained in $R_a(\tau)$ associated with the slower target.

Examining the spectral behavior for the two-target narrowband case provides some interesting insight. First, consider coherently processing the narrowband data using the coherent integration metric $I_c(\dot{R})$ for integration times $T_i = T_{c_1}, 2T_{c_1}, 3T_{c_1},$ and $4T_{c_1}$ where T_{c_1} is chosen to correspond to the fast target. Figure 4.11 depicts the spectrum for the two-target model of figure 4.9 for each integration time. Only for the second case where $T_i = 2T_{c_1}$ does $I_c(\dot{R})$ unambiguously indicate the presence of two targets. However, it is unclear whether one target is rotating at a rate seven times that of the

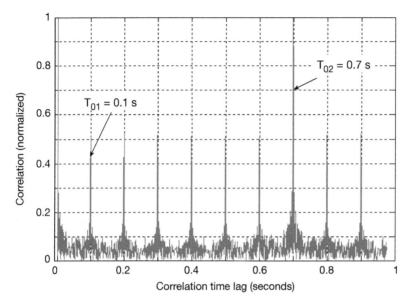

Figure 4.10 Autocorrelation function analogous to figure 4.7 for two targets present in the same narrowband range gate.

other. For integration times $T_i = 3T_{c_1}$ and $4T_{c_1}$ it is not clear whether the spectral spreading is due to multiple targets, or to the overintegration of a single target resulting in a broadening of the spectrum. The difficulty encountered in applying the narrowband metric to try to characterize the rotating motion for each target becomes readily apparent. As the integration time increases, and the range-rate spectrum spreads due to the presence of the two interfering targets, it becomes impossible to determine if the spreading is due to overintegrating the data on a single fast target, or due to the presence of two interfering targets. If one processed the data sequence as it was received and observed the behavior of the spectrum as it evolves in time one could tell what was happening. As the Doppler resolution improves, the two targets would emerge and then the spectrum associated with the higher Doppler target would spread as it as it smears across a number of Doppler channels. Clearly if the targets have different amplitudes, assessing the behavior becomes more difficult. It might be possible to build an adaptive technique to determine how many targets are present. However, the wideband autocorrelation filter concept introduced in the next section provides a better means for overcoming this drawback. By increasing the waveform bandwidth, each target can be isolated to a

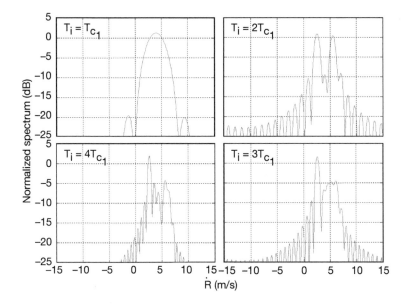

Figure 4.11   Coherent Doppler spectra $I_c(\dot{R})$ analogous to figure 4.8 for two unresolved targets present in the same range gate. The spectrum for the two-target model of figure 4.9 is illustrated for integration times $T_{i_1} = T_{c_1}, 2T_{c_1}, 3T_{c_1}$, and $4T_{c_1}$.

unique, wideband range gate, and the single-target narrowband result then applies individually to the wideband range gate associated with each target.

4.4   The Wideband Range–Autocorrelation Contour Plot

In this section the autocorrelation metric is applied to the wideband case where the bandwidth is sufficient to resolve each target in the compressed pulse. Each pulse in the frequency-time data block is compressed as in the conventional approach discussed in chapter 1 so that individual targets are resolved in range and isolated in separate wideband range gates. Processing across the time sequence in each range gate to estimate $R_a(\tau)$ and plotting it in terms of a color scale as a function of correlation lag as depicted in figure 4.12 defines a range-autocorrelation contour plot. For a block of Q_0 pulses, each range gate consists of a time series of Q_0 complex data samples. Autocorrelation processing can then be applied to the time sequence for each range gate of the Fourier compressed pulse data block. For the two-target example considered in this section, the range gates containing targets will exhibit a gradually decaying autocorrelation measure while other gates that contain only noise will appear uncorrelated as a function of autocorrelation delay. However, for cases where the bandwidth is insufficient to separate two targets and they remain in the same range gate, even though they differ in Doppler, the masking effect encountered in the narrowband example of section 4.3.3 will still occur.

The data organization scheme for the autocorrelation contour concept is illustrated in figure 4.12. Its utility in estimating the relative differences in target motions is demonstrated in the wideband simulation example of the next section.

4.4.1   Wideband Case for Two Targets Separated in Range

In order to compare the information content extracted from a given 2D data block using the range, autocorrelation contour versus that obtained from a range, range-rate image contour, consider the extended data block illustrated in figure 4.13.

By the very nature of estimating the coherence time, the data block size Q_{0_2} used for the coherence estimate must be larger in size than that used for linear imaging. For a given block size of Q_{0_2} pulses, the analyst has the flexibility to process the smaller

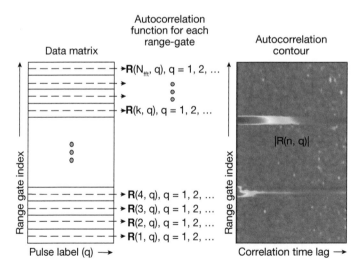

Figure 4.12   Range-autocorrelation contour concept. The autocorrelation function for each range gate is contoured versus the autocorrelation time lag, where N_m is the number of range gates.

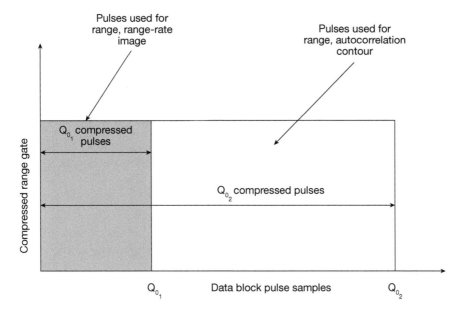

Figure 4.13 Block pulse allocation for the range, range-rate image versus the range, autocorrelation contour.

block of size Q_{0_1} to extract the range, range-rate observable set as developed in chapter 2, and then use the larger data block of size Q_{0_2} to extract the acceleration observable estimate as well as the estimate of the coherence time. Applied to the two-target scenario considered in this section, the smaller data block is used to form a range, range-rate image, while the longer data block is used to form the range-autocorrelation contour. Q_{0_1} is chosen to cover the coherent integration time interval $T_i = T_{c_1}$ of the faster rotating target, and Q_{0_2} is chosen to cover the time interval $2T_{c_2}$ associated with the slower moving target, accounting for the fact that the autocorrelation lag time covers only half the time interval of the data block.

For the wideband simulation the sensor bandwidth is expanded to BW = 1 GHz, adequate to resolve each target into separate range gates. Applying equation (4.21) to the parameters for each target one obtains

$$(T_i)_1 = \frac{T_{0_1}}{2\pi} \sqrt{\frac{\lambda}{2a_1}} = 0.0062 \text{ s} \tag{4.27}$$

$$(T_i)_2 = \frac{T_{0_2}}{2\pi}\sqrt{\frac{\lambda}{2a_2}} = 0.025 \text{ s}. \tag{4.28}$$

Using the same radar sampling parameters as for the narrowband example, $\Delta t = 0.0005$ s for a radar PRF = 2,000, the maximum number of pulses for each target for linear imaging is given by, respectively, $(T_i)_1 = 12.5(\Delta t)$ and $(T_i)_2 = 50(\Delta t)$. The results of table 4.1 indicate that roughly double the number of pulses can be used to form the range-rate image with minimum coherent integration gain loss and spectral distortion. Thus 24 pulses will satisfy the criteria over which both targets appear coherent.

Figure 4.14 illustrates the 24 superimposed wideband Taylor weighted compressed pulse shapes within the image data block, where roughly 18 dB of compressed noise has been added to the simulation for each compressed pulse. Note in the compressed range domain, the targets are separated by roughly 4 m corresponding to the range separation in the two scattering center narrowband example used in section 4.3.2. Each target is isolated in separate range gates, and their spectrum and autocorrelation properties are decoupled relative to the two-target narrowband example.

Figure 4.15 compares the wideband autocorrelation contour plot to that of the conventional wideband range, range-rate image. In figure 4.15(a) the autocorrelation function for each range gate is processed using a data time series of 100 pulses for each range gate. The coherence time for each target is estimated from the autocorrelation filter contour. Comparing the estimates of T_c for each target, it is apparent that target 1 is rotating roughly seven times faster than target 2. Furthermore, comparing figure 4.15(a) to figure 4.10 for the narrowband case, the relative motion of target 2 is estimated using a data sample set covering a much shorter time interval relative to its full rotation period. In the narrowband case, one must wait nearly a full rotation cycle of the slower target for the motion of both targets to separate. For the wideband case, 100 pulses represent only a 0.05 s data time span relative to the 0.7 s slower rotation period.

Comparing figure 4.15(a) to figure 4.15(b), it becomes clear that the correlation time extracted from $R_a(\tau)$ for each range gate provides a measure of the long-duration behavior of the motion, and $I_c(\dot{R})$ extracted from the range, range-rate image characterizes the short duration or instantaneous measure of the scattering centers range-rate at a particular location on the orbit around each target's center-of-rotation.

It should be emphasized that the data set required to assess the coherency of the data using the range, autocorrelation contour must be inherently longer than the data

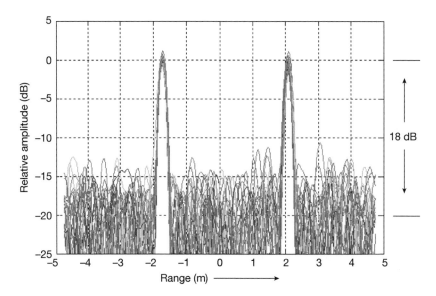

Figure 4.14   Superposition of the 24 compressed wideband pulses defined over the range, range-rate imaging data block for the two-target wideband example.

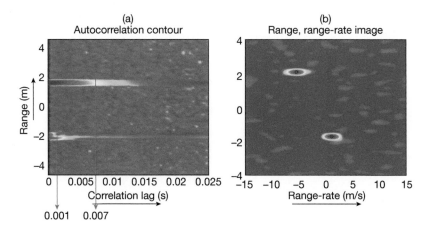

Figure 4.15   Wideband range, autocorrelation contour contrasted to a linear wideband range, range-rate image for the two-target wideband example. (a) Autocorrelation contour using 100 pulses processed for each range gate. (b) Range, range-rate image using 24 pulses processed for each range gate.

set used to form the linear range-Doppler image. To assess coherency using the auto-correlation metric, the data set length must be large enough so the data at the end of the data block is uncorrelated from the data at the beginning of the data block, whereas for optimal linear imaging, all data within the imaging block should be coher-ent. Coupled with this fact is the nature of determining $R_a(\tau)$ using equation (4.11), namely Q_0 pulses of data characterize coherency out to a pulse lag of only $Q_0/2$ pulses. Even with the required increase in block size, the data sample time interval needed is still only a fraction of the rotation period of the motion of the slowest rotating target.

II

The Joint Target-Motion Estimation Problem

5

A Solution Framework for the Joint Target-Motion Estimation Problem

5.1 Introduction

In part I of this book the tools and models that underpin the new framework have been developed and discussed in relation to conventional signal processing techniques. Part II of the text focuses on solving the joint target-motion estimation problem under a variety of conditions and constraints. Solving the joint target-motion estimation problem is a critical step in developing a signature model of the target based on measurement data. Using the geometrical theory of diffraction (GTD) scattering model of the signature defined by equation (2.6), modeling the target signature is accomplished by estimating the number and locations of the scattering centers on the target, as well as the complex amplitude and angular dependence of their respective GTD diffraction coefficients. Field measurements of the target's radar signature contain the combined effects of the target's radar signature as well as the effects produced by the target's motion as described in chapter 1, equation (1.3). As discussed there, the time variation of the radar cross section embedded in the received radar signal is an implicit function of changes in the observation angle versus time. Solutions to the joint target-motion problem result in estimates for the target rotational motion, its scattering center locations, as well as an estimate of each scattering center's diffraction coefficient angular dependence. The scattering center diffraction coefficients are (apart from a phase correction) characterized by the complex pole amplitudes extracted from the measurement data using the sequential processing estimation techniques introduced in chapter 2. The angle dependence of the diffraction coefficients is obtained from the time to angle mapping determined by the joint target-motion solution. A precise methodology for incorporating these into a GTD signature model of the target is developed in chapter 15 and appendix F.

It is important to recognize at the outset that one's ability to achieve a joint solution is strongly dependent on the observability and motion conditions of the target of interest. Many of the techniques discussed in this book were initially developed in order to exploit data taken in a static range radar cross section (RCS) facility with the objective of creating radar signature simulations. Data taken on a target in a static range can be processed in a straightforward manner because the measurement conditions are carefully controlled and precisely known. However, the most valuable data for developing signature simulation models comes from measurements taken on targets during actual field tests and must be handled differently. In addition to data corruption that arises due to poor signal-to-noise conditions, the target's motion is unknown and must be determined from the observations themselves. Thus, the discussion in part II, starting with this chapter, establishes the framework for achieving the book's ultimate goal, a computational physical theory of diffraction (PTD)/GTD radar signature model.

The processing framework introduced in this and the following chapters is subject to specific constraints on the structure of the data. There are two important aspects that need to be carefully addressed in order to analyze the measurement data correctly. The first deals with the relationship between the coordinate system in which the measurements are made and the processing is done, and the coordinate system used to represent the signature model. The second has to do with the reference point on the target used to characterize its motion over the observation time interval. It is useful to summarize the implications of these considerations here as an introduction to the processing framework:

The processing techniques introduced in the following chapters have been developed assuming a target-fixed coordinate system having the target's center of rotation (CoR) located at the origin. They fit nicely into the target-sensor scenario consistent with static range (SR) measurements.

The motion associated with field measurements does not naturally fit into this rotational framework, and data compensation is required before applying the techniques introduced in this chapter. This requires the motion of the target to be decomposed into its translational and rotational components, and the data used for processing registered to the rotational motion. Techniques for removing the translational component of the data are developed in appendix F.

The orientation of the target, as well as the location of the CoR on the target in the measurement system are generally different from those used for the computational model. Field measurements are typically processed by introducing a measurement coordinate system selected for the convenience of the processing.

Proper association of the measurement coordinate system to the computational model coordinate system is required for integration of the measurements with the computational model.

In order to fuse data obtained from a variety of venues and taken under different measurement conditions into the signature model, it is necessary to represent the target model in a local, target-fixed coordinate system. This allows data from various measurement venues (in particular, field measurements and radar static range measurements) to be fused into a single comprehensive signature model. Static range data are handled in a straightforward manner because the measurement conditions are accurately known, and the motion conditions are carefully controlled. Field measurements are a different matter. The observations are taken in a radar-centered inertial frame of reference, may suffer from poor signal-to-noise conditions, and are the result of a mixture of rotational and translational motion. For a target undergoing exoatmospheric torque-free motion, Newton's laws allow the target motion to be decomposed into its translational and rotational components. However, the radar observer, using its own measurements, must estimate what it believes to be the translational and rotational components of the target motion so it can use the rotational motion to "unwind" the motion of the observed scattering centers back to the target-fixed frame of reference. The registration of the data relative to the target's CoR and determining the orientation of the rotational inertial reference frame can be difficult tasks. In cases where the translational motion is not removed adequately, the errors can cause the "unwrapping" process to degrade the location estimates. An extensive and detailed discussion of the coordinate systems and their relationship are provided in appendixes A and F.

Section 5.2 describes the framework using the rotational component of the motion to estimate the scattering center locations in the target-fixed frame. There are certain types of motion that are easier to handle. Under certain motion conditions, constructing a three-dimensional (3D) representation of the target is straightforward. Characterizing these motion condition situations is discussed in more detail in section 5.4.

The second aspect that needs to be understood in order to analyze the measurement data properly deals with the motion conditions that govern the target during the measurement interval. There are three time-dependent observation conditions that apply to how the data can be processed and the amount of information that can be extracted.

Condition A: The observation interval is short enough such that the scattering centers, as viewed from the radar, do not migrate between the radar's range and cross-range resolution cells. This condition applies to the development in part

I in the text and represents the situation for conventional radar imaging techniques. Under this condition the linear signature model introduced in chapter 2 is a valid representation of the data block.

Condition B: The time duration of the data interval allows for significant scattering center migration across resolution cells, but the viewing angle κ between the radar and the target's angular momentum vector, $\underline{J}$, is constant. This condition is illustrated in figure F.5 of appendix F. In this case, the orientation of $\underline{J}$ in inertial space does not have to be known. The methodology here is to solve for the rotational motion parameters introduced in appendix A, and use them to unwrap the scattering center time histories back to a target-fixed coordinate system. This case is the primary focus of this book, and is treated in depth starting with chapter 5.

Condition C: When the radar observes a target over an extended segment of the trajectory, individual observation intervals can have different values of κ as illustrated in figure F.5. This is the most difficult situation to handle and receives only modest attention in the book. In such a case, the observation interval can be divided into smaller time intervals for which κ is constant or nearly so. If the observation intervals are contemporaneous in time, but the data come from different sensors and the sensor locations are known, it is still possible to transform the data sets to a common reference frame and fuse the data. Clearly, if the time history of $\underline{J}$ can be determined, the data sets can be fused. The text does not treat this problem in general. However, fusing estimates of the target scattering center locations obtained from data intervals having different values of κ is made possible using the techniques developed in the text because the location of the target's CoR is fixed, and independent of the time of the observation interval. Thus, the estimates of the target scattering center locations referenced to target-fixed space, apart from scattering center occlusion, are correlated for each individual observation interval, and one must only account for possible differences in the target reference space orientation and possible dimensional scaling due to differences in κ. The noncoherent combination of the scattering center estimates from the individual observation intervals, after rotational adjustment, not only provides more robust location estimates, but also characterizes the diffraction coefficients over a broader rotational viewing angle. This case is discussed in chapters 11 and 14.

In this chapter, the framework is developed for jointly estimating the target's rotational motion and transforming the scattering center's movement as viewed by the radar into the target-fixed coordinate frame of reference. The framework is similar to that introduced in [71, 72], but extended to a broader set of observables. Using the framework, a unique target-motion coupling matrix is developed that is directly related to the range, range-rate and range-acceleration observable sequences extracted from the received radar data. Its unique structure allows the target rotational motion to be separated from the target scattering center locations.

Solutions for the joint target-motion estimation problem are built up in stages in order to develop insight into this complex problem. Three important solution constraints are considered in detail. The remainder of chapter 5, and chapters 6 and 7, consider the constraint where the target motion is known (e.g., as determined by independent analysis or by known constraints on the measurement conditions). Examples illustrating the applications of the framework to estimating the target component locations for specific types of known motion are presented. Chapter 8 examines the case where the target is known, and the motion is to be determined. Such a case might occur in a field experiment where a known target is observed, and the dynamics of the target are to be determined. Finally, in chapters 9 and 10, two solution techniques are developed that are applicable to the case where neither the target nor the motion are known, and a joint solution is required.

It is useful to review the dependence of measurement resolution capability on these three measurement conditions relative to coherent and noncoherent multipulse processing over extended angular regions. The resolution limits for coherent processing can be expressed [73] directly as a function of the radio frequency (RF) wavelength, λ, and RF fractional bandwidth, FBW, where $FBW = BW/f_0$ and $f_0 = c/\lambda$. The range resolution $\Delta\rho_R$ is determined by the RF bandwidth BW:

$$\Delta\rho_R \sim \alpha c/BW = \alpha\lambda/FBW. \tag{5.1a}$$

Assuming arbitrary motion in 3D space, the two cross-range resolution limits are independent of bandwidth, and have resolution limits constrained by

$$\Delta\rho_{\kappa_1} \sim \alpha\lambda/2\Delta\kappa_1, \tag{5.1b}$$

$$\Delta\rho_{\kappa_2} \sim \alpha\lambda/2\Delta\kappa_2, \tag{5.1c}$$

where α is a parameter that depends on the pulse compression weighting and cross-range weighting used for sidelobe reduction in the Fourier transform signal processing, and $\Delta\kappa_1$ and $\Delta\kappa_2$ are two-orthogonal angular intervals covered over the data collection time. Equations (5.1a)–(5.1c) apply to coherent integration processing over a single frequency-time data block satisfying condition A. Consequently, it is assumed the data are collected at a high enough pulse repetition frequency (PRF) consistent with Nyquist angle sampling over the angle sectors defined by $\Delta\kappa_1$ and $\Delta\kappa_2$. Higher resolution estimates of the location of a target's scattering centers can be obtained by decreasing the wavelength, increasing the fractional bandwidth, or incorporating greater angular diversity while observing the target. The cross-range resolution is enhanced as λ decreases, favoring the higher frequency operating regimes. Although the high resolution spectral estimation (HRSE) technique introduced in chapter 2 outperforms the Fourier-based resolution limits, the achievable resolution is still fundamentally dictated by sensor bandwidth, and the extent of the region covered by the viewing angle. Note that equation (5.1a) is independent of equations (5.1b) and (5.1c), that is, resolution in cross range is independent of resolution in range. This fact forms the basis for the narrow-band target characterization framework developed in chapter 6.

Techniques that noncoherently combine multiple scattering center location estimates (obtained using multiple data blocks that span a wide range of viewing aspect angles to the target satisfying condition B) can improve the precision of the location estimates determined under condition A, and are the motivation for developing the composite scattering center extraction techniques introduced later in this chapter. As the extent of the region covered by the viewing angle (θ, φ) increases, a number of estimates of the scattering center locations can be accumulated. For a given observation interval, a number of independent estimates of the range, range-rate observables could be made using nonoverlapping data blocks, where the size of each block satisfies condition A. These independent estimates of the observables could then be processed using the methodology described in section 5.2 to generate independent estimates for the scattering center locations in the target-fixed framework. However, due to the realities of real-world conditions, the fact that scattering centers can disappear because of shadowing or can appear as they emerge from shadowed conditions, and because the set of independent observable estimates may not be correlated, independent data block increments are not used. In practice the observable sequence is obtained by incrementing the size of each data block by a fraction of the maximal length specified by condition A, resulting in better association of the extracted observables over the specified observation interval.

For example, as shown later in section 5.2 [see discussion following equation (5.6)], it is possible to process a set of N_{p_0} sequential range, range-rate observables (covering a small interval within the total set of observables) to develop a single estimate of the N scattering center locations in target-fixed space. For the methodology described there, the size of each data block interval used to estimate a single set of range, range-rate observables is of maximum length that still satisfies condition A. However, the next data block time interval in the sequence does not jump to the next independent set of measurements. Instead, it is incremented further in time by some fraction of the independent data set, thereby taking in a subset of new data while retaining a portion of the previous data set used for the current estimate. In this way the sequence of estimated range, range-rate observables is correlated and can be exploited to associate the range, range-rate estimates of each of the individual scattering centers in the time sequence of observables.[1]

For each set of N_{P_0} sequential observables, an estimate of the N scattering center locations can be obtained. Repeating this process for P times by incrementing the data as described previously results in P estimates for each respective scattering center location. Ideally, in the absence of noise and assuming the scattering center locations are fixed over the processing interval, all P estimates for the location of each scattering center would be identical. In practice, the P estimates have a variance determined by the signal-to-noise ratio and extent of the viewing angle space covered. The composite average of the P noncoherently processed location estimates provide a higher precision estimate relative to each single scattering center location estimate. Note that if the bandwidth is wide enough, some scattering center information can be extracted even if the target were motionless; however, having different viewing angles to the target due to the aspect angle change permits registration of the scattering centers into the target-fixed coordinate system.

A notional scheme for developing a signature model of the target based on the measured signature, emphasizing the dependence on solutions to the joint target-motion estimation problem, is illustrated in figure 5.1. The methodology incorporates the sequential estimation processing scheme developed in chapter 2. The recorded data are organized into data blocks as described in figure 2.6, and B data blocks are processed sequentially to extract the range, range-rate and complex amplitude observables associated with each block, resulting in the observable sequence $(R_n, \dot{R}_n, D_n)_b, b = 1, \ldots, B$ for each of the N scattering centers. This sequence is then used in a joint target-motion space solution framework to obtain the motion solution $\theta(t_b)$, $\varphi(t_b)$, $b = 1, \ldots, B$. [Note that consistent with the discussion in section 2.2, for the development in this chapter, the angles in the coordinate system representing the measurement system are denoted as (θ, φ), and the prime notation is used only when it is necessary to differentiate the measurement system from the computational system.] Given the motion solution, the

---

1. Note that correlating and processing a set of incrementing data blocks satisfying condition A is equivalent to estimating each scattering center acceleration observable $\ddot{R}_n$ as discussed in chapter 3 and again in section 5.6.

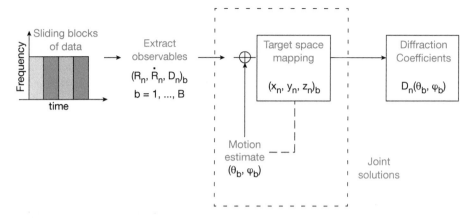

Figure 5.1   Target modeling methodology.

observable sequence is mapped into a fixed target space coordinate system using the framework developed in section 5.2, yielding the scattering center location estimates. These scattering center locations are defined relative to the measurement coordinate system having origin at the CoR of the target. The complex amplitudes for each of the N extracted poles, $D_n (\theta_b, \varphi_b)$, $b = 1, \ldots, B$, viewed as a function of angle, characterize the magnitude of the diffraction coefficient $D_n(\theta_b, \varphi_b)$ averaged over the frequency bandwidth (assuming $FBW \le 10\%$). These coefficients contain information that also help to identify the scattering center types (described later in chapter 6). The scattering center locations on the target are time invariant (except for their disappearance caused by shadowing), except for a special class of time varying scattering center locations (known as slipping scattering centers caused by edges and spherical shapes), and collectively provide a measure of the target's topology [72, 74]. The estimate of the target topology is essential for incorporating the estimated target model parameters into a PTD-based computational electromagnetic scattering code to characterize those non-fixed-point scattering centers, as discussed in chapter 15.

The target radar signature model also allows for extrapolating the target's signature outside the measurement observation region. Estimating the target model parameters from the data and then using a computational signature model to extrapolate observations outside the observation region provides a robust complement to the data extrapolation techniques developed in chapter 12.

Measurement data used to develop signature models typically come either from an RCS static range or from observation of an actual flight test in the field. Static range measurements are taken on a standard radar cross section range having controlled known motion and measurement conditions. The data can be processed in a straightforward manner because the measurement system is highly calibrated, and the sensor is referenced to a constant orientation in the plane of rotation. Figure 5.2 illustrates two types of angle sampling trajectories for two methods commonly used in static range data collections. (Figure 15.9 illustrates two static range data collection schemes applicable to 3D target characterization.) For a *sting-like* platform on the static range, the aspect angle is fixed at an angle to the target reference z-axis, and the target is rotated about the azimuthal axis defining the roll angle. For a *roll-cut measurement*, implemented by fixing the target atop the pylon and rotating the pylon 360°, the roll angle defined relative to the target-centered coordinate system is fixed and the aspect angle is characterized by the pylon rotation angle defined relative to the z-axis of the target. Because static range measurements are much less expensive than the cost of a field test measurement, many roll-cut measurements can be taken for a fraction of the cost of a field test. However, for each roll cut, the target must be physically displaced in discrete roll angles, which slightly changes the CoR and adds a rotationally dependent phase uncertainty to the data when attempting to coherently combine one roll-cut data set to another. Furthermore, complete roll characterization of the signature requires Nyquist angle samples in roll, which typically requires closely spaced angle samples, resulting in a prohibitive number of roll cuts to cover the entire observation space. Consequently, coherent processing of multiple pulses of static range data is typically limited to a single aspect or roll-angle cut.

The static range measurement types illustrated in figure 5.2 are compared to a typical angle sampling trajectory obtained from field measurements. During a field measurement both θ and φ can vary over a significant angular spread but are not completely under the observer's control. In general, field measurements allow one to obtain a coherent set of Nyquist sampled observation angle data over a more robust set of trajectories in observation space. The field experiment offers sophisticated motion changes on the target both in aspect and roll, as illustrated in figure 5.2 for a target undergoing spin and precession motion, having spin frequency much greater than precession frequency. For torque-induced motion, such as would occur for aircraft undergoing aerodynamic and propulsion force, the angle space trajectories would exhibit a more general behavior.

There are other, primarily practical reasons for desiring measurements under flight test conditions. For example, sting-type static range measurements on large targets, because of size and weight considerations, become impractical and prohibit a rear view of the target. As an alternative, one typically resorts to approximating the roll angle data in

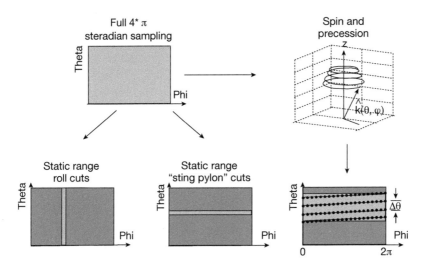

Figure 5.2 Motion observation space: Trajectories for Nyquist angle sampling for static range data versus field data.

increments, collecting data from a rotating pylon using multiple aspect cuts, fixing the target at discrete roll angles.

5.2   The Joint Target-Motion Coupling Framework

The joint target-motion estimation framework presented in this chapter is similar in structure as that introduced in [71, 72, 75]. To develop the framework, consider the range, range-rate observables $(R_n, \dot{R}_n)_b$, $n = 1, \ldots, N$ extracted from a single wideband data block centered at $t = t_b$. Equation (2.17) defines the R_n, $\dot{R}_n$ observable mapping to target space that is used to estimate the location of the nth scattering center located at $\underline{r}_n = (x_n, y_n, z_n)$ in the target-fixed coordinate system. For the data block centered at $t = t_b$, the mapping is given by

$$\hat{\underline{k}}_b \cdot \underline{r}_n = (R_n)_b \tag{5.2}$$

$$\dot{\hat{\underline{k}}}_b \cdot \underline{r}_n = (\dot{R}_n)_b \tag{5.3}$$

where $\hat{\underline{k}}_b$ and $\dot{\hat{\underline{k}}}_b$ characterize the motion of the target at $t = t_b$ according to equations (2.4) and (2.5).

Equations (5.2) and (5.3) can be structured more generally to emphasize the coupling between the target scattering center locations and the motion. Assume that B blocks of sensor data are processed sequentially and the range, range-rate observables $(R_n, \dot{R}_n)_b$, $n = 1, \ldots, N$, $b = 1, \ldots, B$ are extracted from these B data blocks. Next assume that as the data blocks are processed sequentially, the range, range-rate observables corresponding to a given scattering center can be associated in time; that is, $(R_n, \dot{R}_n)_{b_1}$ and $(R_n, \dot{R}_n)_{b_2}$ at times t_{b_1} and t_{b_2} correspond to the same scattering center. Express the transformation defined by equations (5.2) and (5.3) for a single data block centered at $t = t_{b_1}$ in matrix form:

$$
M_{b_1} \cdot \begin{bmatrix} x_n \\ y_n \\ z_n \end{bmatrix} = \begin{bmatrix} R_n \\ \dot{R}_n \end{bmatrix}_{b_1}
\tag{5.4}
$$

where M_{b_1} is a 2×3 transformation matrix given by

$$
M_{b_1} = \begin{bmatrix} \hat{k}_{b_1} \\ \dot{\hat{k}}_{b_1} \end{bmatrix}
\tag{5.5}
$$

and $\hat{k}_b$ and $\dot{\hat{k}}_b$ are expressed as row vectors. Equation (5.4) defines the mapping at $M = M_{b_1}$. Accumulate the observables at t_{b_1} and t_{b_2} in the form

$$
\begin{bmatrix} M_{b_1} \\ M_{b_2} \end{bmatrix} \cdot \begin{pmatrix} x_n \\ y_n \\ z_n \end{pmatrix} = \begin{bmatrix} \begin{pmatrix} R_n \\ \dot{R}_n \end{pmatrix}_{b_1} \\ \begin{pmatrix} R_n \\ \dot{R}_n \end{pmatrix}_{b_2} \end{bmatrix}
\tag{5.6}
$$

As additional observables are obtained, they can be appended to equation (5.6). Denote the number of accumulations as N_{p_0}. After N_{p_0} accumulations, the appended version of equation (5.6) is inverted to provide an estimate of the N scattering center locations, (x_n, y_n, z_n), $n = 1, \ldots, N$. As time progresses, additional estimates of the scattering center locations are obtained by applying equation (5.6) to a new set of N_{p_0} observables.

The appended version of equation (5.6) can be expressed more generally in the form

$$M \cdot \begin{pmatrix} x_n \\ y_n \\ z_n \end{pmatrix} = O \qquad (5.7)$$

where O denotes the observable matrix and M denotes the motion matrix, both accumulated over $B = N_{p_0}$ blocks of radar data.

Equation (5.7) illustrates the general decoupling of the motion and target:

$$[Motion] \cdot [Target] = [Observables] \qquad (5.8)$$

Viewed in the target coordinate system, because the [Target] matrix is constant for the fixed-point scattering centers, the time evolution of the observable sequence is driven directly by the motion mapping of the (fixed) target scattering locations into observable space. The general case illustrating the matrix decoupling is illustrated pictorially in figure 5.3, assuming only the range and range-rate observables are extracted from each data block. The motion matrix consists of B rows of block motion matrices M_b. The target matrix assumes N scattering centering centers, and the observable matrix now has dimension $2 * B \times N$.

For the special case where the motion of the target is characterized by torque-free Euler motion, the motion matrix can be described by the Euler variables introduced in appendix A. For targets that are nearly rotationally symmetric about the target's z-axis, that is, having nearly symmetric moments of inertia $I_x \approx I_y$, as discussed in section 2.5, the rotational motion is characterized by the six parameters θ_p, f_p, α_p, f_s, α_s, and κ, the radar look angle to the target's angular momentum vector. In this case joint solutions to equation (5.7) for both motion and target can be reduced to a $6 + 3(N) + 1$ variable search space, where the additional variable (+1) corresponds to a range bias estimate that can result from imperfectly removing the translational motion from the received signature data, as discussed in appendix F.

The nature and coupling of the general motion–target estimation problem defined by equation (5.8) can be generalized as the observable matrix is extended to include more observables, such as the differential phase associated with the complex amplitudes $(D_n)_b$ and $(D_n)_{b + b_0}$ extracted at times $t = t_b$ and at a later time $t = t_{b+b_0}$ (see chapter 7 on Interferometric inverse synthetic aperture radar (ISAR) processing) or the acceleration $\ddot{R}_n$ of each scattering center (see section 5.6).

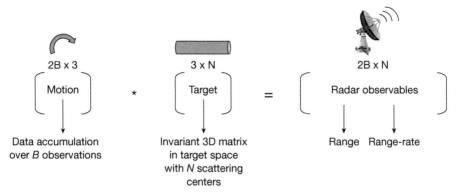

Figure 5.3   The joint target-motion estimation coupling composed of B paired range, range-rate observables and N scattering centers.

Special case solution techniques for estimating the motion and scattering center locations are developed in subsequent chapters: In chapters 5, 6, and 7 for the special case where the motion is known, for which equation (5.7) can be inverted to solve for the scattering center locations; in chapter 8 for the special case assuming the target is known and equation (5.7) is used to estimate the target motion; and chapters 9 and 10 develop more general case solutions for which both motion and target are unknown.

5.3   Estimating the 3D Location of Scattering Centers: Motion Known

Various solution techniques for the joint target-motion estimation problem for known motion have been developed in the literature, and results presented in this chapter are based on the work in [75]. The remainder of this chapter exploits the joint target-motion estimation framework introduced in section 5.2 to develop two-dimensional (2D) and 3D solution techniques for estimating the scattering center locations imposing this constraint. Traditional techniques for 3D scattering center location estimation for known motion are based on conventional Fourier-based processing and use the concept of correlation imaging developed in [73]. The correlation imaging technique estimates the scattering center locations by extracting the peak values from a 3D correlation image function $I(x, y, z)$, defined relative to a fixed 3D target space coordinate system. The correlation image function $I(x, y, z)$ introduced in [73] can be expressed using the notation used in this text in the form

$$I(x, y, z) = \iiint \sigma_c(f, \theta, \varphi) e^{j\frac{4\pi f}{c}(x\sin\theta\cos\varphi + y\sin\theta\sin\varphi + z\cos\theta)} df d\theta d\varphi \qquad (5.9)$$

Applied to the measurement data, σ_c in equation (5.9) represents the measured signature, and the limits of integration are taken over the measurement observation space corresponding to condition B as discussed previously.

The image correlation function $I(x, y, z)$ corresponds to coherent processing of the complex signature data. Reference [76] introduces the concept of extended coherent processing of the complex signature data for the special case of targets exhibiting uniform rotational motion about the angular momentum vector $\underline{J}$. For this special case, the integration in equation (5.9) can be decomposed into piecewise segments consisting of blocks of range-Doppler images, properly scaled in cross range and rotated using the known rotational motion. These images can be coherently summed to provide a 2D estimate $I(x, y)$ of the scattering center locations. Coherent processing of the complex radar signature data in this manner is advantageous for fixed-point scattering centers having constant amplitude as a function of rotation angle. However, it tends to suppress the contribution of the more practical case of scattering centers where the amplitude and phase of the scattering center changes with rotation angle.

As an alternative to extracting the scattering location estimates from the correlation imaging function, the target-motion matrix coupling developed in section 5.2 based on equation (5.7) is applied to the observable sequence. As the extent of the rotational viewing angle space increases, still satisfying condition B over the time duration of the processing, a number of estimates of the scattering center locations can be accumulated. By processing P subintervals of angle data (each subinterval corresponding to the accumulation of N_{p_0} sequential observables), P estimates of the scattering center locations can be obtained. Applying this sequential processing technique to estimate the scattering center locations offers numerous advantages over the conventional Fourier-based 3D correlation imaging technique:

1. For each inversion of equation (5.7), it provides for a direct 3D HRSE estimate of the location of the dominant scattering centers on the target visible to the radar relative to the radar line-of-sight to the target. This is contrasted to correlation image processing for which the scattering centers must be extracted from a 3D correlation image averaged over the entire processing observation space.

2. The sequential nature of the processing provides an estimate of each scattering center amplitude and phase as a function of look angle to the target.

3. It produces 3D location estimates using a minimum number of observables. Because it requires only a limited set of observables, it is particularly applicable to 3D fusion of sectors of sparse angle data, such as might be obtained using periodically spaced bursts of high PRF radar pulses.

4. It gracefully reduces to a range-only (e.g., wideband radar, low PRF) and Doppler-only (e.g., narrowband radar, high PRF) 3D scattering center location estimate.

5. By implementing range, range-rate association applied to a sequence of observables, one can isolate the 3D motion of any specific scattering center and extract the scattering center component amplitude and phase associated with the diffraction coefficient $D_n(\theta_b, \varphi_b)$ as a function of observation angle. Correlating the measured GTD diffraction coefficients to hypothesized characteristics of each scattering center component provides the potential for parametric modeling of each component diffraction coefficient. These concepts are exploited further in chapters 14, 15, and 16.

The rank of the motion matrix M and hence the solutions for (x_n, y_n, z_n) depend on numerous variables: the sparsity of the motion of the target, the number of data blocks N_{p_0}, the time spacing between data blocks, and the signal-to-noise ratio present in the data. A detailed analysis of these variables relative to their effect on the resultant image is not considered in this text. Rather, the results to follow serve more as a baseline study of the general technique.

Examination of equations (2.4) and (2.5) defining $\hat{k}$ and $\dot{\hat{k}}$ indicate that, for arbitrary motion, the motion matrix M_{b_1} defined in equation (5.4) has a complex time variation. However, the form of M_{b_1} for some special cases for specific types of motion commonly observed in static range and field data collections simplifies considerably, and are considered in the following section.

5.4   The Target-Motion Mapping for Some Special Cases

It is useful to consider the motion to target space mapping for some special cases commonly observed in static range and field data collections:

5.4.1   Constant Aspect Angle ($\theta = \theta_0, \dot{\theta} = 0$)

This case is most indicative of static range data collected using a static range *sting* pylon measurement (see figure 15.9) where the target is mounted on the pylon such that the target's z-axis is oriented at fixed aspect angle $\theta = \theta_0$ to the incident wavefront. Data are collected as a function of rotation angle as the target is rotated $360°$ in roll about its z-axis. Because the data are recorded as a function of angle, and the framework in this text assumes a time sequence of single pulse data, it is convenient to map angle to time assuming a rotation rate $\dot{\varphi} = 1$, $(T = 2\pi)$, so that the target rotates $360°$ over a complete rotation period. The roll angle is characterized in increments $\Delta\varphi = \dot{\varphi}\Delta t$. The target is effectively in spin motion and $\dot{\hat{k}}$, as defined in equation (2.5), takes a particularly simple form

$$\dot{\hat{k}} = \dot{\varphi}\sin\theta_0\hat{\varphi} = \dot{\varphi}\sin\theta_0(-\sin\varphi\hat{x} + \cos\varphi\hat{y}). \tag{5.10}$$

The transformation defined by equation (5.4) can then be written in the form

$$\begin{bmatrix} \cos\varphi & \sin\varphi & 1 \\ -\dot{\varphi}\sin\varphi & \dot{\varphi}\cos\varphi & 0 \end{bmatrix} \cdot \begin{pmatrix} x'_n \\ y'_n \\ z'_n \end{pmatrix} = \begin{bmatrix} R_n \\ \dot{R}_n \end{bmatrix} \tag{5.11}$$

where the apparent coordinates x'_n, y'_n, z'_n are defined as $x'_n = \sin\theta_0 x_n$, $y'_n = \sin\theta_0 y_n$, $z'_n = \cos\theta_0 z_n$. Equation (5.11) takes the general form indicated by equation (5.4), and as a result additional observables must be accumulated to obtain scattering center location estimates.

5.4.2   Constant Roll Angle ($\varphi = \varphi_0$, $\dot{\varphi} = 0$)

This case is indicative of static range measurement data where the target is placed on the pylon oriented at a fixed roll angle relative to the target axis and the pylon is rotated over $360°$ in aspect. The target is effectively in tumble motion as viewed from a Euler motion perspective and the radar line of sight (LoS) to the target lies in the tumble plane. Similar to the previous discussion in section 5.4.1, the angle data are scaled to time by increments $\Delta\theta = \dot{\theta}\Delta t$, where $\dot{\theta}$ is typically chosen as $\dot{\theta} = 1$, $(T = 2\pi)$, so that the target rotates $360°$ over a complete rotation period. The look angle $\hat{k}$ is determined using $\varphi = \varphi_0$ and θ becomes the rotational angle. The time derivative of $\hat{k}$ for this special case is given by

$$\dot{\hat{k}} = \dot{\theta}\hat{\theta} = \dot{\theta}[\cos\theta\cos\varphi_0\hat{x} + \cos\theta\sin\varphi_0\hat{y} - \sin\theta\hat{z}]. \tag{5.12}$$

The transformation mapping can be expressed in the simple 2D form

$$\begin{pmatrix} \sin\theta & \cos\theta \\ \cos\theta & -\sin\theta \end{pmatrix} \begin{pmatrix} x'_n + y'_n \\ z_n \end{pmatrix} = \begin{pmatrix} R_n \\ \dot{R}_n / \dot{\theta} \end{pmatrix}, \tag{5.13}$$

where a new set of apparent coordinates are defined as

$$x'_n = \cos\varphi_0 x_n, \; y'_n = \sin\varphi_0 y_n. \tag{5.14}$$

Define $\rho_n = x'_n + y'_n$. Note that a 2D solution for (ρ_n, z_n) is now obtained from a single range, range-rate observable set, and association from block to block is not required to obtain a 2D scattering center location estimate. One can show that ρ_n corresponds to the radial projection of the 3D scattering center location onto the plane of rotation.

It should be noted that for this special data collection scheme, the angular extent of the spherical coordinate θ is extended to cover the region $0 \le \theta \le 2\pi$. The more general case of tumble motion, and the associated use of spherical coordinates to model the motion for this case, is treated in chapter 9.

5.4.3   Range-Only Data (e.g., Wideband Radar, Low PRF)

This special case is applicable to a low PRF sensor where the time (angle) samples are not spaced closely enough (below Nyquist sampling) to unambiguously extract the range-rate of each scattering center. However, the bandwidth is sufficient to isolate the scattering centers in range using sequential single pulse processing. The *range-only* data can still be used to extract the scattering center locations. In this case a minimum of three wideband range observables must be processed to obtain a single estimate of (x_n, y_n, z_n). The single wideband pulse 3D mapping equation takes the form

$$[\sin\theta\cos\varphi, \sin\theta\sin\varphi, \cos\theta] \cdot \begin{pmatrix} x_n \\ y_n \\ z_n \end{pmatrix} \equiv \underline{T} \cdot \begin{pmatrix} x_n \\ y_n \\ z_n \end{pmatrix} = [R_n] \tag{5.15}$$

Combining three wideband pulses collected at times $t = t_{q_1}$, $t = t_{q_2}$, and $t = t_{q_3}$

$$
\begin{pmatrix} \underline{T}_{q_1} \\ \underline{T}_{q_2} \\ \underline{T}_{q_3} \end{pmatrix} \cdot \begin{pmatrix} x_n \\ y_n \\ z_n \end{pmatrix} = \begin{pmatrix} (R_n)_{q_1} \\ (R_n)_{q_2} \\ (R_n)_{q_3} \end{pmatrix} \tag{5.16}
$$

A sequential range association tracking technique is required to properly associate the range time samples for this special case. The reader is referred to chapter 10 for further discussion on range-only processing.

5.4.4   Range-Rate-Only (Doppler) Data

This case is the dual of equation (5.15) and corresponds to a narrowband sensor operating at high PRF where only the range-rate can be unambiguously determined. The transformation mapping to target space takes the form

$$
[\dot{\theta}\hat{\theta} + \dot{\varphi}\sin\theta\hat{\varphi}] \cdot \begin{pmatrix} x_n \\ y_n \\ z_n \end{pmatrix} \equiv \underline{T} \cdot \begin{pmatrix} x_n \\ y_n \\ z_n \end{pmatrix} = [\dot{R}_n]. \tag{5.17}
$$

Combining the range-rate observables extracted using three data blocks centered at block times $t = t_{b_1}$, $t = t_{b_2}$, and $t = t_{b_3}$ yields the dual of the mapping defined in equation (5.16). The reader is referred to chapter 6 for further discussion on narrowband processing techniques.

5.4.5   A Known Surface Constraint

This is an important special case where the target topology is known, and the objective is to determine and characterize the dominant scattering centers on the target. This is the case for example when the target has scattering components that are difficult to model using existing electromagnetic codes. In this case the shape of the target is known, and one desires to isolate the components and extract the scattering centers diffraction coefficients for use in developing the target signature model. The measurement might take place on a radar cross-section range (known motion) or during a field test (unknown motion). For this purpose, it is useful to complement equations (5.2) and (5.3) with a third equation that expresses the known surface constraint

$$
S(x_n, y_n, z_n) = 0, \tag{5.18}
$$

where $S(x, y, z)$ denotes the target surface topology. For a given data block, imposing the constraint $S(x_n, y_n, z_n) = 0$ provides an additional equation yielding 3D solutions for extracting the (x_n, y_n, z_n) locations from the measurement data. The associated diffraction coefficients $D_n(\theta, \varphi)$, $n = 1, \ldots, N$ can be obtained by applying the motion solution mapping to angle space. A detailed analysis of this technique applied to several canonical target surface constraints is presented in [77].

The correct association of the accumulated range, range-rate observables as a function of increasing block time is essential for 3D target characterization and emphasizes the importance of an $(R_n, \dot{R}_n)_b$ association technique, particularly in low signal-to-noise ratio environments. A simple nearest-neighbor technique that associates the scattering centers in range, range-rate observable space is developed in the following section.

5.5   Correlation and Cumulative Processing

Essential to the application of equation (5.7) is that, as time progresses, the trajectories of each scattering center described by the set of observables $(R_n, \dot{R}_n)_b, n = 1, \ldots, N, b = 1, \ldots, B$ are correctly associated. Thus the sets $(R_n, \dot{R}_n)_{b_1}$ and $(R_n, \dot{R}_n)_{b_2}$, for two times $t = t_{b_1}$ and $t = t_{b_2}$ must be associated before applying equation (5.7) to the N_{p_0} accumulated observables. The association is required only over the time span of each set of the N_{p_0} observables. The persistence of the scattering center location estimates obtained by applying P sequential inversions of equation (5.7) effectively acts as the longtime scattering center location correlation agent.

The nearest-neighbor association technique used in this text exploits the time evolution of the $(R_n, \dot{R}_n)_b$ trajectories viewed parametrically versus time in range, range-rate space as discussed in section 2.6 and illustrated in figure 2.12. Consider, for example, the application of equation (5.7) to the example of the spinning cylinder introduced in chapter 2. Figure 2.11, for convenience replicated as figure 5.4 in this chapter, illustrates the range and range-rate sequence extracted from the data over a complete spin cycle using the simulation parameters discussed in section 2.5.

Examination of figure 5.4 present an unambiguous visual of the time evolution of each specific scattering center observable with a unique trajectory. As discussed in the simulation example introduced in section 2.6, although the time evolution of the observables associated with a specific scattering center can be visually associated over the longer time span, the output of the HRSE spectral estimator does not consistently associate (as denoted by the color of the plotted values) the $(R_n, \dot{R}_n)_b$ observables from data block to data block. The values associated with each observable pair are

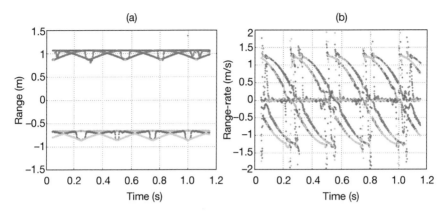

Figure 5.4   Range, range-rate observable sequence for the spinning cylinder example in chapter 2 : $N = 6$, $Q_0 = 16$ pulses in block, $n_f = 48$ frequency samples, 2 GHz bandwidth at X-band. (a) Range sequence versus time. (b) Range-rate sequence versus time.

correct; only the labeling (color) of the plotted value might change. Thus, the trajectories for the range and range-rate extractions for a given scattering center often change color as a function of time due to the inconsistency of the labeling associated with the outputs of the HRSE estimator. Hence, a technique is required to associate the range, range-rate observables that belong to individual scattering centers as time progresses.

There are other reasons it may be difficult for a correlation technique to correctly associate the scattering center observables over a long time span. In particular, scattering centers may lack persistence over time due to shadowing, low scattering amplitudes versus viewing aspect, as well as a number of other factors. Figure 2.12, replicated here as figure 5.5, illustrates the time evolution of the $(R_n, \dot{R}_n)_b$ trajectories viewed in range, range-rate space defined over a much shorter time span of 50 sequential observables.

Observe from figure 5.5 that over short time spans, but typically long enough in duration to accumulate observables for application of equation (5.7), the time evolution of the $(R_n, \dot{R}_n)_b$ trajectories viewed in range, range-rate space remain closely spaced. This suggests that a simple correlation technique, referred to as the nearest-neighbor association scheme that tracks the evolution of each scattering center trajectory in range, range-rate space can be used to satisfy the short time span requirement. The scheme used in the framework is developed in the following section.

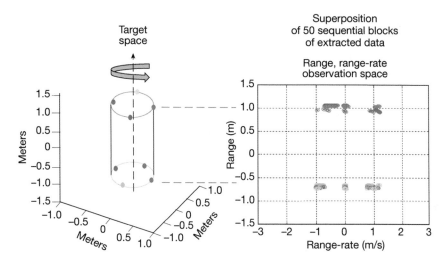

Figure 5.5 Sequential range, range-rate observable pairs superimposed in 2D range, range-rate space. Each data block incremented by one pulse. Fifty sequential range, range-rate observables are displayed.

5.5.1   The Nearest-Neighbor Correlation Technique

Consider the problem of associating the observables extracted from two sequential data blocks. Denote the observables corresponding to the range, range-rate of the scattering center extracted from the data block at time t_{b_1} as $\{R_n, \dot{R}_n\}_{b_1}$. Denote the observables extracted from the data block at t_{b_2} as $\{R_m, \dot{R}_m\}_{b_2}$. If the extracted range, range-rate values $(R_m, \dot{R}_m)$ from the data block at time t_{b_2} are such that $|R_m - R_n| < \Delta_1$ and $|\dot{R}_m - \dot{R}_n| < \Delta_2$ (corresponding to a small change Δ_1, Δ_2 for the two estimates) the observables are deemed to come from the same scattering center. To properly associate the two observables, define the vectors:

$$\underline{d}_n = \begin{bmatrix} R_n \\ \dot{R}_n/\beta \end{bmatrix}_{b_1}, \quad \underline{d}_m = \begin{bmatrix} R_m \\ \dot{R}_m/\beta \end{bmatrix}_{b_2}, \tag{5.19}$$

where the scale factor β is introduced to scale and weight the $\dot{R}_n$ measurement into the same units and magnitude as R_n. Consider the distance $|\underline{d}_n - \underline{d}_m|$. In order to combine the distance measure d_n with the distance measure d_m, the scale factor β must be appropriately chosen so that the range-rate measurement contributes to the distance measure. If $\{R_m, \dot{R}_m\}_{b_2}$ is properly associated with $\{R_n, \dot{R}_n\}_{b_1}$, the distance $|\underline{d}_n - \underline{d}_m|$ will be small. In other words, when viewed in 2D range, range-rate space, the two vectors are nearly colocated, differing only by the small motion changes of the target over the time interval $t_{q_2} - t_{q_1}$. The nearest-neighbor correlation technique searches the matrix $\underline{\underline{\Delta}}$ defined by

$$\Delta_{n,m} = \left[|\underline{d}_n - \underline{d}_m|^2\right]_{n,m} \tag{5.20}$$

for the N minimum values for N scattering centers extracted from each data block. The corresponding indices of $\Delta_{n,m}$ corresponding to the N minimum values of the matrix lead to the proper association of the range, range-rate observables from block to block. Typically, the association defined using this technique works well over short data intervals, but long enough that equation (5.7) applied to N_{p_0} data blocks can be inverted for a solution for (x_n, y_n, z_n) for each of the N scattering centering centers.

5.5.2   A Correlation Example: The Spinning Cylinder

Consider the application of the nearest-neighbor technique to the set of range, range-rate observables obtained over the time interval (0.1–0.2 s) for the spinning cylinder example illustrated in figure 5.6. The sequence consists of roughly 100 time-indexed observables. Recall from chapter 2 that for this example, only six of the 10 scattering centers are visible at any given time due to the shadowing constraint imposed by equation (2.21). Figure 5.6 compares the six range-rate observable sequences versus time both before and after correlation. The solid lines denote the simulation values and the dotted points the spectral estimator output. Note that the two cylindrical end scattering centers are slipping scattering centers and always visible to the sensor. Slipping scattering centers always exhibit zero range-rate relative to changes in roll angle, and are plotted atop each other as illustrated in figures 5.6(a) and 5.6(b). In figure 5.6 the color code indicates a scattering center ID, n, assigned to each estimated observable pair, $\{R_n, \dot{R}_n\}_b$, by the HRSE spectral estimator. The estimator does not attempt to correlate the scattering centers, but simply outputs an ordered list of paired observable estimates. One can see from figure 5.6(a) that for this case the ID assignment is inconsistent; the colors corresponding to the HRSE outputs along a true scattering center trajectory (solid lines) are changing due to changes

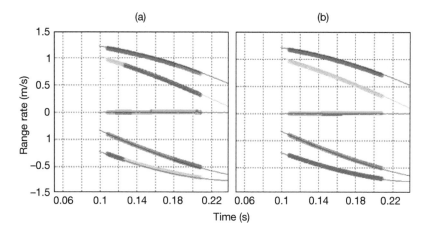

Figure 5.6   Range-rate sequence over the time interval (0.1–0.2 s) (a) before, and (b) after correlation. The color ID of each observable after correlation indicates the scattering center with which it is associated.

in the ordering of the observable sequence output from the spectral estimator. Some examples, especially for low signal-to-noise ratio cases, can be significantly worse. Application of the nearest-neighbor technique clears up the ambiguities. After correlation, the results of figure 5.6(b) show the correct association over the correlation time interval. Typically, for other than noise-free simulation data, the correlated observables $\{R_n, \dot{R}_n\}_b$ exhibit variances due to noise, and a simple polynomial fit to the correlated time sequence before application of equation (5.6) is applied to smooth the data.

5.5.3   3D Scattering Center Location Estimates: The Spinning Cylinder Example

In order to illustrate the flexibility of applications of equation (5.7), consider an example having bursts of high PRF data over widely spaced sectors of look angle data to the target. Assume four short, noncontiguous, equally spaced time intervals, each of duration 0.1 s, to which the correlation technique is applied to the extracted observables over these time intervals. Because of shadowing, only six of the 10 scattering centers for this target are visible at any given time. The four time intervals are spaced to correspond to four orthogonal quadrant views of the target. The HRSE extracted range and range-rate observables are correlated using the nearest-neighbor technique, and the associated observable estimates are illustrated in figure 5.7. In the figure, they are overlaid on solid lines that represent the true trajectories of each of the scattering centers for the assumed spinning motion.

For each inversion of equation (5.7), $N_{P_0} = 6$ correlated sequential $(b,\ b+1, \dots, b+N_{P_0}-1)$ range, range-rate observables are accumulated and applied to equation (5.7) consistent with the methodology described in section 5.3. Each inversion of equation (5.7) results in an estimate of each $(x_n, y_n, z_n)_p$ for each scattering center location. The accumulation of location estimates $(x_n, y_n, z_n)_p$ for $p = 1, \dots, P$ inversions of equation (5.7) is plotted in 3D target space and shown in figure 5.8. A total of $P = 68$ scattering center location estimates are plotted. Note that for this noise-free example, there is only a slight variation in the location estimates. The variation in the estimates for the z-coordinate z_n versus inversion index p is illustrated in figure 5.9. The consistency of location estimates as the solutions for the scattering center locations are accumulated in target space is very good, and they agree very closely with the scattering center locations assumed for the simulation target.

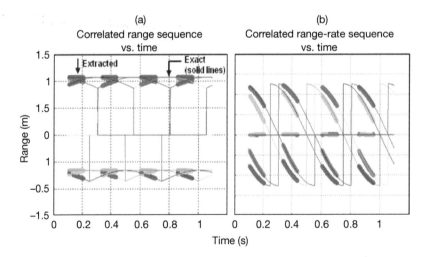

Figure 5.7   Range, range-rate observable sequence assuming four short, noncontiguous, equally spaced time intervals, each of duration 0.1 s, after applying the nearest neighbor correlation technique to the extracted observables. (a) Range sequence versus time. (b) Range-rate sequence versus time

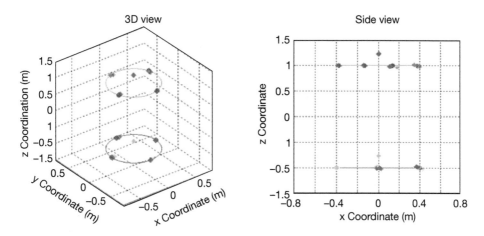

Figure 5.8   The 3D scattering location estimates extracted using sequential processing of each set of $N_{p_0} = 6$ observable pairs illustrated in figure 5.7. A total of $P = 68$ scattering center location estimates are superimposed on the figure.

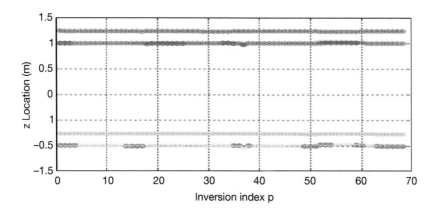

Figure 5.9   Estimates of the z-coordinate locations versus inversion index p.

Observe that the location of the two slipping scattering centers corresponding to the edges of the cylinder at either end project to a location on the z-axis independent of roll angle. This is due to the assumed zero precession motion of the target, resulting in a constant aspect angle $\theta = \theta_0$ as time evolves. The $(R_n, \dot{R}_n)_b$ estimates corresponding to the slipping scattering centers exhibit a constant range variation versus time, having zero range-rate. Slipping scattering center locations are projected to the z-axis, with locations determined by the range values. This is a general characteristic of slipping scattering centers: their location in target space can only be determined from changes in aspect angle, independent of roll angle. To show this, consider equation (5.11) for fixed $\theta = \theta_0$ applied to two successive slipping scattering center observables at roll angles φ_1 and φ_2. The range, range-rate observables for these two data points are given by $(R_n)_{b_1} = (R_n)_{b_2}$ and $\dot{R}_n = 0$ for the slipping scattering centers. Applying equation (5.11) to these two successive observables results in the solution

$$(x_n, y_n, z_n) = (0, 0, R_n/\cos\theta_0), \tag{5.21}$$

lying on the z-axis.

For this example, the time intervals used for processing were intentionally selected to avoid the noisier data evident in the complete set of observables illustrated in figure 5.4. Applying the nearest-neighbor correlation technique over the entire $(0, 1.2 \text{ s})$ interval becomes problematic when including the noisier regions. The need to filter the input observables to avoid such situations is eliminated using the target space persistence filter introduced in the next section.[2] Furthermore, the need for scattering center correlation in observable space is removed in section 5.6 by the inclusion of the acceleration observable $\ddot{R}_n$ to complement equations (5.2) and (5.3). The incorporation of the acceleration observable is equivalent to correlation in the sense that a sequence of correlated range-rate observables could in principle be used to obtain an acceleration estimate. Thus, accumulation of N_{p_0} observables in the form of equation (5.7) can be thought of as being equivalent to applying a tracking filter to the N_{p_0} range-rate observables to obtain an acceleration estimate for each scattering center. (This, in fact, forms the basis for the polynomial filter technique used to estimate $\ddot{R}_n$ developed in appendix G.) Accumulating the N_{p_0} range, range-rate observable pairs is consistent with the accumulation of sequences of range-Doppler images to obtain a 3D scattering center acceleration estimate discussed in [74], except only the essential ingredients of each range-Doppler image data block are used.

---

2. An alternate approach incorporating a target space filter is developed in appendix D.

5.6   Location Estimation Using Scattering Center Acceleration as an Observable

5.6.1   Persistence Filtering in Target Space

Unfortunately, when applied to field data from complex targets, association in observable space can be difficult. The simple nearest-neighbor association technique developed in section 5.5 may become intractable when characterizing complex targets having a large number of scattering centers that are not clearly resolved. As discussed earlier, an alternative approach to the correlation (association) problem is to append equation (5.4) with a third equation that incorporates the scattering center acceleration observable $\ddot{R}_n$. The resulting three equations can be solved for each of the three scattering component locations (x_n, y_n, z_n) using the range, range-rate and range-acceleration observables extracted over a single, expanded data block time and referenced to the center block time, eliminating the need for block-to-block correlation in observable space. Because the estimate of $\ddot{R}_n$ developed in chapter 3 results from using data over an extended data block characterizing the nonlinear signal model, using the larger data block effectively results in scattering center correlation imposed by the pole pairing applied to the nonlinear signal model. However, as can be seen from the simulation results of chapter 3, the estimates of $\ddot{R}_n$ using the techniques developed in chapter 3 can be somewhat noisy, particularly in regions where two scattering centers are sufficiently close in range or range-rate so as to confuse the estimate of $\ddot{R}_n$. To circumvent this problem the concept of the target space persistence filter is introduced. This filter relies on the behavior that, when mapped into target space, for $p = 1, \ldots, P$ scattering center location estimates, only the location estimates that correspond to actual scattering centers will exhibit persistence over an extended processing time. By subdividing target space into 3D bins, one can identify those bins that are most frequently occupied (persist) as the sequence of observables are processed. By establishing a simple a bin count threshold, the estimates caused by noisy observable data can be rejected. Incorporation of the target space filter along with the augmentation of equation (5.4) with the acceleration estimate allows for all the extracted observables to be processed sequentially, without association, and leads to a fairly robust technique for estimating the target scattering center locations.

5.6.2   Location Estimation and the Target-Space Persistence Filter

By including the acceleration variable $\ddot{R}_n$ in the scattering center location estimation equations, a set of three independent equations is obtained, allowing for a direct inversion of the target-motion coupling equations using observables extracted from a single frequency-time data block centered at $t = t_b$. When the acceleration term $\ddot{R}_n$ is included, the matrix M_{b_1} in equation (5.4) is now expanded to include $\ddot{k}$

$$
\begin{bmatrix} \hat{k}_{b_1} \\ \dot{k}_{b_1} \\ \ddot{k}_{b_1} \end{bmatrix} \cdot \begin{pmatrix} x_n \\ y_n \\ z_n \end{pmatrix} \equiv M_{b_1} \cdot \begin{bmatrix} x_n \\ y_n \\ z_n \end{bmatrix} = \begin{bmatrix} R_n \\ \dot{R}_n \\ \ddot{R}_n \end{bmatrix}_{b_1}.
\tag{5.22}
$$

where $\ddot{k}$ is given by equation (3.39). Equation (5.22) is similar in form to that introduced in [74] used as a first step in estimating the topology of a target. Because the motion is assumed known, the vectors $\hat{k}_{b_1}$, $\dot{k}_{b_1}$, and $\ddot{k}_{b_1}$ are known and equation (5.22) can be solved to obtain the scattering center location (x_n, y_n, z_n) estimates. In practice, in the presence of noisy data, the triplet $(R_n, \dot{R}_n, \ddot{R}_n)_b$ results in a noisy estimate of (x_n, y_n, z_n), and the target space persistence filter concept can be used to eliminate the false location estimates.

To illustrate the utility of using equation (5.22) coupled with the target-space persistence filter, consider its application to the spinning cylinder example treated in section 5.5. The extracted sequence of the triplets $(R_n, \dot{R}_n, \ddot{R}_n)_b$ versus sliding block time $t = t_b$ over the complete $(0, 1.2 \text{ s})$ interval, were examined for this example in section 3.5 and are illustrated in figure 3.8. For each extracted triplet $(R_n, \dot{R}_n, \ddot{R}_n)_b$, equation (5.22) is used to estimate the scattering center locations $(x_n, y_n, z_n)_b$, $n = 1, \ldots, N$ associated with the observables at $t = t_b$. The observable sequence is then incremented to $(R_n, \dot{R}_n, \ddot{R}_n)_{b+1}$ and the inversion of equation (5.22) is again repeated. This process continues sequentially to cover the complete observation time interval. Thus an estimate of $(x_n, y_n, z_n)_b$ is obtained from the observables extracted from each extended frequency-time data block.

Two cases are considered. The first, considered in section 5.5, applies equation (5.22) using the $(\ddot{R}_n)_b$ extracted from data taken only over the four-quadrant time intervals illustrated in figure 5.7, in which the observable sequence is well behaved. The second processes all the observables over the complete $(0, 1.2 \text{ s})$ time interval. In both cases, no association is applied to the observable sequence.

First consider the four-quadrant time interval example illustrated in figures 5.7 and 5.8, but now using the uncorrelated range, range-rate and range-acceleration

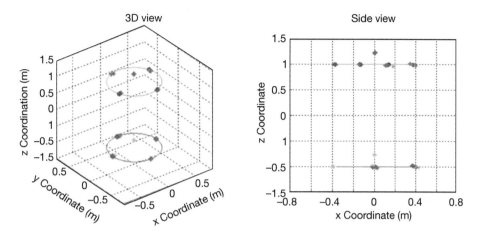

Figure 5.10 Target-space scattering center location estimates obtained using equation (5.22) incorporating the addition of the acceleration estimate, using the uncorrelated observables over the time intervals illustrated in figure 5.7.

sequences. The 3D scattering center location estimates obtained using equation (5.22), including the $(\ddot{R}_n)$ sequence superimposed for each $(x_n, y_n, z_n)_b$ estimate, are illustrated in figure 5.10. The results are virtually identical to those illustrated in figure 5.8, except in this case no correlation was required from one observable set to the next.

This result is satisfying to some extent, but one still had to do some prefiltering of the input data observable sequence to assure that the $(R_n, \dot{R}_n, \ddot{R}_n)_b$ sequence applied to the location estimate is of good quality [i.e., those regions where the $(R_n, \dot{R}_n, \ddot{R}_n)_b$ estimate are somewhat noisy have been excluded], and is therefore still analyst intensive. For the second case, all of the $(R_n, \dot{R}_n, \ddot{R}_n)_b$ observables estimated over the (0, 1.2 s) time interval illustrated in figure 3.8 are processed sequentially, without association, and any erroneous estimates are filtered in target space using the persistence target space filter. Figures 5.11 and 5.12 illustrate the results obtained by directly processing all the observables over the (0, 1.2 s) time interval without target-space filtering. Figure 5.11 illustrates the $(z_n)_b$ and radial coordinates $(\rho_n = sqrt(x_n^2 + y_n^2))_b$ estimates resulting from inverting equation (5.22) as a function of sliding block sequence number. Observe that where the estimates for $\ddot{R}_n$ are poor, there is considerable variation in the (ρ_n, z_n) estimates. The resulting 3D location

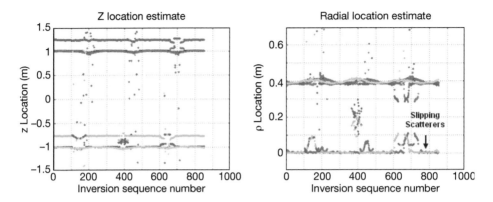

Figure 5.11   Radial and z-location estimates for the spinning, canonical cylinder incorporating the acceleration observables into the solution framework versus observable sequence number (no correlation). All observables illustrated in figure 3.8 were applied sequentially to equations 5.22.

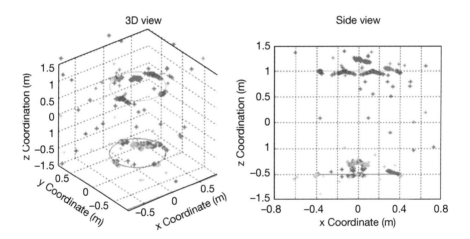

Figure 5.12   The 3D location estimates incorporating acceleration term into equations 5.22 without target-space filter.

estimates are illustrated in figure 5.12. Although the majority of the location estimates are correct (identical estimates are accumulated at the same 3D location so that erroneous estimates appear overly predominant on the figure), the resulting variance in the $(x_n, y_n, z_n)_b$ distribution in target space considerably degrades the characterization of the scattering center location estimates.

Now consider the application of the target-space filter to the sequential block $(x_n, y_n, z_n)_b$ estimates. The target-space filter accumulates the number of times (bin counts) a location estimate falls into a particular 3D bin in target space. For this example, define a target-space grid according to the increment

$$x_y = -1:0.025:+1$$
$$y_y = -1:0.025:+1$$
$$z_y = -1.5:.05:+1.5$$

There are $81 \times 81 \times 61 = 400{,}221$ total bins for this example. Figure 5.13 illustrates the bin count occupancy versus number of bins for the totality of estimated scattering center locations, from which it is clear that statistically only a fraction of the bins are occupied. For example, if a threshold bin count greater than 25 is used to filter the

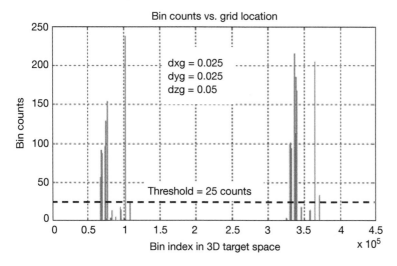

Figure 5.13 Bin counts versus grid label applied to target-space filter.

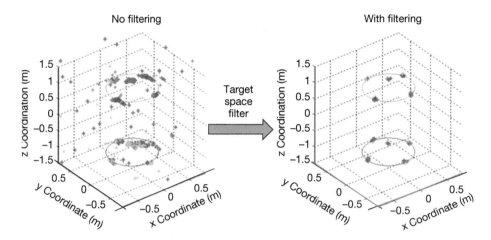

Figure 5.14 The 3D scattering center location estimates after the target-space persistence filter in figure 5.13 is applied. A threshold of 25 bin counts applied to target space filter.

spurious scattering center locations estimates, the composite target-space location estimates of the scattering center positions appear as illustrated in figure 5.14, resulting in scattering center location estimates essentially identical to those illustrated in figure 5.10.

The utility of implementing the simple persistence filter in target space described in this section can be generalized to include more elaborate target-space filtering schemes, and examples of such filters are described in appendix D.

6

Narrowband Signature Modeling Techniques

6.1   Introduction

Modeling the signature of a target from radar measurements using the wideband techniques introduced in chapter 5 requires one to estimate the number and locations of the dominant scattering centers on the target, as well as their geometrical theory of diffraction (GTD)-based diffraction coefficients. As developed in chapter 5 in equation (5.7), joint solutions for the motion and scattering center locations require processing a number of two-dimensional (2D) frequency-time data blocks leading up to solutions to equation (5.7). However, a single wideband range-rate image obtained using a single 2D frequency-time data block can often provide useful information about the general nature of the target and the types of scattering centers located on the target. For example, as discussed in chapter 1 (see figure 1.12), the wideband range, range-rate image of the canonical spinning cylinder provides information about the range extent of the target as viewed along the line of sight (LoS) of the radar, as well as an estimate of the number of scattering centers required to define the target signature model. Because the image range-rate axis is not scaled to target physical dimensions, the cross-range dimensions of the target cannot be determined from a single image. The Doppler spread does provide some information relative to the instantaneous motion of the target, but not a useful motion solution for estimating the scattering center locations defined relative to a target-space model of the target. Additional isolated 2D data block range, range-rate images exhibiting various scattering component phenomena are illustrated in figures 6.1, 6.2, 6.16, and 6.20 and discussed in later sections.

In some cases it is either not possible or desirable to apply the wideband techniques of chapter 5 to estimate the scattering center locations. This can occur for two specific cases: (1) when the sensor is narrowband with a bandwidth for which the range resolution limit governed by equation (5.1a) encompasses the entire target; or (2) for wideband sensors when the target has scattering centers that exhibit delayed returns (figures 6.16 and 6.20). In the first case, scattering centers can still be extracted by exploiting the change in look angle to the target that takes place during the observation interval. Scattering center location estimation using narrowband measurement data as a function of look angle has resolution limits characterized by equations (5.1b) and (5.1c), and govern the location estimate limits for narrowband processing. In the second case, delay is inherently a wideband phenomenon, and the application of the wideband location estimation techniques introduced in chapter 5 to targets having delayed returns can result in erroneous location estimates. In fact, the problem is magnified in the sense that, when mapping range, range-rate observables from multiple wide-angle sectors into target space, delayed returns will cause erroneous scattering center location estimates to appear not just as single points, but rather as *trajectories* in target space. Such trajectories do provide useful information, but do not characterize fixed-point scattering centers. To mitigate this effect, one approach used to characterize these types of scattering centers employs a combination of narrowband and wideband techniques. The range information is discarded and only the range-rate and range-acceleration observables are used to estimate the scattering center locations (e.g., as in section 5.4.4). The wideband extracted amplitudes and delayed returns associated with each scattering center are then used to model the diffraction coefficient associated with the scattering center and the delayed return.

The scattering center location estimation techniques developed in this chapter are referred to as narrowband and/or Doppler-only location estimation techniques. The term *Doppler-only* is used when wideband data are available, but only the range-rate and/or range-acceleration observables are used to estimate the scattering center locations. One property of the Doppler-only processed wideband data is that higher resolution is achieved when compared to the Doppler observables extracted from narrowband data that covers the same angular measurement extent. An example would be two scattering centers located such that they have nearly the same Doppler relative to the radar look angle to the target's rotational motion, but occur in different range gates of a wideband range-Doppler image. The wideband Doppler resolution achieved for this example is much improved over the narrowband Doppler resolution. Thus, it is always advantageous to use wideband data when available.

Another property of the scattering center location estimates obtained using narrowband or Doppler-only processing is that the location estimates cannot fall outside a *bounding volume* enclosing the target, thus eliminating the problem caused by delayed returns. The concept of using narrowband data to obtain estimates of the scattering center locations is not new, and it is discussed in some detail in [78]. In fact, noting the angular resolution bounds governed by equations (5.1b) and (5.1c), it is shown in [78] that by coherently processing narrowband data over a complete 2π angular circumference, a theoretical resolution of $\lambda/2$ is achievable on an isolated scattering center, considerably better than the range resolution characterized by equation (5.1a).

The attributes of the narrowband formulation developed in this chapter are based on a fundamental theorem of electromagnetic scattering theory [79]: the spherical harmonic modal basis functions characterizing the single frequency ($f = f_0$) scattered field contains radiation modes that decay rapidly for

$$N_0 > \frac{2\pi a}{\lambda_0} \tag{6.1}$$

where N_0 denotes the spherical harmonic mode index for which equation (6.1) is satisfied, a is the bounding spherical radius that just encloses the target, and λ_0 is the wavelength $= c/f_0$. In the context of this textbook, the spherical radius a enclosing the target is referred to as the bounding volume.

The bound imposed by equation (6.1) is inherently a single frequency, continuous wave constraint. Implicit in the application of equation (6.1) is that the bounding volume encloses the target, and excludes the undesirable artifacts associated with wideband processing. Two important corollaries to this narrowband attribute should be emphasized:

1.   The range-rate observables are necessarily bounded in extent and contained within a range-rate observable interval dependent on the target size and instantaneous motion.

2.   Estimating the scattering center locations using narrowband or Doppler-only techniques constrains the location estimates to lie inside the bounding volume.

Techniques for estimating the bounding volume using the modal constraint imposed by equation (6.1) are developed in [80]. The techniques introduced in this chapter focus on estimating the scattering center locations on the target.

Physically, the constraint defined by equation (6.1) is inherently related to physical limits imposed on energy stored within this bounded volume, which electrical engineers denote as the Q of the scattering object [79]. Typically, components or targets that have a high Q are highly resonant and exhibit strong delayed returns. This phenomena can be representative of the target (e.g., small in size, but resonant) or of components on the target (e.g., a cavity or resonant slot).

In this chapter, two techniques for estimating the locations of the scattering centers contained within the bounding volume are examined: first, the narrowband formulation developed in section 5.4.4 is extended to incorporate both the sequential range-rate and range-acceleration observable sequence; second, an alternate narrowband scattering center location estimation technique that utilizes only the sequential range-rate estimates is developed. Both techniques are combined with a target space filter as described in section 5.6.

This chapter is divided into three parts:

1. Section 6.2 discusses some wideband scattering mechanisms that produce delayed returns.

2. Section 6.3 applies the narrowband and wideband scattering center location estimation techniques developed in chapter 5 to a two-point scattering center model that includes single- and double-bounce mutual coupling terms producing wideband delayed returns. The implications of mutual coupling and delay on the respective narrowband and wideband location estimates are discussed.

3. In section 6.4 three canonical targets are introduced for which either static range or electromagnetic computational model data are available and the narrowband scattering center location estimation techniques are applied to each of these targets. Two of the targets have relatively small dimensions and have open and closed ends, characterized by wideband cavity-like delayed returns.

The narrowband signature model is based on the high-frequency formulation of the geometrical theory of diffraction introduced in chapter 2. As discussed in chapter 2, the GTD model is inherently a high-frequency scattering model but is extendable to the low-frequency transition region by introducing the concept of mutual coupling between isolated scattering centers located on the target. It is useful to consider this transition to lower frequencies starting with a general scattering model characterized by the target *mutual admittance matrix* introduced in [81], which characterizes the ultra-wideband scattering from the target as discussed in chapter 2. The general relationship characterizing the scattered electric field, $\underline{E}$ is given by [81]:

$$\underline{E} \sim \underline{V}_i^T \cdot Y \cdot \underline{V}_i \qquad (6.2)$$

where $\underline{V}$ represents the voltage induced on the target by the incident field at each point on the target surface, and Y is the body mutual admittance matrix. From the viewpoint of GTD, in the high-frequency limit, Y transitions to a diagonal matrix for targets exhibiting no delay artifacts. Wideband delay is observed when single and/or multiple bounce mutual coupling between true scattering centers occurs, in which case non-zero off-diagonal elements occur in Y. In the transition to lower frequency, Y becomes dense as many locations on the target become mutually coupled. As the scattering centers become closely coupled relative to a wavelength, the target begins to exhibit large delay responses and various resonance phenomena as dictated by the off-diagonal elements of Y. In the limit of very low frequencies, the off-diagonal elements of Y becomes fully filled with non-zero values.

6.2   Wideband Scattering Mechanisms

Many different physics-based scattering phenomena can be represented within Y including:

a. Specular reflection

b. Creeping waves

c. Traveling and surface waves

d. Edge diffraction

e. Surface discontinuities

f. Resonating gaps and cavities

g. Multipath mutual coupling

More detail can be found in [82]. Each physical mechanism listed previously can be labeled according to differing types:

Type 0: Scattering from a component on the target directly back to the radar

Type 1: Scattering that couples to various parts of the target, but scatters back to the radar *emanating from the original point of incidence.*

Type 2: Scattering that couples to various parts of the body but scatters back to the radar *emanating from a location on the target different from the original point of incidence.*

Examples of type 1 scattering are mechanisms *a* and *f*. Examples of type 2 scattering are mechanisms *b*, *c*, and *g*. For a Fourier-based wideband range, range-rate image, type 1 mutual coupling leads to a wideband delayed return that images as a delay emanating from the location of the original point of incidence on the target component causing the delay, contained within the same Doppler resolution cell. Type 2 mutual coupling leads to a wideband delayed return that images as emanating from a fictitious location either inside or outside the target's bounding volume. Specular scattering is represented by range, range-rate poles distributed along the target surface as viewed perpendicular to the radar LoS.

Figures 6.1, 6.2, 6.16, and 6.20 illustrate range-Doppler images representative of the various types of scattering centers. The image axis scales as depicted are notional only, as the images are meant to convey only the nature of the delayed return as it appears on the image. Figure 6.1 illustrates type 1 scattering using static range data collected on a circumferential slot positioned on a conic section. The wideband range-Doppler image exhibits the characteristics indicative of a resonant scattering component. Note the presence of a strong delayed return that emanates from the location of the slot on the target and decays along the radar LoS.

Figure 6.2 illustrates both type 1 and type 2 scattering from targets depicting two types of creeping waves, and their corresponding notional range-Doppler images. Note, as expected for type 2 scattering, that the wideband delayed return appears in the image as a scattering center located behind the physical target relative to the radar LoS. In the next section, a simple two-point scattering center example, including mutual coupling, is introduced and used to illustrate the effects of type 1 and type 2 delay mechanisms on mapping the range, range-rate and range-acceleration observables to target space.

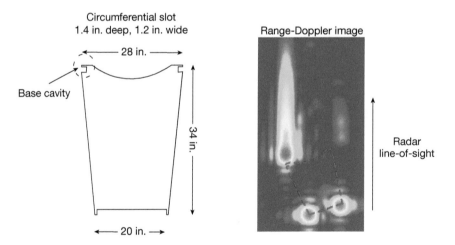

Figure 6.1 Type 1 scattering mechanism for a circumferential slot at X-band, 1 GHz bandwidth with associated range, Doppler image (notional scale). *Source*: Courtesy of Dr. M. L. Burrows.

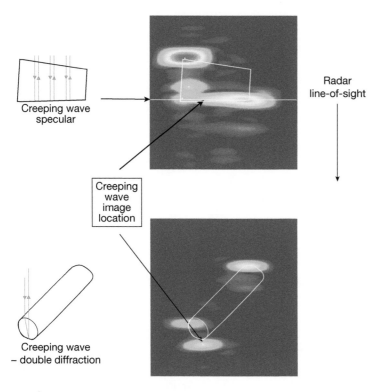

Figure 6.2   Notional type 2 scattering mechanism for canonical target shapes. Range-Doppler image with notional Doppler scale. Both type 1 and type 2 scattering depicting two types of creeping waves are illustrated.

6.3   On the Relationship between Wideband and Narrowband Scattering Phenomena

6.3.1   Two-Point Scattering Center Model with Mutual Coupling

In this section a two-point scattering center model is introduced to characterize the type 1 and type 2 scattering centers discussed in section 6.2, and is used to compare the scattering center location estimates obtained by the narrowband and wideband processing techniques. Each of the scattering mechanisms listed in section 6.2 (other than specular scattering) can be modeled by introducing one-way and two-way mutual

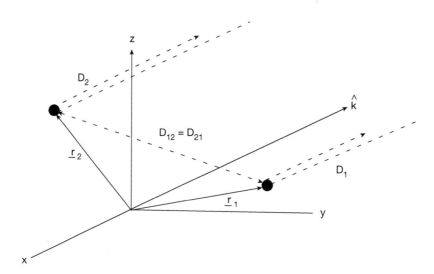

Figure 6.3   Two-point scattering center example having mutual coupling $D_{12} = D_{21}$ exhibiting type 1 and type 2 scattering mechanisms.

coupling between the two scattering centers. Consider the geometry illustrated in figure 6.3. Two-point scattering centers located at positions r_1 and r_2 are coupled by mutual coupling coefficients $D_{12} = D_{21}$. The direct return from each scattering center has amplitudes given by D_1 and D_2. The double-bounce returns are representative of type 1 scattering, characterized by a two-way transit of the mutual coupling path between r_1 and r_2, having amplitude $D_{12}D_{21}$. The double-bounce return emanates back to the radar from the original point of incidence after experiencing the two-way transit delay. The one-way single bounce returns are representative of type 2 scattering. The wavefront path originates at one location and exits back to the radar from the second location after experiencing a one-way transit delay.

Using the GTD scattering model introduced in chapter 2, the field scattered from this "target", denoted by $E(f, t)$, including the direct, single, and double bounces, can be obtained using simple ray-tracing analysis techniques. The scattered field takes the form

$$
\begin{aligned}
E(f,t) \sim D_1 e^{-j\frac{4\pi f}{c}\hat{k}\cdot r_1} &+ D_2 e^{-j\frac{4\pi f}{c}\hat{k}\cdot r_2} + D_{12}e^{-j\psi_{12}(f,\,t)} \\
&+ D_{21}e^{-j\psi_{21}(f,\,t)} + D_{12}D_{21}e^{-j\psi_{22}(f,\,t)} + D_{21}D_{12}e^{-j\psi_{11}(f,\,t)},
\end{aligned}
\tag{6.3}
$$

where the phase functions ψ_{12}, ψ_{21}, ψ_{11}, ψ_{22} are given by

$$\psi_{12}(f,t) = \frac{2\pi f}{c}\hat{k}\cdot\underline{r}_1 + \frac{2\pi f}{c}\hat{k}\cdot\underline{r}_2 + \frac{2\pi f}{c}|\underline{r}_2 - \underline{r}_1|, \tag{6.4}$$

$$\psi_{21} = \psi_{12}, \tag{6.5}$$

$$\psi_{22}(f,t) = \frac{4\pi f}{c}\hat{k}\cdot\underline{r}_2 + \frac{4\pi f}{c}|\underline{r}_2 - \underline{r}_1|, \tag{6.6}$$

$$\psi_{11}(f,t) = \frac{4\pi f}{c}\hat{k}\cdot\underline{r}_1 + \frac{4\pi f}{c}|\underline{r}_2 - \underline{r}_1|. \tag{6.7}$$

Equation (6.4) can be reformulated as if the scattering originates from an apparent position $(\underline{r}_1 + \underline{r}_2)/2$ between the scattering centers with a one-way path length time delay $|\underline{r}_2 - \underline{r}_1|/2c$:

$$\psi_{12}(f,t) = \frac{4\pi f}{c}\hat{k}\cdot\left(\frac{\underline{r}_1 + \underline{r}_2}{2}\right) + \frac{4\pi f}{c}|\underline{r}_2 - \underline{r}_1|/2. \tag{6.8}$$

Note that the second term on the right side of equation (6.8) is independent of observation angle. Equation (6.6) represents a wideband delayed return emanating from the original point of incidence at location $\underline{r}_2$, but time delayed by the round-trip path length time $|\underline{r}_2 - \underline{r}_1|/c$ between the two scattering centers. It is similar for equation (6.7), but corresponds to a wideband round-trip delayed return emanating from location $\underline{r}_1$.

Note that the phase contributions $\frac{4\pi f}{c}|\underline{r}_2 - \underline{r}_1|$ and $\frac{4\pi f}{c}|\underline{r}_2 - \underline{r}_1|/2$ are independent of changes in look angle, $\hat{k}$, corresponding to delay only. Their Doppler exhibits no change with variation in look angle to the target.

The wideband observables $(R_n, \dot{R}_n)$ characterized by $E(f,t)$ can be extracted directly from equations (6.3)–(6.8) using equations (2.15)–(2.18), yielding five range, range-rate observables (the range, range-rate pairs corresponding to phase terms ψ_{12} and ψ_{21} are identical)

$$R_1 = \hat{k}\cdot\underline{r}_1 \qquad \dot{R}_1 = \hat{k}\cdot\underline{\dot{r}}_1 \tag{6.9}$$

$$R_2 = \hat{k}\cdot\underline{r}_2 \qquad \dot{R}_2 = \hat{k}\cdot\underline{\dot{r}}_2 \tag{6.10}$$

$$R_3 = \hat{k} \cdot \left(\frac{r_1 + r_2}{2} \right) + |r_2 - r_1|/2 \qquad \dot{R}_3 = \hat{k} \cdot \left(\frac{r_1 + r_2}{2} \right) \tag{6.11}$$

$$R_4 = \hat{k} \cdot r_1 + |r_2 - r_1| \qquad \dot{R}_4 = \hat{k} \cdot r_1 \tag{6.12}$$

$$R_5 = \hat{k} \cdot r_2 + |r_2 - r_1| \qquad \dot{R}_5 = \hat{k} \cdot r_2. \tag{6.13}$$

Note in particular that there are five independent range observables corresponding to the direct and delayed returns, but only three independent range-rate observables, because $\dot{R}_1 = \dot{R}_4$ and $\dot{R}_2 = \dot{R}_5$. Consequently, if only the range-rate observables are used to estimate the scattering center locations, only locations contained within the bounding volume will appear. Note also that the location of the fictitious scattering center for the range-rate pole of equation (6.11) emanates from the center of the line connecting the two scatterers.

Consider the two-point scattering center example from the viewpoint of forming a GTD-based scattering model based on the measurement data. The key ingredients to the model are the parameters defining the target model of figure 6.3: r_1, r_2, and the associated mutual coupling diffraction coefficient parameters. Essential to the modeling process is a correct estimate of r_1 and r_2, and the fact that only two actual scattering locations are used to characterize the target. However, as will be demonstrated, applying the wideband target-space mapping equations defined by equations (5.2) and (5.3) to the five range, range-rate observable pairs yields multiple scattering center location estimates, some of which are time varying, and only two correspond to r_1 and r_2. Thus, it is important to examine how one can identify the true scattering centers from the totality of estimates. Furthermore, it is important to identify how the errors introduced to these location estimates affect the resulting target model. Section 6.3.3 examines the wideband mapping defined by equations (5.2) and (5.3) and section 6.3.4 compares the wideband mapping location estimates to the narrowband mapping location estimates obtained using equation (5.22) applied to the range-rate and range-acceleration observables $(\dot{R}_n, \ddot{R}_n)_b$.

6.3.2   The Wideband Target Space Mapping

Consider the scattering center location estimates obtained using the five wideband range, range-rate observables defined by equations (6.9)–(6.13) in the observables to target space mapping defined by equations (5.2) and (5.3). Equations (5.2) and (5.3) are functions of the motion defined by $\hat{k}$ and $\dot{\hat{k}}$. Assume the scattering centers are constrained to the (x, y) plane $\underline{r}_1 = (x_1, y_1)$ and $\underline{r}_2 = (x_2, y_2)$. Further assume a pure rotational motion, so that $\varphi(t)$ is linear in time, characterized by $\varphi(t) = 2\pi(t)/T_p$, where T_p denotes the rotation period, and the radar views the target in the rotational plane. For this special motion, equations (5.2) and (5.3) reduce to the form

$$\hat{k} \cdot \underline{r}_n = x_n \cos\varphi + y_n \sin\varphi = R_n$$
$$\dot{\hat{k}} \cdot \underline{r}_n = \dot{\varphi}(-x_n \sin\varphi + y_n \cos\varphi) = \dot{R}_n. \tag{6.14}$$

Denote $\underline{r}'_n = (x'_n, y'_n)$, $n = 1, \ldots, 5$ as the five locations determined using equation (6.14) applied to each of the five wideband range, range-rate observable pairs. It is useful to compare the five location estimates $\underline{r}'_n$ to the actual locations. The $\underline{r}_1 = (x_1, y_1)$ and $\underline{r}_2 = (x_2, y_2)$ are the locations of the two scattering centers that define the target. Using each of the two range, range-rate observables in equations (6.9) and (6.10) directly in equation (6.14), it follows that each corresponds to a scattering center location given by $\underline{r}_1$ and $\underline{r}_2$.

$$\underline{r}'_1 = \underline{r}_1,$$
$$\underline{r}'_2 = \underline{r}_2,$$

The locations $\underline{r}'_1$ and $\underline{r}'_2$ correspond to the true scattering center locations.

For the remaining range, range-rate observables, $n = 3, 4, 5$, the fixed delay independent of look angle introduces errors in the mapping to target space relative to the true scattering locations. To estimate this error, reference each location estimate relative to the true point scatter locations (x_1, y_1) and (x_2, y_2), as well as an apparent location (x_3, y_3) located midway between (x_1, y_1) and (x_2, y_2). Using equation (6.14), it follows that each (x'_n, y'_n) must satisfy the range and range-rate constraint equations

$$\cos\varphi(x'_n - x_k) + \sin\varphi(y'_n - y_k) = \Delta R_n$$
$$-\sin\varphi(x'_n - x_k) + \cos\varphi(y'_n - y_k) = 0, \tag{6.15}$$

where $\Delta R_3 = \Delta R_4 = |\underline{r}_2 - \underline{r}_1|$ for the two type 1 scattering centers, and $\Delta R_5 = |\underline{r}_2 - \underline{r}_1|/2$ for the type 2 scattering center. In equation (6.15), the correspondence between the indices n and k is represented as $(n = 3, 4)$ corresponds to $k = (1, 2)$ and $n = 5$ corresponds to $k = 3$. Solutions to equation (6.15) are readily determined to be

$$x'_n - x_k = \Delta R_n \cos\varphi(t)$$
$$y'_n - y_k = \Delta R_n \sin\varphi(t),$$

(6.16)

so that

$$(x'_n - x_k)^2 + (y'_n - y_k)^2 = \Delta R_n^2.$$

(6.17)

Thus, as time evolves, $\varphi(t)$ covers the full 360° rotational angle and the wideband location estimates (x'_n, y'_n), $n = 3, 4$, which result from the delayed returns, will trace out circles centered at each of the (x_1, y_1) and (x_2, y_2) scattering locations having radius $\Delta R_1 = \Delta R_2 = |\underline{r}_2 - \underline{r}_1|$. Similarly, the fictitious scattering center corresponding to $n = 5$ will trace out a circle of radius $\Delta R_3 = |\underline{r}_2 - \underline{r}_1|/2$ centered midway between the two real scattering centers with apparent location (x_3, y_3).

To summarize, the composite wideband target space scattering center location estimates using this two-point scattering center example with mutual coupling delay yields five scattering center location estimates—two fixed, and three time varying—where the latter evolve over time into circles enclosing either the two true scattering center locations or the fictitious scattering center located between the two. In the next section a simulation for this two-point scattering center example that contrasts the wideband (see figure 6.6) versus the narrowband (see figure 6.7) scattering center location estimates is presented.

6.3.3   Two-Point Scattering Center: Simulation

Consider a simulation of the two-point scattering example illustrated in figure 6.3 using equation (6.3) to simulate the frequency-time data matrix $E(f, t)$ using the parameters illustrated in table 6.1.

Table 6.1   Simulation parameters for two-point scattering center example in figure 6.3

	Scatterer 1	Scatterer 2
Locations (m) (x, y)	$(0, -0.5)$	$(0, +0.75)$
Mutual Coupling	$D_{11} = D_{22} = 0$ dB	
	$D_{12} = D_{21} = -10$ dB	
f_0, BW, PRF	10 GHz, 1 GHz, 1,000 Hz	
Motion	$\varphi(t) = \dfrac{2\pi}{T_p} t - \pi/2,\ T_p = 1$ s	

The two-point target is oriented in the $z = 0$ plane on the y-axis, and the look angle to the target at $t = 0$ is along the direction aligned with the two scattering centers ($\varphi = -\pi/2$). The two scattering centers are chosen to have unequal distances relative to the target center of rotation at $(0, 0)$ in order to emphasize the impact of the delayed returns. Because of this asymmetry in positioning, the target exhibits symmetry versus viewing angle only at $\varphi = 0$ ($t = 0.25$ s) and $\varphi = \pi(t = 0.75$ s).

The wideband case

The target geometry is illustrated in figure 6.4. A typical wideband range, range-rate image is illustrated in figure 6.4 for a 45° look angle to the target. Note that there are five well-defined peaks that show up in the image: two peaks that correspond to the true scattering center locations, two peaks that are delayed in range from the true locations, and a fifth peak that is delayed in range relative to a scattering center location located midway between the two true locations. Observe that the range delay for the peaks relative to the true locations is double (two-way transit mutual coupling) that of the delay relative to the location midway between the true locations (one-way transit mutual coupling). Note also that each of the delayed returns lies in the same range-rate bin as its parent scattering center, consistent with the five range, range-rate pairs determined analytically using equations (6.9)–(6.13).

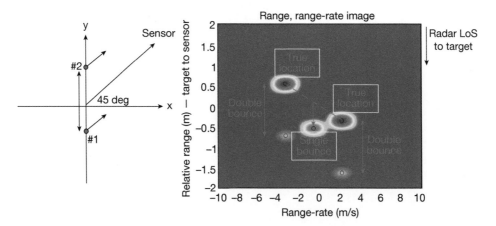

Figure 6.4 Typical range, range-rate image for the two-point scattering center simulation example having type 1 and type 2 mutual coupling.

Figure 6.5 illustrates the sequential estimates of the observables $(R_n, \dot{R}_n, \ddot{R}_n)_b$ for block time t_b, as a 2D data block (using $Q_0 = 32$ pulses per block) is sequenced through the simulation data over a complete $360°$ motion period (in block time increments $\Delta t = 1/PRF$). Note as per the development shown previously, there are five distinct range observables extracted from the simulation data shown in figure 6.5(a), but only three distinct range-rate and range-acceleration observables shown in figures 6.5(b) and 6.5(c). On examination of figure 6.5(a), the double and single bounce delays

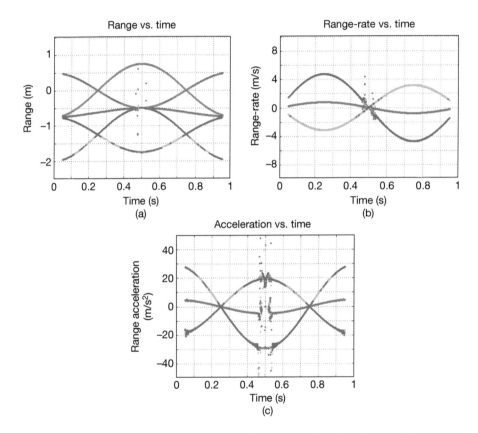

Figure 6.5   Estimated range, range-rate and range-acceleration observable sequences for the two-point scattering center wideband delay example. (a) Scattering center range observable estimates versus time ($t = 0$ corresponds to a look angle $\varphi = -\pi/2$ along the line connecting the scattering centers, and $t = 0.5$ s, $\varphi = +\pi/2$, has a directly opposite viewing angle). (b) Range-rate observable estimates versus time. (c) Range acceleration observable estimates versus time.

indicative of type 1 and type 2 delay mechanisms result in range observables having a fixed range delay offset from the range values corresponding to the true scattering centers, and therefore have the same range-rate value for each parent scattering center.

Because the motion is known, and the scattering centers each lie in the x, y-plane, the wideband target space mapping equations characterizing each scattering center [using equations (5.2) and (5.3) with $\theta = \pi/2$] are given by

$$\cos\varphi x_n + \sin\varphi y_n = R_n$$
$$-\sin\varphi x_n + \cos\varphi y_n = \dot{R}_n/\dot{\varphi}. \qquad (6.18)$$

where $\dot{\varphi} = 2\pi/T_p$ and the scattering center locations (x_n, y_n) in equation (6.18) are determined by inverting equation (6.18) using the time progression of each of the five range, range-rate pole pairs.

The composite wideband mapping of the range, range-rate observables to target space is illustrated in figure 6.6(b) and compared in range scale to the two-point scatter geometry in figure 6.6(a). Clearly, as predicted by equation (6.17), the location estimation error for the type 1 scattering centers forms a circle (blue/green) of radius ΔR about the true location estimates, and the type 2 scattering center location estimate forms a trajectory (red) in target space that encircles the midpoint of the two true scattering

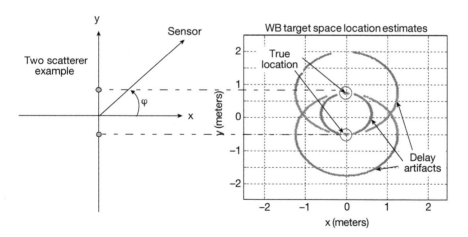

Figure 6.6 Wideband scattering center location estimates using the wideband target space mapping for the two-point simulation example having type 1 and type 2 mutual coupling. The mutual coupling delays exhibit a circular nature centered about the true and midpoint (fictitious) scattering centers.

center locations at one-half the radius of the type 2 delayed returns. [Recall that the value of ΔR applicable to equation (6.17) is given by $|r_2 + r_1|$ for each of the type 1 scattering centers and by $(r_1 + r_2)/2$ for the type 2 midpoint scattering center.]

The narrowband correction

Applying the mapping from observable space to target space using narrowband processing eliminates the location estimate ambiguities obtained using the wideband target space mapping equations. For the narrowband case, using equation (5.22) applied to each range-rate, range-acceleration pair, the narrowband target space mapping equations take the form

$$\dot{\hat{k}} \cdot \underline{r}_n = \dot{R}_n$$
$$\ddot{\hat{k}} \cdot \underline{r}_n = \ddot{R}_n$$

(6.19)

Now apply equation (6.19) to the two-point scattering center example. The vector $\dot{\hat{k}}$ is characterized by equation (2.5) repeated here for convenience:

$$\dot{\hat{k}} = \dot{\theta}\hat{\theta} + \dot{\varphi}\sin\theta\hat{\varphi}$$

(6.20)

For the assumed geometry where the scattering centers lie in the x, y-plane, it follows that the spherical coordinate $\theta = \pi/2$ and $\dot{\theta} = 0$. For the assumed motion $\varphi(t) = \dfrac{2\pi}{T_p}t - \pi/2$, equation (6.20) reduces to the simple form

$$\dot{\hat{k}} = \dot{\varphi}(-\sin\varphi\hat{x} + \cos\varphi\hat{y})$$

(6.21)

where $\dot{\varphi} = \dfrac{2\pi}{T_p}$. Using equation (6.21),

$$\ddot{\hat{k}} = -(\dot{\varphi})^2(\cos\varphi\hat{x} + \sin\varphi\hat{y})$$

(6.22)

Thus, the coupled equations for estimating the scattering center locations (x_n, y_n) reduce to

$$-\sin\varphi x_n + \cos\varphi y_n = \dot{R}_n/\dot{\varphi}$$
$$\cos\varphi x_n + \sin\varphi y_n = -\ddot{R}_n/(\dot{\varphi})^2.$$

(6.23)

Using the sequential values for the observables $(\dot{R}_n, \ddot{R}_n)$ illustrated in figure 6.5, but excluding the time interval where the $\ddot{R}_n$ estimate is poor (note that one could include this interval by adding a target space filter to eliminate the spurious points similar to that introduced in chapter 5), the resulting location estimates using equations 6.23 are illustrated in figure 6.7, superimposed as a function of time over (0, 1 s). When comparing figure 6.7 using narrowband processing to figure 6.6 using wideband processing, the erroneous wideband location estimate artifacts outside the bounding volume enclosing the target are removed. The effect of the delay mechanisms using the narrowband target space mapping equations result only in the addition of a single fictitious scattering center, located precisely at the center of the mutual coupling path, that is, at $(\underline{r}_1 + \underline{r}_2)/2$.

6.3.4   The Bounding Volume

In this section a simple technique is developed to characterize a *bounding volume* that encloses the target, analogous to the development of [80], which uses a modal analysis based on the discussion in section 6.1. Because the range-rate observables are necessarily bounded within an interval dependent on the target size and instantaneous motion, the bounds associated with the observable sequence can be used to estimate the bounding

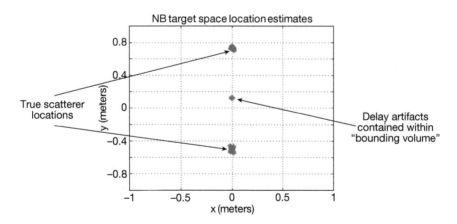

Figure 6.7   Narrowband scattering center location estimates using the narrowband target space mapping for the two-point simulation example having type 1 and type 2 mutual coupling. False scattering center delay artifact caused by type 2 mutual coupling located midway between the true scattering center locations.

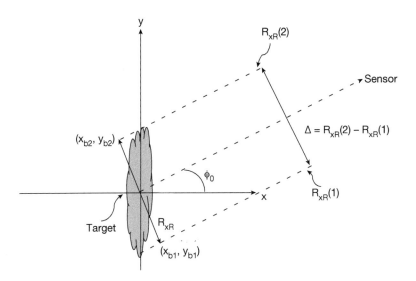

Figure 6.8   Target cross-range bounds using maximum and minimum cross-range observable estimates defined by $R_{xR}(1)$ and $R_{xR}(2)$, where $R_{xR}(1)$ and $R_{xR}(2)$ are determined for each range-rate observable set using the maximum and minimum range-rate estimates at each observation angle, scaled according to equation (6.25).

volume. Consider the diagram in figure 6.8 for the case where the target motion is constrained to lie in the x, y-plane. For radar angle of incidence φ_0, let Δ represent the cross-range extent of the target. For this look angle, the target can be no wider than the cross-range extent measure Δ. Translating the width Δ to the origin of the target, the points $(x_{b1},\, y_{b1})$ and $(x_{b2},\, y_{b2})$ at the end points of Δ are determined from the equations

$$x_{b1} = R_{xR}(1)\sin\varphi_0,\; y_{b1} = R_{xR}(1)\cos\varphi_0$$
$$x_{b2} = R_{xR}(2)\sin\varphi_0,\; y_{b2} = R_{xR}(2)\cos\varphi_0. \tag{6.24}$$

$$R_{xR} = \dot{R}/\dot{\varphi}. \tag{6.25}$$

It follows that $R_{xR}(1) = \min_n(\dot{R}_n/\dot{\varphi})$ and $R_{xR}(2) = \max_n(\dot{R}_n/\dot{\varphi})$. For targets in pure rotation, a bounding enclosure that envelops the target is obtained by graphically plotting the coordinates $(x_{b1},\, y_{b1})$ and $(x_{b2},\, y_{b2})$ as φ_0 rotates over the full 360°. This result

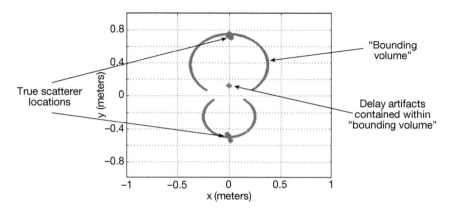

Figure 6.9   Narrowband target bounding volume estimate applied to the two-point scattering center example of figure 6.3. Scattering center location estimates from figure 6.7.

is illustrated in figure 6.9 using the range-rate extractions illustrated in figure 6.5(b). Note that if one views this two-point scattering center example as determining the end points of a thin wire, where the length is such that the end points are mutually coupled, then the target bound obtained using equation (6.24) is loose for the thin dimension and tight for the long dimension. Note the gap in the boundary occurs because the angular region of φ for which the acceleration estimate was poor has been excluded. One anticipates that the bounding volume realized would be much tighter for a realistic target having a more typical length to width (diameter) ratio.

6.4   Narrowband Scattering Center Location Techniques

Due to the nature of the scattering center acceleration estimation technique, the range-acceleration observable estimates are often noise-like in regions where the range or range-rate observable sequences coincide. Thus the narrowband and/or Doppler-only scattering center location estimation techniques developed in section 6.3 using both the range-rate and range-acceleration observables are usually applied using a target space filter as described in section 5.6.2. In this section, an alternate target scattering center location estimation technique is developed that exploits the bounded nature of the range-rate observable sequence. This technique is based on the special case solution for range-rate scattering center estimation described in section 5.4.4, based on sequential applications of equation (5.17). Sequential applications of equation (5.17) typically require sequential association of the estimated range-rate observables as time evolves. The technique described here eliminates the need for sequential range-rate correlation and uses a target space *bin-count filter* to eliminate spurious scattering center location estimates. The technique is referred to as the cross-range line integration technique and is fully developed in [82]. It incorporates a 2D target space mapping, and is therefore primarily applicable to static range data collections obtained for a fixed roll-cut orientation on the target. It is also applicable to field measurements for targets undergoing torque-free rotational motion as discussed in chapter 9. The technique is based on the fact that the range-rate poles extracted from a given data block isolate the locations of the scattering centers in cross range (i.e., perpendicular to the radar LoS), when scaled to cross range using equation (6.25), without a determination in range. Thus, for a given look angle to the target, the Doppler returns can be visualized by forming lines in target space, separated in cross range. When viewed from a number of different angles, the intersection of these lines determines the location of the scattering centers in target space. A procedure to implement this technique fills all the target space bins along the radar LoS for each of the cross-range offsets, and accumulates the bin amplitudes (or bin counts) for every observation angle. The persistence of bin counts in a given range, cross-range bin determines the scattering center locations.

Figure 6.10 illustrates a comparison of the two techniques used to estimate the scattering center locations using narrowband processing. Technique 1 uses the range-rate, range-acceleration observable mapping to target space characterized by equation (6.23) applied to B blocks of sequential observable estimates. Technique 2 uses the cross-range line integration technique described previously, and is represented graphically. The cross-range line integration target space mapping is illustrated in figure 6.10 for three separate look angles to the target. For each look angle, the cross-range amplitudes (or bin counts) are accumulated in the bins in target space for every range bin along the radar LoS for that specific cross-range offset. The accumulation (persistence) of amplitudes in the target space bins corresponds to the location of the true scattering centers. As the target rotates, successive cross-range line intersections add noncoherently, with the resulting integrated image indicating the scattering center locations. Typically, to avoid the occurrence of large specular returns dominating the line integration, the amplitudes are either limited to a maximum value, or alternately, simply set to unity. In the later case, the line integration reduces to a bin count in target space. This is a reasonable approach because the resultant target space map is used only for location

Technique 1

Use range-rate and range acceleration
observables to estimate the SC locations

$$\hat{k} \cdot \underline{r}_n = -\dot{\varphi} \sin \varphi x_n + \dot{\varphi} \cos \varphi y_n = \dot{R}_n$$

$$\hat{k} \cdot \underline{\ddot{r}}_n = -(\dot{\varphi})^2 (\cos \varphi x_n + \sin \varphi y_n) = \ddot{R}_n$$

Location estimate for each of
B sequential data blocks
Motion matrix

$$\begin{pmatrix} x_n \\ y_n \end{pmatrix}_b = -\begin{pmatrix} \sin \varphi & -\cos \varphi \\ \cos \varphi & \sin \varphi \end{pmatrix}_b^{-1} \begin{pmatrix} \dot{R}_n / \dot{\varphi} \\ \ddot{R}_n / (\dot{\varphi})^2 \end{pmatrix}_b$$

Target space filter output
$$(x'_n, y'_n)_b, \, n = 1,...,N, \, b = 1,...,B'$$

Technique 2

Cross-range line integration technique

(Noncoherent integration of
amplitudes or bin count in target space)

Use range-rate observables to characterize
SC "cross-range" lines and integrate

$$\hat{k} \cdot \underline{\dot{r}}_n = -\dot{\varphi} \sin \varphi x_n + \dot{\varphi} \cos \varphi y_n = \dot{R}_n$$

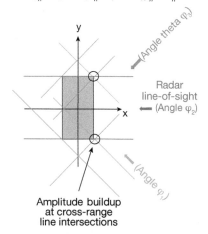

Radar
line-of-sight
(Angle φ_2)

Amplitude buildup
at cross-range
line intersections

Figure 6.10 Contrast of the two techniques for narrowband processing. Technique 1 illustrates equations (6.23) using the range-rate, range-acceleration observable sequence. Technique 2 illustrates the target space cross-range line integration technique. Intersecting line amplitudes add noncoherently in target space, and spurious estimates are filtered using an appropriate bin threshold as described in chapter 5. Cross-range lines are illustrated for three example look angles.

estimation, not scattering center range cross section (RCS) determination. Spurious estimates are filtered using an appropriate target space bin threshold as discussed in chapter 5.

For the examples to follow, these techniques are applied to three targets having more realistic features and delay components compared to the two-point scattering model used in section 6.3. Both static range and simulation data examples are considered. The three example targets are illustrated in figure 6.11. Specifically, the three targets and data types examined are as follows:

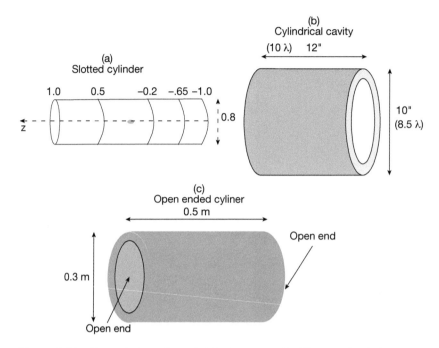

Figure 6.11   Example targets and data types used to illustrate both the cross-range line integration and NB range-rate, range-acceleration observable scattering center location estimation techniques. (a) GTD simulation data using the cylindrical target introduced in chapter 2 with azimuthally symmetric slots added and having no fixed-point scattering centers. (b) Static range data on a cylindrical cavity open at one end. (c) High-fidelity electromagnetic code data simulation on a cylinder open at both ends.

1. A slotted cylindrical target. The canonical cylindrical target introduced in chapter 2 is modified with three azimuthally symmetric slots added to the cylinder sides, and the fixed-point scattering centers are omitted. A GTD-based scattering model incorporating each of the slots and cylinder ends is used to develop a set of synthetic static range measurement data. The target is placed in rotational motion as if positioned on a static range pylon and the GTD computational model is used to compute the data samples.

2. A closed-end cylindrical cavity, which is open at one end, has extended wideband delayed returns, and uses static range data. Note the dimensions of this target are small relative to wavelength at X-band (10 GHz).

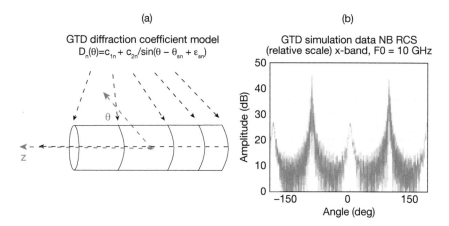

Figure 6.12 (a) Illustration of the slotted cylindrical target having axially symmetric slots modeled using the parameter-based diffraction coefficient in equation (14.6), introduced in chapter 14. (b) Simulated RCS (relative scale) of the target plotted versus rotation angle at 10 GHz using the GTD scattering model.

3. An open-ended cylindrical cavity, which has wideband delayed returns, and uses high-fidelity electromagnetic code simulation data representative of static range data.

6.4.1   GTD-Based Scattering Model Applied to a Slotted Cylindrical Target

Figure 6.12(a) illustrates the target, as well as the functional form of the diffraction coefficients used to model each of the target components. The target consists of three azimuthally uniform slots introduced along the midportion of the cylinder as well as the two cylinder ends.

A GTD-based scattering model is used to model the scattering from the target, which is placed in rotational motion to emulate a static range data collection. As discussed earlier in section 2.2, it is often necessary to introduce two different coordinate systems: one for characterizing the computational model and one for processing the measurement data. For the computational model, a standard spherical coordinate system is used where the target is oriented in an (x, y, z) system characterized by spherical coordinate angles (θ, φ) as illustrated in figure 6.13(a), and the diffraction coefficients are defined in this coordinate system.

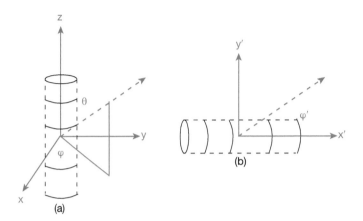

Figure 6.13   (a) The (x, y, z) coordinate system used for the GTD computational model. (b) The (x′, y′) coordinate system used to process the synthetic static range measurement data.

To process the static range measurements, it is most convenient to use the 2D measurement coordinate system (x', y') characterized by rotational angle φ' defined relative to the x'-axis. Having generated the synthetic signature data representative of a static range data collection using the computational model, the coordinate system in figure 6.13(b) is used to process the data. (for additional discussion, refer also to figures 9.1(a) and 9.1(b) in chapter 9)

In order to properly model the scattering from the axially symmetric target, the computational model z-axis must be oriented along the cylinder axis and the roll angle φ characterizes the roll-symmetric nature of the scattering from the azimuthally symmetric slots. For a given angle of incidence to the target z-axis, accounting for shadowing, six scattering centers are visible to the radar: the two forward (relative to the radar LoS) cylinder end scattering centers, the three slots and the rear portion of the cylinder visible to the radar. Each of these six scattering centers can be modeled as a slipping scattering center as described by equation (2.20). The scattering by the forward visible portion of the cylinder requires two slipping scattering centers, such that equation (2.20) is applied to each, with $\varphi \rightarrow \varphi + \pi$ for the lower portion of the visible edge. To model the target motion as if placed on a static range pylon, the following Euler parameters are chosen

$$[\kappa,\ \theta_p,\ T_p, f_s,\ \alpha_p,\ \alpha_s] = [\pi/2,\ \pi/2,\ 2\pi,\ 0,\ \pi/2,\ 0]$$

For the target model, each end of the cylinder is assumed characterized by the diffraction coefficient for a perfectly conducting 90° wedge angle, characterized parametrically using equation (14.6) as discussed in chapter 14. For simplicity, the slots are assumed to have the same general scattering behavior as the cylinder ends, characterized by the same functional form with different parameter choices.

Applying the functional form of equation (14.6), the assumed diffraction coefficient for each respective scattering center diffraction is expressed as

$$D_n(\theta) = c_{1n} + c_{2n}/\sin(\theta - \theta_{sn} + \varepsilon_{sn})$$

where the index n ranges over the target components. The parameter ε_{sn} has a small complex value and is introduced into the diffraction coefficient model to control the amplitude of the specular for both the forward, rear, and side speculars. The following parameter choices for c_{1n}, c_{2n}, θ_{sn} and ε_{sn} are used in the simulation:

$$c_{1n} = [.01, .01, .01, .01, .01, .01]$$

$$c_{2n} = [0.75, 1, 0.5, 0.9, 0.25, 0.8]$$

$$\theta_{sn} = [0, \pi/2, \pi/2, \pi/2, \pi/2, \pi/2]$$

$$\varepsilon_{sn} = j*[.05, .02, .02, .02, .02, .02]$$

For a given specular angle θ_{sn}, specular scattering using the expression for $D_n(\theta)$ will occur at angles θ_{sn} and $\theta_{sn} + \pi$. Shadowing of each of the three slots and rear portion of the cylinder end not visible referenced to the radar LoS is included in the simulation.

Narrowband data samples were calculated from the GTD signature model at roughly 0.1° rotational angle increments simulating a static range data collection over the 2π angular region. As discussed further in chapter 7, for a rotating target of length L, Nyquist sampling of the scattered field requires samples spaced no larger than $\lambda/(2L)$, which is approximately 0.5° for the 2 m length simulation target. Thus the 0.1° simulated model data are oversampled by a factor of five relative to Nyquist sampling at X-band. The narrowband RCS (10 GHz) of the target versus aspect angle is illustrated in figure 6.12(b). Note, in particular, the forward (0°), side (±90°), and rear (180°) specular responses evident in the data.

Figure 6.14 illustrates the narrowband range-rate, range-acceleration observables, $(\dot{R}_n, \ddot{R}_n)_b$, $b = 1, \ldots.B$, extracted from $B = 900$ data blocks as a function of time assuming the rotation period of 2π s used to simulate the static range pylon motion. The observables were extracted using the HRSE nonlinear signal model acceleration

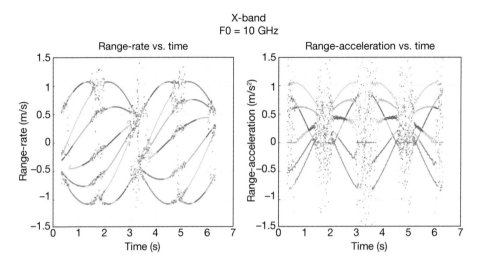

Figure 6.14   Narrowband range-rate, range-acceleration observable sequences (order $N = 6$) extracted from simulation data for the slotted cylindrical target at 10 GHz. A rotation period of 2π s is used for the motion to scale the range-rate observable.

estimation technique developed in chapter 3 and described in appendix E. (Note: the narrowband version of the 2D HRSE signal model described in appendix E is implemented incorporating two frequencies per pulse, closely spaced about the center frequency.) The discontinuities in each of the observable trajectories versus time are the result of shadowing of the scattering centers as the target rotates. The parameters used for estimating the observables, referenced in figure 3.7, were chosen as follows: for estimating a single expanded block observable pair $(\dot{R}_n, \ddot{R}_n)_b$, the initial block size used was 64 pulses, along with three iterations each increasing the block size by $N_0 = 40$ pulses. This results in a total of 304 data samples covering the entire expanded data block. As discussed previously, the angle change per pulse for the simulation data is roughly 0.1° per sample, corresponding to five pulses per Nyquist sampling interval. The entire expanded data block for each observable extraction then covers an angle sector of approximately 30°.

Using range-rate, range-acceleration observable sequences applied to the narrowband observable space to target mapping technique 1 described in figure 6.10 results in the narrowband target scattering location estimates illustrated in the composite target space mapping figure 6.15(a). The measurement coordinate system illustrated in

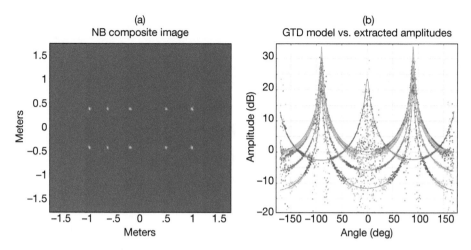

Figure 6.15   (a) Narrowband target scattering center location estimates using the target space mapping technique 1 in figure 6-10 . (b) HRSE amplitudes extracted from the simulation data (dots) versus GTD model diffraction coefficients assumed for the data simulation (solid lines).

figure 6.13(b) was used to carry out the mapping of the observables to target space, for which the observation angle φ' covers the observation angle $0 < \varphi' \le 2\pi$. An observable sampling rate of four (every fourth observable) was used in the mapping to target space and a grid size of 0.025×0.025 m was used to cover the region of target space illustrated in the figure. The resultant narrowband target space composite mapping is well capable of identifying the locations of the scattering centers on the target. Figure 6.15(b) illustrates the HRSE amplitudes extracted from the data compared to the GTD model diffraction coefficients used in the signature model. The results illustrate the capability to extract the component scattering behavior versus angle using strictly narrowband data. The correlation of the extracted amplitudes to the GTD component characterization is discussed in considerable detail in chapter 16.

6.4.2   Static Range Data: Cylindrical Cavity

The cylindrical cavity exhibits significant delayed returns, due to the resonances induced within the cavity. It has been analyzed extensively by [83]. Of particular interest here, along with the presence of delayed returns, are the relatively small dimensions of the

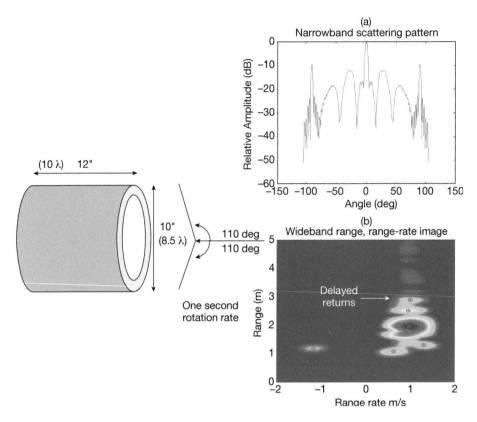

Figure 6.16   Scattering from cylindrical cavity. Data over ±110°. (a) Narrowband RCS versus angle at X-band. (b) Typical wideband range, range-rate image (1 GHz *BW*) illustrating extended delayed returns.

target measuring only $8.5\lambda \times 10\lambda$ at X-band. However, because it is highly resonant, it exhibits range delayed returns far exceeding the target dimensions. Yet the narrowband cross-range line integration technique is capable of identifying features on the target as well as its basic dimensions. Figure 6.16 illustrates the narrowband scattering pattern and a typical wideband range, range-rate image. For this target, static range data were only available from (−20° to +110°) relative to the forward (open end) view of the target. By symmetry, we have extended this interval to ±110°. A 1 s rotation rate is used to provide scale to the range, range-rate image. Figure 6.17 illustrates the wideband (1 GHz bandwidth at X-band) estimated range, range-rate observables versus

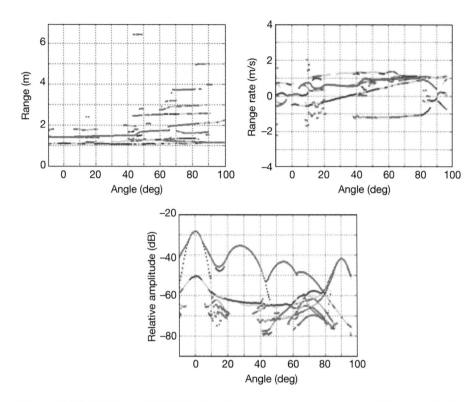

Figure 6.17   Wideband (1 GHz) estimated range, range-rate, and relative amplitude sequential observables using the HRSE spectral estimation technique applied to the measured data over the (–20, 110) degree interval.

rotation angle using the initial (–20° to +110°) data set. Note the large range extent of the delayed returns (~4 m in extent at 75° aspect for a 0.3 m target) as the look angle covers regions which strongly couple into the cavity. Also note the bounded nature of the range-rate estimates. Figure 6.18 illustrates the cross-range and amplitude estimates obtained using the 1D HRSE spectral estimation technique described in appendix B applied to the single frequency narrowband data using the data set symmetrically extended over ±110°. Note the bounded extent of the cross-range observables corresponds well to the dimensions of the target as it rotates. Figure 6.19 shows the narrowband composite target space mapping. For comparison, the target outline has been superimposed over the scattering center location estimates. Note especially the accurate identification of target dimensions, as well as the fictitious scattering center location

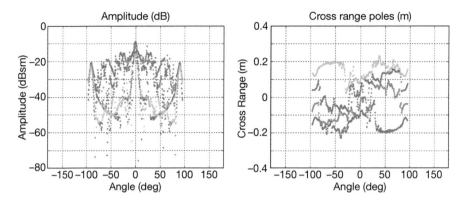

Figure 6.18   Narrowband (10 Ghz) cross-range poles (scaled to meters) and associated pole amplitudes using the measurement data extended symmetrically over (–110, 110) degrees.

estimates caused by type 2 delay mechanisms occurring within the target bounding volume.

Because the scattering center location estimates are obtained using data only over the forward look direction, the limited angular extent of the static range data prevents the target space mapping from imaging the rear end of the target. This is the reason the outside rear edges of the target are not prominent.

6.4.3   Simulation Data: Cylinder Open at Both Ends

To emphasize the occurrence of delayed returns, monostatic data at X-band from an open-ended cylinder having dimensions somewhat larger than the previous example was simulated using traditional high fidelity electromagnetic codes (acknowledgment to Audrey Dumanian). To minimize computation time resulting from applying the computational codes at the X-band wavelength, the scattering was computed only over a limited region of (–10, 90) degrees (0° corresponding to the cylinder axis), and the target symmetry was used to extrapolate this data over the complete 360° angular region of (–270, 90) degrees. Figures 6.20–6.23 present the results applying the narrowband cross-range line integration technique in a manner parallel to figures 6.16–6.19 of the previous example. Figure 6.20 illustrates a typical wideband range, range-rate image assuming a 1 s rotation rate, as well as the narrowband scattering pattern at X-band, as a function of look angle $-270° < \varphi < 90°$, defined relative to the cylinder axis.

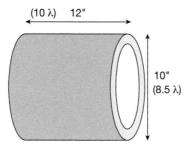

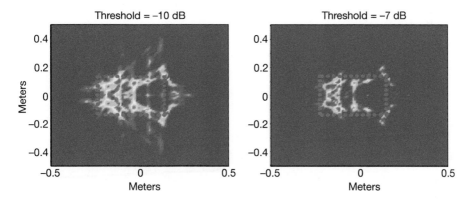

Figure 6.19   Narrowband (10 Ghz) target space composite mapping obtained using technique 2 illustrated in figure 6.10. Red dashed outline illustrates cylindrical cavity outer boundary.

Figure 6.21 illustrates the wideband range and range-rate sequential observables estimated using the HRSE spectral estimation technique applied to the simulation data interval of (0–90) degrees. The delayed returns are enhanced at near 60° indicating maximum coupling into the cavity at these look angles. Note the large range extent of the wideband delayed returns and the bounded nature of the range-rate poles. Figure 6.22 indicates the narrowband range-rate (scaled to cross range) and amplitude observable sequence estimated using the 1D HRSE technique described in appendix B applied to the narrowband data as extended to the complete 360° rotation cycle. Note the maximum and minimum spread in cross-range corresponds directly to the physical dimensions of the target. Figure 6.23 illustrates the composite target space mapping using

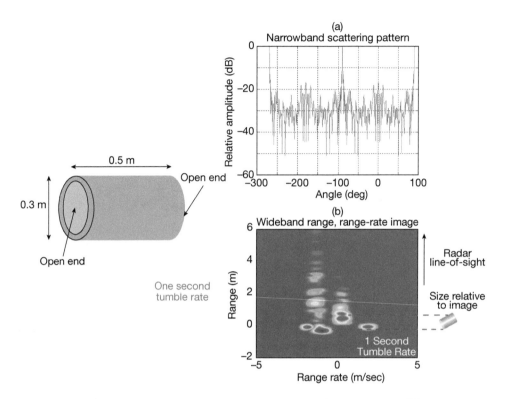

Figure 6.20   Scattering from cylinder open at both ends. (a) Narrowband RCS scattering pattern as a function of look angle $-270° < \varphi < 90°$. (b) Typical wideband range, range-rate image (1 GHz *BW*) illustrating long delayed return.

technique 2 for the target scattering center location estimates, for two differing thresholds applied to the target space filter. The full details of the target are now visible, and the scattering center location estimates are clearly bounded within the target physical boundaries. Note also that the fictitious scattering centers located internal to the cylinder at the midway between the cylinder openings indicate the presence of type 2 transit delays through the open cylinder, consistent with the two-point scattering center example treated in section 6.3.3.

Figure 6.24 illustrates the narrowband target space composite mapping resulting from applying both techniques 1 and 2 to the open-ended cylinder data at 10 GHz.

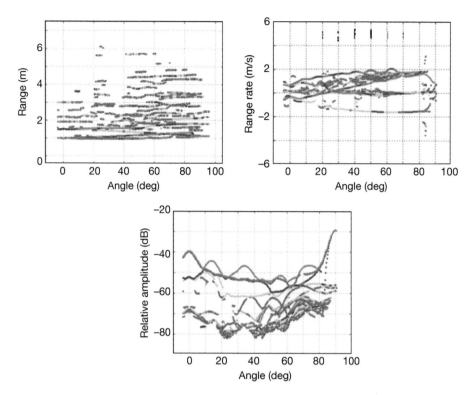

Figure 6.21 Estimated wideband (1 GHz) scattering center range, range-rate, and relative amplitude sequential observables obtained using the measurement data set over the interval (−10, 90) degrees.

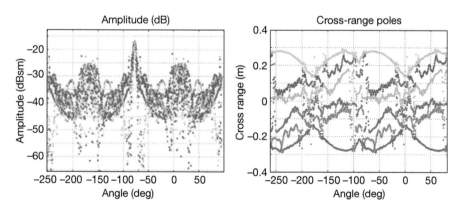

Figure 6.22 Narrowband estimated cross-range and amplitude observables extracted using the measurement data set symmetrically extended over (−270, 90) degrees. Note the maximum and minimum spread in cross-range corresponds directly to the physical dimensions of the target.

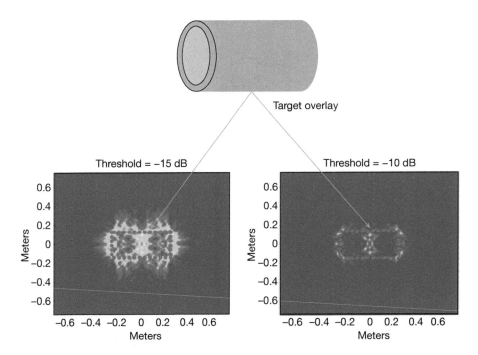

Figure 6.23   Cylinder open at both ends. Narrowband target space composite mapping using narrowband cross-range estimates applied using technique 2 illustrated in figure 6.10. Red dashed outline illustrates open cylinder outer boundary.

Both techniques provide a detailed view of the internal mechanisms associated with the delayed returns transitioning through the inner portion of the open-ended cylinder.

6.5   Limitations on Narrowband Processing for Scattering Center Location Estimation

There is one special case where narrowband processing gives degraded results relative to 3D scattering center location estimation: $\dot{\theta} = 0$. This special situation can be quantified by examination of the expression for $\hat{k}$ and $\dot{\hat{k}}$, repeated here as

$$\hat{k} = \sin\theta\cos\varphi\hat{x} + \sin\theta\sin\varphi\hat{y} + \cos\theta\hat{z} \qquad (6.26)$$

$$\dot{\hat{k}} = \dot{\theta}\hat{\theta} + \dot{\varphi}\sin\theta\hat{\varphi} \qquad (6.27)$$

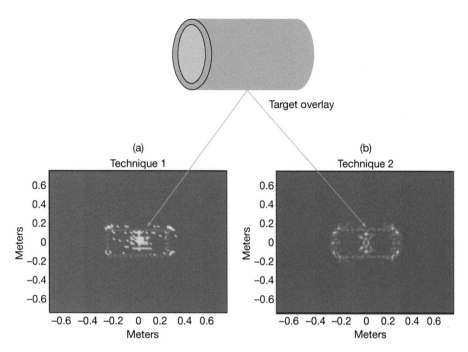

Target overlay

Figure 6.24   Cylinder open at both ends. Narrowband target space composite mapping technique 1 versus technique 2. (a) Composite mapping formed using technique 1 described in figure 6.10. (b) Composite mapping formed using technique 2 described in figure 6.10.

Note that the unit vectors $\hat{\theta}$ and $\hat{\phi}$ are given by the expression

$$\hat{\theta} = \cos\theta\cos\varphi\hat{x} + \cos\theta\sin\varphi\hat{y} - \sin\theta\hat{z}$$
$$\hat{\phi} = -\sin\varphi\hat{x} + \cos\varphi\hat{y}$$

(6.28)

When $\dot{\theta} = 0$, the narrowband equation $\dot{\hat{k}} \cdot \underline{r}_n = \dot{R}_n$ reduces to $\dot{\varphi}\sin\theta\hat{\phi} \cdot \underline{r}_n = \dot{R}_n$ and the dependence on z_n is lost. Thus, for this special case no *height* information is obtainable.

7

Interferrometric ISAR

7.1 Introduction

Chapters 5 and 6 present techniques for estimating a target's scattering center locations using the wideband and narrowband techniques applied to data received by a single sensor. In this chapter, the joint target-motion coupling framework is applied to the two-sensor problem for the case where the motion solution is known. One sensor is monostatic and the other bistatic, and they are closely spaced, but have independent viewing angles of the target. The technique, referred to as interferometric-inverse synthetic aperture radar (IF-ISAR) relies on extracting the phase difference between the field scattered from the target received by each of the sensors. For some limited motion solutions, the techniques developed can also be applied using a single monostatic sensor, where the different viewing angles to the target are obtained at different times and are a result of target motion.

Three-dimensional (3D) synthetic aperture radar (SAR) imaging of ground reflecting surfaces using interferometric processing has shown much promise using synthetic aperture radars [84]. A key factor in achieving such 3D images is precise control over the observation angle (motion) used by the radar sensor. A particular advance in this research area is the technique referred to as interferometric SAR (IF-SAR). For this case, a unique observation angle sampling grid is generated that allows overlaying of nearly identical two-dimensional (2D) range-Doppler images and uses phase differences between these images, each collected from closely spaced look angles to the target and calibrated for sensor motion, to estimate the out-of-plane height information, from which a 3D image is developed. References [84–86] provide a good discussion of the basic principles and associated radar calibration requirements. Various schemes for collecting the SAR data are discussed, each focused on the peculiar hardware calibration requirements inherent in applying these techniques.

In this chapter, the IF-SAR interferometric concept is integrated into the high-resolution 3D target signature modeling framework developed in chapter 5, incorporating results presented in the literature [87]. The essence of the 3D scattering center location estimation framework is to apply the target space mapping from radar observable space to 3D target space to estimate the target scattering center locations, referenced to a fixed target space coordinate system. In section 5.4, several solution constraints were imposed on the target space mapping when only the range and range-rate observable sequence is available. For this case, a single observable pair $(R_n, \dot{R}_n)_b$ at block time $t = t_b$ provides only two equations for the corresponding (x_n, y_n, z_n) target space mapping equation, and several techniques for augmenting these two equations with additional constraint equations (e.g., using sequential correlation) were developed. Subsequently, the observable space was extended to three dimensions, using the triplet of observables $(R_n, \dot{R}_n, \ddot{R}_n)_b$, providing a direct estimate of (x_n, y_n, z_n) from a single expanded data block, and eliminating the need for correlation over the sequence of observables. In this chapter, by incorporating interferometric processing techniques, a new observable triplet $(R_n, \dot{R}_n, -\Delta_n(phase))_b$ using the relative phase, $\Delta_n(phase)$, between two sensors is introduced and used to augment the range, range-rate equations. This new observable triplet is extracted using two sensors having differing look angles to the target.

The most common methods for measuring the target radar signature separate into two general categories: radar cross section static range measurements and field measurements. The development in this chapter addresses the estimation of the scattering center locations on a target using field data measurements obtained by both monostatic and bistatic sensors. Two sensor configurations are considered: monostatic and bistatic (two sensors) and monostatic (one sensor). When applied to a single sensor monostatic case, the bistatic calibration problems are eliminated and replaced by the need for an accurate motion solution, but not all motions will qualify. However, the set of allowable motion solutions that are applicable to the monostatic case is robust enough to warrant discussion. Section 7.2 presents the bistatic development and the monostatic case is discussed in section 7.3. Applications to static range data are discussed in chapter 15.

7.2   The Bistatic Case

Consider the generic bistatic geometry illustrated in figure 7.1, where sensor 1 (transmit/receive) and sensor 2 (bistatic/receive) are separated in viewing angle by $\Delta\theta_B$, referenced to the target coordinate system. Because interferometric processing utilizes

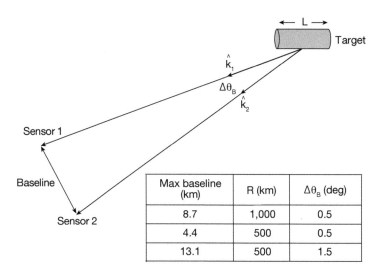

Figure 7.1 showing sensor geometry and the following table:

Max baseline (km)	R (km)	$\Delta\theta_B$ (deg)
8.7	1,000	0.5
4.4	500	0.5
13.1	500	1.5

Figure 7.1   Maximum baseline for bistatic collections for $L = 1$ m at X-band, as a function of range to the target.

the phase difference between the signals received from sensors 1 and 2, the allowable separation angle $\Delta\theta_B$ must satisfy certain phase ambiguity constraints specified later in this section. These constraints require $\Delta\theta_B$ be small, placing certain requirements on the baseline separation distance between the sensors as a function of the range from the sensor to the target. Tolerable values for $\Delta\theta_B$ are dependent on what are referred to as Nyquist angle sampling of the bistatic scattering pattern of the target, and are strongly dependent on operational frequency. For example, for a target of characteristic dimension L, the lobing structure of the bistatic scattering pattern occurs at angle intervals $\lambda/(2L)$, and this is often taken as the upper limit of Nyquist angle sampling. For angles $\Delta\theta_B$ spaced larger than Nyquist, the phase difference between sensors becomes ambiguous relative to 2π phase changes, and erroneous location estimates occur. Some typical values of $\Delta\theta_B$ at X-band for $L \approx 1$ m are given in figure 7.1 as a function of range to the target, along with the baseline separation limits. Note in particular that for targets at long ranges from the sensor, the separation limits are indeed large compared to traditional phased array element placement separation values used for coherent processing. Because monostatic data received at sensor 1 and bistatic data received at sensor 2 are eventually combined, the fusion processing is accomplished using the data recorded by each sensor, after registering the data to the rotational motion of the target.

7.2.1   Development

Assume sensor 1 both transmits and receives the monostatic scattered signal, and sensor 2 receives the bistatic signal scattered from the target resulting from the incident field of sensor 1. The monostatic scattered field in the direction of sensor 1 (direction vector $\hat{k}_1$) is characterized by equation (2.6), except now a one-way propagation time delay τ_1 from sensor 1 to the target is included for comparison to the propagation path time delay relative to the bistatic sensor. The monostatic scattered field received at sensor 1 takes the form

$$E_1 \sim e^{-j\omega(2\tau_1)} \sum_n D_n e^{-j\frac{4\pi}{\lambda}(\hat{k}_1 \cdot \underline{r}_n)}. \tag{7.1}$$

Typically, D_n is a relatively slow varying function of (f, θ, φ) relative to the phase changes in equation (7.1) and characterizes the geometrical theory of diffraction (GTD) coefficient of the nth scattering center located on the target. In a similar manner, the bistatic field received by sensor 2, located in the direction $\hat{k}_2$ and having one-way propagation time delay τ_2 from the target to sensor 2 is given by

$$E_2 \sim e^{-j\omega(\tau_1 + \tau_2)} \sum_n D_n' e^{-j\frac{2\pi}{\lambda}(\hat{k}_1 + \hat{k}_2) \cdot \underline{r}_n}. \tag{7.2}$$

Assume the bistatic angle $\Delta\theta_B$ is small. It follows that for $\hat{k}_1$ close to $\hat{k}_2$, because D_n varies slowly with angle changes, $D_n' \approx D_n$. Expressing $\hat{k}_2$ in the form $\hat{k}_2 = \hat{k}_1 + \Delta\underline{k}$, equation (7.2) can be written as

$$E_2 \sim e^{-j\omega(2\tau_1)} \sum_n D_n \left[e^{-j\frac{2\pi\Delta R}{\lambda}} e^{-j\frac{2\pi}{\lambda}\Delta k \cdot \underline{r}_n} \right] e^{-j\frac{4\pi}{\lambda}\hat{k}_1 \cdot \underline{r}_n}, \tag{7.3}$$

where $c = 3 \times 10^8$ m/s and the path length difference ΔR is defined as

$$\Delta R = c(\tau_2 - \tau_1). \tag{7.4}$$

Comparing equations (7.3) and (7.1), the phase difference between the nth scattering center (assumed to be at the same physical location on the target) as viewed by each sensor is given by

$$\Delta_n(phase) = -\frac{2\pi}{\lambda}(\Delta R) - \frac{2\pi}{\lambda}(\hat{k}_2 - \hat{k}_1) \cdot \underline{r}_n. \tag{7.5}$$

Note the occurrence of the phase factor $(2\pi/\lambda)$ in equation (7.5) versus $(4\pi/\lambda)$ that is conventionally encountered for monostatic scatter. Equation (7.5) can be used to augment the target space mapping equations of equations (5.2) and (5.3). However, the difference in path length ΔR must be estimated. To accomplish this, one can either correct the data received by sensor 2 by applying a range shift ΔR to the received data, or explicitly use ΔR in equation (7.5). Techniques for estimating ΔR are discussed subsequently and additionally developed in chapter 13.

Assuming the correction for ΔR is applied, the three equations for estimating the scattering center locations (x_n, y_n, z_n) can now be summarized in the form

$$\hat{k}_1 \cdot \underline{r}_n = R_n, \tag{7.6a}$$

$$\dot{\hat{k}}_1 \cdot \underline{r}_n = \dot{R}_n, \tag{7.6b}$$

$$\frac{2\pi}{\lambda}(\hat{k}_2 - \hat{k}_1) \cdot \underline{r}_n = -\Delta_n(phase). \tag{7.6c}$$

The angular separation between sensors, $\Delta\theta_B$, is assumed small enough that the phase Δ_n lies in the interval $-\pi < \Delta_n < \pi$. The phase difference between sensors can be determined either by forming the range, range-rate images from dual 2D data blocks of wideband data received at each sensor (having the same center block time) and extracting the differential phase at each image peak (Fourier-based resolution), or extracting the phase difference between the complex weight amplitudes obtained using the high resolution 2D spectral estimation technique applied to each dual data block. For either case, the observable sequence $(R_n, \dot{R}_n)_b$ is replaced by the triplet $(R_n, \dot{R}_n, -\Delta_n(phase))_b$ for each of the N scattering centers present. Applying the concepts of sequential estimation processing introduced in chapter 2, a sequence of B observables, $(R_n, \dot{R}_n, -\Delta_n phase)_b, b = 1, \ldots, B$ leads to B scattering center location estimates, $(x_n, y_n, z_n)_b, b = 1, \ldots, B$.

Implicit in equation (7.6) is that the motion is known, so that, referenced to the target-centered coordinate system introduced in chapter 2, the two motion solutions $(\theta_1(t), \varphi_1(t))$ and $(\theta_2(t), \varphi_2(t))$, defined relative to each sensor, are known. For example, if the motion solution relative to sensor 1 is known, so that $\hat{k}_1(t)$ is known, and the location of sensor 2 relative to sensor 1 is such that $\theta_2(t) = \theta_1(t) + \Delta\theta_B$, and $\varphi_2(t) = \varphi_1(t) + \Delta\phi_B$, then the separation angles $(\Delta\theta_B, \Delta\phi_B)$ are sufficient to characterize the motion $\hat{k}_2(t)$.

7.2.2   Bistatic Simulation

The example of the spinning cylinder having no precession (chapter 5, section 5.5.3) is now used to illustrate the application of 3D interferometric processing. Recall from the example that for pure spin motion, the 3D component location equations developed in chapter 5 applied to a monostatic sensor are unable to determine the radius of the two-cylinder end slipping scattering centers, which are projected to the z-axis when mapping into target space. This occurs because of the lack of independent views of the target relative to changes in motion of the slipping scattering center. For a constant aspect angle $(\theta = \theta_0)$, as the target rotates about the z-axis, the slipping scattering center does not exhibit any phase change versus roll angle and consequently it is not possible to determine its radius using a single sensor. If a bistatic receiver is displaced in look angle in the theta direction relative to the z-axis, two independent theta angle views of the slipping scattering center are obtained as it rotates about the z-axis, allowing a determination of the slipping scattering center radius. A more rigorous development of this independence property is provided in section 7.2.2.

For the simulation example, consider a bistatic sensor located in the plane of the monostatic sensor at an angular separation $(\Delta\theta_B, \Delta\varphi_B) = (0.5°, 0)$ from the monostatic sensor using the same target and motion considered in section 5.5.3. The bistatic geometry is illustrated in figure 7.2.

Consider the application of equation (7.6) to a typical set of dual 2D data blocks, each referenced to block time $t = t_b$ processed by each of the sensors, at a particular roll angle view of the target. Each data block provides a simultaneous view of the target relative to each sensor's viewing angle. Figures 7.3(a) and 7.3(b) depict the range, range-rate images as determined at each of the two receivers, compensated so that $\Delta R = 0$. The locations of the peak scattering centers are indicated on each image, as well as the locations of the shadowed scattering centers not present in the data. Note that because $\hat{k}_1$ is close to $\hat{k}_2$, the amplitudes of each range, range-rate image look virtually identical. Not obvious from the image contours is that the phase associated with each image pixel has changed considerably. The image phase-difference contour plot is illustrated in figure 7.3(c), and depicts the locations of each of the scattering center peaks shown in figure 7.3(a) and (b). Note that the phases extracted from the image contour are directly correlated to the location of the peaks. Upon extracting the locations $(R_n, \dot{R}_n)_b$ and $\Delta_n(phase)_b$ from the corresponding images, equation (7.6) can be solved for the $(x_n, y_n, z_n)_b$ scattering center locations associated with block time $t = t_b$. The solutions for the N scattering centers $(x_n, y_n, z_n)_b$ obtained from a single set of dual data blocks are illustrated in figure 7.4. The estimated locations are indicated as red

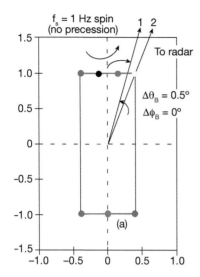

Figure 7.2 Bistatic geometry and sensor locations using the example target and motion of figure 2.8.

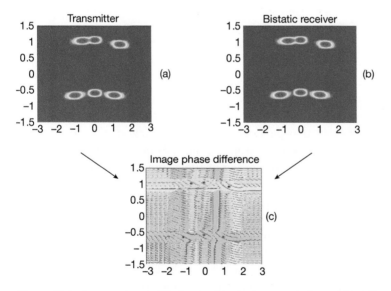

Figure 7.3 Image phase difference of typical monostatic and bistatic range, range-rate images. (a): Monostatic range, range-rate image. (b) Bistatic range, range-rate image. (c) Image phase difference contour.

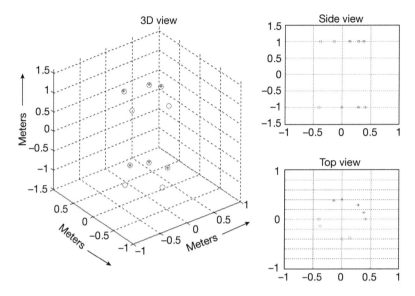

Figure 7.4   The 3D target space scattering center location estimates for block time t_b used in figure 7.3 obtained using IF-ISAR processing. Red dots are the estimated scattering center locations. Blue circles are the true scattering center locations, including those shadowed at the specified radar line of sight (LoS) to the target.

dots, and the true scattering center locations are indicated by blue circles. The empty circles represent the scattering center locations shadowed [according to equation (2.21), for which, according to this simplistic shadowing constraint, the rear scattering centers on the top portion of the cylinder are also discarded] to the sensors at this viewing angle. Note, in particular, that on examining the top-view of the 3D location estimates, the slipping scattering centers (top and bottom of the cylinder) are now located along their proper position on the ring of the cylinder edges at the assumed roll angle view of the target, even though there is no precession. Thus, the addition of the second sensor in the theta plane provides an additional look at the target in a direction orthogonal to the spin direction. This additional degree of freedom generally provides a more robust 3D location estimation.

Figure 7.5 illustrates the extracted 3D target location estimates obtained by sequentially processing the dual sets of monostatic and bistatic data blocks over an entire spin period, using dual data block pairs incremented by 30 pulses. The locations of the fixed and slipping scattering centers are clearly determined, and the slipping scattering

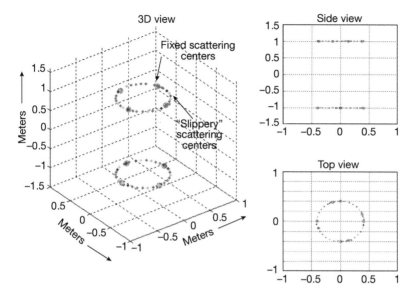

Figure 7.5 Composite of 3D scattering center location estimates for the spinning cylinder using IF-ISAR processing applied to a sequence of 2D dual wideband data blocks at times $t = t_b$, $b = 1, \ldots, B$ for B dual data block pairs covering a complete 360° roll angle cycle.

centers are properly located along the rim of the cylinder edges at spacings corresponding to the 30-pulse data block jump angle changes.

7.2.3   Linear Independence of Solutions for Bistatic Looks

It is instructive to examine the structure of equations (7.6) for a target undergoing pure spin motion and no precession when the bistatic sensor is located in the theta plane, perpendicular to the spin direction. First, consider equation (7.6a). The expression for $\hat{k}$ is given by equation (2.4), so that equation (7.6a) reduces to

$$\sin\theta\cos\varphi x_n + \sin\theta\sin\varphi y_n + \cos\theta z_n = R_n \tag{7.7}$$

The general expression for $\dot{\hat{k}}$ is given by equation (2.5). Using equation (2.5) with $\dot{\theta} = 0$, equation (7.6b) reduces to the simpler form:

$$\dot{\varphi}\sin\theta\hat{\varphi}\cdot\underline{r}_n = \dot{R}_n. \tag{7.8}$$

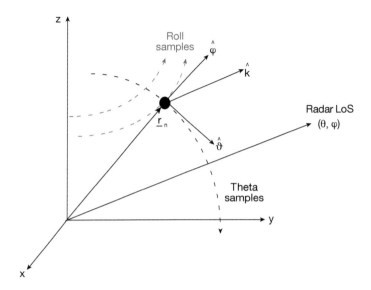

Figure 7.6 Localized orthogonal coordinate system for 3D phase-enhanced precessing.

Now, examine equation (7.6c). Because the φ coordinates for each sensor are the same and $\theta_2 = \theta_1 + \Delta\theta_B$, it can be shown that because $\Delta\theta_B$ is small,

$$\frac{2\pi}{\lambda}(\hat{k}_2 - \hat{k}_1)\cdot\underline{r}_n \approx \frac{2\pi}{\lambda}\Delta\theta_B\hat{\theta}\cdot\underline{r}_n. \tag{7.9}$$

Equations (7.7), (7.8), and (7.9) have an interesting graphical interpretation that illustrates the basic orthogonality (and consequent linear independence) of the equation set. The associated orthogonal coordinate system is illustrated in figure 7.6. Recall that $\hat{k}$ is a unit vector in the direction of the radar line of sight. Because $\hat{k}$ is perpendicular to both $\hat{\theta}$ and $\hat{\varphi}$, the three unit vectors $(\hat{k}, \hat{\varphi}, \hat{\theta})$ form an orthogonal set, defining a localized 3D coordinate axis set that changes with look angle to the target, but always remain orthogonal. Equations (7.7)–(7.9) can be written in the form

$$\hat{k}\cdot\underline{r}_n = R_n, \tag{7.10a}$$

$$\hat{\varphi}\cdot\underline{r}_n = \dot{R}_n/(\dot{\varphi}\sin\theta), \tag{7.10b}$$

$$\hat{\theta} \cdot \underline{r}_n = -\frac{\lambda}{2\pi} \cdot \Delta_n(phase)/\Delta\theta_B.$$ (7.10c)

Equations (7.10a)–(7.10c) clearly delineate the projections of the scattering center location vector $\underline{r}_n$ onto the axes of the localized 3D orthogonal coordinate system.

Using this geometrical interpretation of the linear independence of equation (7.6), one can compare the bistatic sampling space to the monostatic sampling grids applied to equation (5.7) in chapter 5 for the spinning cylinder. The result is depicted in figure 7.7.

For the pure spin example and the monostatic sensor, observe in figure 7.7(a) that as the data block is progressively incremented, data are collected only along the spin direction (roll angle φ), for θ taking on a constant value. Even though $\hat{k}_1$ is close to $\hat{k}_2$ for increments along the roll angle sampling grid, the cylinder edge slipping scattering center exhibits no change in phase over this sampling trajectory so its radius cannot be determined. The fixed-point scattering center locations for this motion are correctly estimated, because the phase for these scattering center types does change over the sampling trajectory. For the bistatic case, two different sampling trajectories are used, as illustrated in figure 7.7(b). These different views of the target provide changes in viewing angle in the $\hat{\theta}$ plane as well as the $\hat{\varphi}$ plane, resulting in complete 3D characterization of the target for the slipping as well as the fixed-point scattering centers.

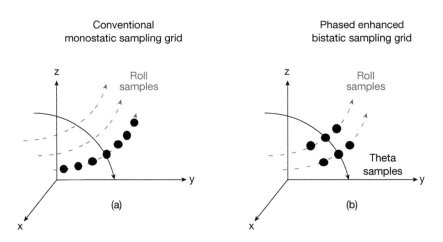

Figure 7.7 Data sampling grids for a target undergoing pure spin motion using (a) monostatic sensor samples, and (b) bistatic and monostatic sensor samples.

Although equations (7.10a)–(7.10c) are directly invertible to solve for $(x_n, y_n, z_n)_b$ using observables extracted at a single block time $t = t_b$, it is still useful to apply a cumulative approach that provides smoothing in the presence of noisy data. Rewrite equation (7.6) in the form

$$\underline{\underline{T}}_b \cdot \underline{R}_n = \begin{bmatrix} R_n \\ \dot{R}_n \\ -\Delta_n(\text{phase}) \end{bmatrix}_b, \tag{7.11}$$

where $\underline{\underline{T}}$ is the 3×3 motion matrix defined by

$$\underline{\underline{T}} = \begin{bmatrix} \hat{k} \\ \dot{\varphi} \sin \theta \hat{\varphi} \\ \dfrac{2\pi}{\lambda} \Delta \theta_B \hat{\theta} \end{bmatrix}. \tag{7.12}$$

Accumulate observable samples at $t = t_{b_1}$ and $t = t_{b_2}$ in a manner analogous to the technique used in chapter 5:

$$\begin{bmatrix} \underline{\underline{T}}_{b_1} \\ \underline{\underline{T}}_{b_2} \end{bmatrix} \cdot \begin{pmatrix} x_n \\ y_n \\ z_n \end{pmatrix} = \begin{bmatrix} \begin{pmatrix} R_n \\ \dot{R}_n \\ -\Delta_n(\text{phase}) \end{pmatrix}_{b_1} \\ \begin{pmatrix} R_n \\ \dot{R}_n \\ -\Delta_n(\text{phase}) \end{pmatrix}_{b_2} \end{bmatrix}. \tag{7.13}$$

By appending additional observables, solutions to equation (7.13) in the least-squares sense provide a more robust estimate of each scattering location (x_n, y_n, z_n).

7.3   The Monostatic Case

Typically, one would not consider using interferometric processing techniques for strictly monostatic processing because, by observation of the bistatic development, it is clear one requires two closely spaced views of the target constrained by the fact that these views must provide orthogonal (independent) information. However, for some specialized motion types, one can look ahead in time (using the recorded data) and search for a look angle to satisfy this condition. A specific condition where this is met is when the roll frequency f_φ is much greater than the precession frequency f_p. In this situation, as the roll angle $\varphi(t)$ increases by 2π increments, $\theta(t_2)$ sampled at the later time t_2 typically undergoes only small changes in look angle, emulating the bistatic case. At this later time the look angle $\hat{k}_2$ is close to $\hat{k}_1$. One can develop correlation search criteria that, assuming the motion is known, check whether or not the independence condition is met, and if met, apply interferometric processing techniques. In this case the data block at the later time effectively emulates that of a bistatic sensor placed at the same look angle, the difference being that the two look angles to the target are not simultaneous. A second instance where the look-ahead condition can be utilized is applicable to static range measurements and discussed in section 15.6. For this case the look-ahead condition can be simulated by the addition of a second roll-cut measurement, for which the target is physically displaced in roll angle by $\Delta\theta_B$, repositioned on the pylon, and the static range measurement repeated. The two roll-cut measurements are then processed coherently using the technique developed in this section. However as noted in appendix F, care must be taken to align the second roll cut measurement to the same CoR for the target as placed on the pylon. Otherwise data compensation as discussed in appendix F must be applied to the second roll cut data set.

7.3.1   Development

The development of the basic equation set used to estimate the N scattering locations (x_n, y_n, z_n) for the case of a monostatic sensor parallels the development for the bistatic case in section 7.2, with minor modifications. The starting point for the development begins with the expression for the monostatic scattering from the target given by

$$E_1 = \sum_n D_n e^{-j\frac{4\pi}{\lambda}\hat{k}_1 \cdot r_n}, \tag{7.14}$$

where the set $\{r_n\}$, $n=1,\ldots,N$ denotes the N dominant scattering centers on the target, located at (x_n, y_n, z_n), having complex amplitude D_n; and $\hat{k}$ denotes the radar LoS unit vector. Let $\hat{k}=\hat{k}_1$ at time $t=t_1$, and $\hat{k}=\hat{k}_2$ at $t=t_2$ where $t_2 \gg t_1$. Assume that at the later time, $t=t_2$, the scattered field is sampled in a direction $\hat{k}_2$, where $\hat{k}_2$ is close to $\hat{k}_1$. Express $\hat{k}_2$ in the form

$$\hat{k}_2 \approx \hat{k}_1 + \underline{\Delta k}. \tag{7.15}$$

Using equation (7.15), the scattered field E_2 at $t=t_2$ can be expressed in the form

$$E_2 = \sum \left[D'_n e^{-j\frac{4\pi}{\lambda}(\Delta k \cdot r_n)} \right] e^{-j\frac{4\pi}{\lambda}\hat{k}_1 \cdot r_n}, \tag{7.16}$$

where, as in the bistatic case, $D'_n \approx D_n$. Because of the proximity of $\hat{k}_2$ to $\hat{k}_1$, the range, range-rate images extracted from each 2D data block centered at times $t=t_1$ and $t=t_2$ respectively, are nearly identical, so that the individual scattering centers in each image can be identified and correlated. Extracting the differential phase from each scattering center in the dual range, range-rate images, equations (7.14) and (7.16) yields

$$Phase(I_{2n}) - Phase(I_{1n}) = -\frac{4\pi}{\lambda}(\hat{k}_2 - \hat{k}_1) \cdot r_n, \tag{7.17}$$

where I_{1n} and I_{1n} denote the complex amplitude of the image pixel of the nth scattering center for each image. Equation (7.17) can be written in a form analogous to equation (7.4),

$$\frac{4\pi}{\lambda}(\hat{k}_2 - \hat{k}_1) \cdot r_n = -\Delta_n(phase), \tag{7.18}$$

except the term containing ΔR in equation (7.4) is no longer present. The essence of the difference between the monostatic and bistatic cases is threefold:

1. The factor $\dfrac{2\pi}{\lambda}$ is replaced by the factor $\dfrac{4\pi}{\lambda}$ when applying the target space mapping.
2. The difference in look angles, $\hat{k}_2 - \hat{k}_1$ is obtained from a later time (vs. simultaneous) look angle to the target.

3. The need for the bistatic estimate of ΔR resulting from the LoS separation between sensors is replaced by the need for an accurate motion solution.

The second criteria assumes that, given a look angle to the target at $t = t_1$, a second look at the target at $t = t_2$ can be found where the look angle difference, $\hat{k}_2 - \hat{k}_1$, provides an *independent* differential look angle to the target. Independence essentially means out-of-plane of the general directional motion characterized by $\hat{k}_1$. Given that the motion solution is known, one can search ahead in time for a direction $\Delta \underline{k}$ projecting out of the $\hat{k}_1, \dot{\hat{k}}_1$ plane—or in the direction of $\hat{k}_1 \times \dot{\hat{k}}_1$ where $\times$ denotes the vector cross-product operation. For the pure spin example treated in chapter 5, imposing this criterion prohibits selecting successive roll angle samples at incremental times $t_2 = t_1 + b\Delta t$ for which $\hat{k}_2$ is contained in the plane determined by $\hat{k}_1, \dot{\hat{k}}_1$.

Define the correlation lag function $C(t_1, t_2 - t_1)$ according to

$$C(t_1, t_2 - t_1) = \frac{4\pi}{\lambda} |\hat{k}_2 - \hat{k}_1| \cdot z_m, \tag{7.19}$$

where z_m represents a maximum target dimension, the subscripts 1 and 2 refer to the times t_1 and t_2 respectively, and assume $t_2 \gg t_1$. Define the lag $\Delta \equiv t_2 - t_1$, and search the motion solution for the time t_2 such that the condition $C(t_1, \Delta) < \pi$ provides an unambiguous phase estimate. Choosing z_m to be larger than the maximum dimension of the composite of all the target scattering centers assures that when the phase condition $C(t_1, t_2 - t_1) < \pi$ is met, it holds for each scattering center. To eliminate those conditions for which $\hat{k}_2 - \hat{k}_1$ is in the plane of $\hat{k}_1, \dot{\hat{k}}_1$, impose a second criteria where the matrix T_m used in the mapping to target space, defined by

$$T_m \equiv \begin{bmatrix} \hat{k}_1 \\ \dot{\hat{k}}_1 \\ -\dfrac{4\pi}{\lambda}(\hat{k}_2 - \hat{k}_1) \end{bmatrix}, \tag{7.20}$$

is well conditioned from the viewpoint of matrix inversion. To implement the criteria, apply the MATLAB "cond" function, defined as the ratio of the largest to the smallest singular value of T_m. Large condition numbers indicate a nearly singular matrix relative to matrix inversion.

Table 7.1   Target and motion parameters applied to cylindrical precessing target of figure 7.2

T_p	1 s
T_s	100 s
κ	35°
θ_p	5°
$\alpha_p = \alpha_s$	0°
z_m	1 m
cond(T)	<5

Consider a specific example. First examine a small precession case having target and motion parameters defined in table 7.1 applied to the cylindrical target considered in section 7.2. The radar parameters are assumed the same as defined earlier for this example.

Figure 7.8 illustrates the assumed motion [figure 7.8(a)] and correlation time-lag [figure 7.8(b)] contour for this example. The correlation lag $C(t_1, t_2 - t_1)$ is plotted in figure 7.8(b) as a function $C(b_1, b_2)$ relative to the start data block time $t_1 = b_1 \Delta t$ (vertical axis) versus future data block lag time $t_2 - t_1 = b_2 \Delta t$ (horizontal axis, b_2). Brighter red regions represent large values of $C(b_1, b_2)$, and blue regions the lowest values indicating that in these regions the look angle difference $\hat{k}_2 - \hat{k}_1$ is small. Note that for the assumed motion parameters used in the simulation, for each block time sample b_1 there exists a lag b_2 $(t_2 \gg t_1)$ for which $C(b_1, b_2) < \pi$, a condition which characterizes the maximum possible phase difference for a target of length $z_m = 1$ m. (Note the red line overlays in this blue region of the contour plot are used only to elucidate those regions on the contour for which $C(b_1, b_2) < \pi$, and do not denote amplitude.) A magnified view of the lag b_2 for which this condition is satisfied is shown in figure 7.8(c). However, not all values of lag that satisfy the differential phase constraint lead to a well-conditioned T_m matrix.

Consider the application of the target space mapping equations defined by equation (7.11) where T_b is replaced by T_m defined in equation (7.20). The processing begins by selecting a data block time at $t = t_1$ and also a data block at the later time $t = t_2$ dictated by the condition on $C(t_1, t_2 - t_1) < \pi$. These two data blocks, referred to as dual data block pairs, are used to develop and solve the modified version of equation (7.11) (where T_b is replaced by T_m) for the scattering center location (x_n, y_n, z_n) estimates. For each set of dual data blocks, the two range, range-rate images are used to

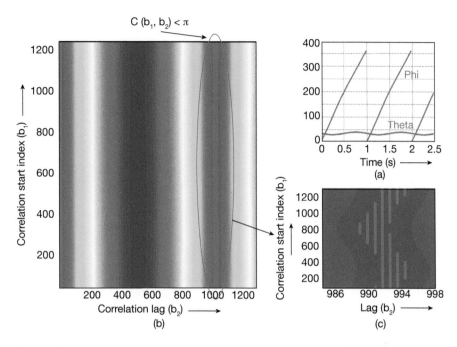

Figure 7.8   Motion versus time and correlation lag contour. (a) $[\theta(t), \varphi(t)]$ for 5° precession motion. (b) Correlation lag contour. (c) Expanded region of correlation lag contour satisfying differential phase solution constraint $C(b_1, b_2) < \pi$.

extract the observables $(R_n, \dot{R}_n)$ and $\Delta_n(phase)$ for each scattering center. The solution space can be expanded by selecting subsequent dual data block pairs sequentially and accumulating the sequential (x_n, y_n, z_n) estimates in target space.

Figure 7.9(b) illustrates the condition number for T_m versus sliding dual data block increments for this small precession example, where the red asterisk overlay indicates those occurrences for $\mathrm{cond}(T_m)$ having condition numbers less than five, for which a location estimate is calculated. Figure 7.9(a) illustrates the corresponding differential angles $\theta_2 - \theta_1$ (blue) versus sliding data block that meet the criteria $C(b_1, b_2) < \pi$ The red asterisk overlay in figure 7.9(a) indicates those regions where the condition number is greater than five and are excluded from the solution space. Observe that when the condition number is large θ_2 is very close to θ_1 and because $\varphi_2 \sim \varphi_1 + 2\pi$, the lag look angle is nearly in-plane with the initial look angle, meeting the differential phase condition but not meeting the condition number constraint. Figure 7.10 illustrates

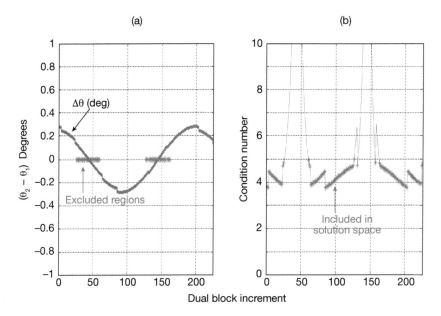

Figure 7.9 Solution robustness versus differential look angle for 5° precession motion. (a) Differential angle $\theta_2 - \theta_1$ (blue) versus sliding data block. Excluded block times that meet the criteria $C(b_1, b_2) > \pi$ indicated in red. (b) Condition number for T_m versus sliding dual data block increments. Condition numbers less than five are indicated in red.

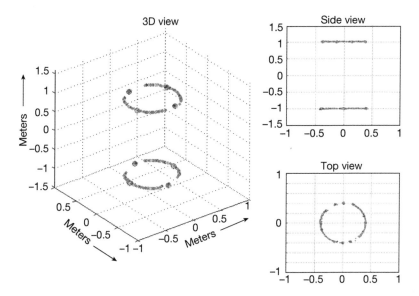

Figure 7.10 The 3D scattering center location estimates for cylindrical target illustrated in figure 7.2. Monostatic sensor, target motion precession $\vartheta_p = 5°$.

the resulting scattering center location estimates as the processing time $t = t_1$ sequences over the entire roll cycle, choosing only those time $t_2 > t_1$ having look angles for which the condition number is less than five. The results show a good correspondence to the actual scattering center locations on the target, and the slipping scattering centers on each ring are properly located indicating the true radius of the cylinder. The gaps in the slipping rings in target space occur because of the exclusion of the data due to the condition number constraint, and correspond to the excluded regions in figure 7.9(a).

Now consider the same motion example, but increase the precession angle to 10°, keeping the other motion parameters the same. The results are depicted in figures 7.11, 7.12, and 7.13 compared to figures 7.8, 7.9, and 7.10, respectively, for $\theta_p = 5°$. Some specific comments related to the comparison are as follows:

Note that examining figures 7.11(b) and 7.11(c), the look-ahead regions where $C(b_1, b_2) < \pi$ becomes sparse, and when combined with the criteria that cond

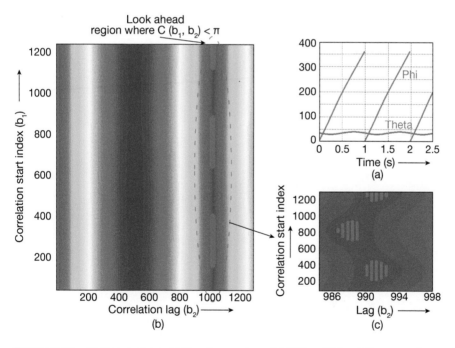

Figure 7.11 Motion and correlation lag contour. (a) $[\theta(t), \varphi(t)]$ for 10° precession motion. (b) Correlation lag contour. (c) Expanded region of correlation lag contour satisfying differential phase solution constraint.

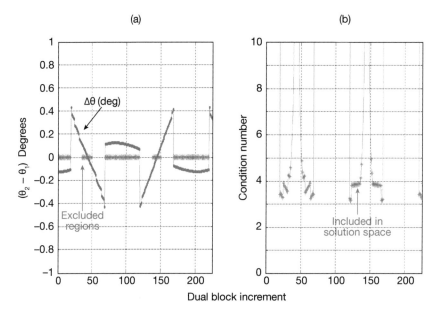

Figure 7.12 Solution robustness versus differential look angle for 10° precession motion. (a) Differential angle $\theta_2 - \theta_1$ (blue) versus sliding data block. Excluded block times that meet the criteria $C(b_1, b_2) > \pi$ indicated in red. (b) Condition number for T_m versus sliding dual data block increments. Condition numbers less than five indicated in red.

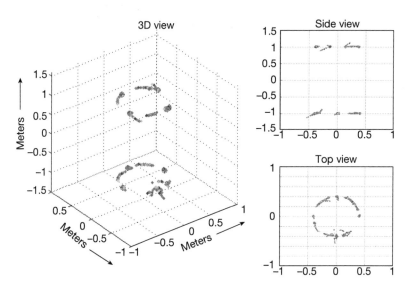

Figure 7.13 3D scattering center location estimates for cylindrical target illustrated in figure 7.2. Monostatic sensor, target motion precession $\vartheta_p = 10°$.

(T_m) is less than five, results in fewer solution occurrences for estimates of (x_n, y_n, z_n). This results in gaps in the slipping scattering center location estimates on the cylinder ring edge, although the fixed-point scattering center estimates are simply duplicated fewer times.

Although the condition number criteria is satisfied less often [compare figure 7.12(a) to figure 7.9(a)], the 3D location estimates accumulated in figure 7.13 offer a reasonable estimate of the overall locations of the scatterers on the target, albeit somewhat noisy when compared to the small precession case.

Increasing the precession angle even further results in an increasingly poor solution space. Thus it is clear that applications of interferometric processing for the monostatic sensor case is restricted to special classes of motion.

8

Motion Estimation Techniques

8.1 Introduction

In chapters 5–7 the problem of estimating various target specific characteristics such as the locations of the scattering centers and their associated geometrical theory of diffraction (GTD) coefficients using either wideband or narrowband signature data was addressed assuming the motion of the target was known. This is one special case of the more general joint target-motion estimation problem depicted in figure 5.3. In this chapter the complementing assumption of estimating the rotational motion of the target using a limited set of a priori target geometric information is considered. An application of this methodology would be to analyze a field test experiment where one has knowledge of the target under test and must extract the motion of the target from the test data collection. For example, if the target composition is complex, and static range measurements on the target are limited, one might design a field test to collect signature measurements on the target over trajectories not realizable on the static range. The techniques are based on fixing the target in its localized target space reference frame, and using the spherical coordinate look angles $[\theta(t), \varphi(t)]$ as motion parameters. This allows one to reference and correlate the measured signature of the target, defined relative to the motion parameters $[\theta(t), \varphi(t)]$ estimated from the flight test, to electromagnetic code computations using the target configuration as well as static range measurements made on the target as input.

The origin of the target-centered coordinate system is assumed located at the target's center of rotation, and the rotational motion of the target is defined relative to the origin of this system. The target space mapping defined by equation (5.22) relating the rotational motion of the target, its scattering centers, and the observable sequence $(R_n, \dot{R}_n, \ddot{R}_n)_b, n = 1, \ldots, N, b = 1, \ldots, B$ must be properly aligned in this system. It is assumed that the radar data are registered to the rotational motion defined relative to the CoR of the target as described in appendix F.

The chapter begins by assuming the target topology is known to some limited degree, and gradually relaxes these a priori assumptions as the development progresses through the chapter, eventually jointly estimating the motion along with some basic information relative to the topology of the target, for example, its length and/or width. Incorporating the surface topology into the solution methodology is particularly important in extracting the diffraction coefficients of components on the target for incorporation into a measurements-based modeling computational model of the target. The motion solution techniques in this chapter serve as a precursor to the more general solution framework of jointly estimating target and motion discussed in the subsequent chapters 9 and 10. Similar to the techniques introduced in chapters 5–7, the mathematical framework is based on using the range and range-rate observable sequence $(R_n, \dot{R}_n)_b, n = 1, \ldots, N, b = 1, \ldots, B$ for N scattering centers extracted from B data blocks centered at block times $t = t_b, b = 1, \ldots, B$.

Although the framework is generally valid for all motion types, this chapter focuses on the important special class of motion for targets undergoing torque-free motion governed by Euler's equations of motion [88, 89]. The important attributes of this type of motion are developed in appendix A and previously summarized in chapter 2, section 2.5. For torque-free Euler motion, the rotational motion $[\theta(t), \varphi(t)]$ is fully determined by the functional mapping, $F\{\cdot\}$, of the Euler dynamical parameters defined in equations (2.22), (2.23), and (2.24) and repeated here for convenience.

$$F\{\kappa, \theta_p, \phi_p, \psi\} \to [\theta(t), \varphi(t)] \tag{2.22}$$

where κ is the localized aspect angle from the radar to the target-motion angular momentum vector, θ_p is the precession cone angle relative to target-motion angular momentum vector, θ_p is the precession rotation angle, and ψ is the spin-angle.

The aspect and roll angles $[\theta(t), \varphi(t)]$ are given by equations (2.23) and (2.24):

$$\cos\theta = \cos\kappa \cos\theta_p + \sin\kappa \sin\theta_p \sin\phi_p \tag{2.23}$$

$$\varphi = -\psi + \tan^{-1}\left(\frac{\cos\kappa \sin\theta_p - \sin\kappa \sin\phi_p \cos\theta_p}{\sin\kappa \cos\phi_p}\right) \tag{2.24}$$

Equations (2.23), and (2.24) represent a general characterization of the rotational motion for torque-free Euler motion. As discussed in appendix A, for target types nearly rotationally symmetric about the target's z-axis, the precession angle θ_p is nearly constant and the motion can be further simplified by introducing the precession frequency and phase, f_p, α_p, and the spin frequency and phase f_s, α_s. The Euler angles θ_p and ψ reduce to the linear time variation $\varphi_p = 2\pi f_p t + \alpha_p$ and $\psi = 2\pi f_s t + \alpha_s$. In this case, the rotational motion is characterized by the five Euler parameters $\theta_p, f_p, \alpha_p, f_s, \alpha_s$, together with the sensor look angle κ to the target's angular momentum vector. The rotational motion is completely characterized by these six parameters and estimating the rotational motion is equivalent to estimating these six parameters. In this special case solutions to the joint target-motion estimation problem defined by equation (5.7) can be reduced to a $6 + 3(N) + 1$ parameter estimation problem, where N denotes the number of scattering centers on the target and the additional parameter $(+1)$ corresponds to the range bias estimate R_{bias}. For the techniques developed in this chapter, the five Euler parameters and κ are estimated using the range, range-rate observable sequences coupled with specific a priori target information.

The problem of estimating Euler motion from a time sequence of observation data has been discussed by numerous authors. References [90, 91, 92] discuss the estimation problem of determining Euler motion from a time sequence of images, and the object under motion is unknown. Typically, the time sequence considered in these references is relatively short in duration. Reference [93] incorporates a time sequence of specific features extracted from the image over a somewhat longer time sequence. Reference [94] presents a comprehensive review of the various approaches and techniques associated with time sequences of features extracted from camera images. The formulation developed in [95] is most representative of the approach used in this chapter. The time sequence of a particular observable feature—a magnetometer mounted on the target—is measured for an extended time, of a duration long enough that the harmonic nature of the motion of the object becomes readily manifested in the observation data. A parameter-based model of the motion is developed, and a nonlinear optimization of the fit of the model to the data is used to estimate the Euler parameters. In this chapter, the target topology is assumed known to some degree, and this a priori target knowledge is used to simplify the solution methodology.

8.2 Formulation

Using the GTD scattering model as discussed in chapter 2, the net scattering from the target can be decomposed into scattering from discrete scattering centers, located at (x_n, y_n, z_n), $n = 1, \ldots, N$. Relative to the target-centered coordinate system, these positions are either fixed or slipping. The range, range-rate radar observable $(R_n(t_b), \dot{R}_n(t_b))$ extracted from a block of radar data centered at time $t = t_b$ corresponding to the scattering center located at (x_n, y_n, z_n) is characterized by equations (5.2) and (5.3), expressed here in the form

$$R_n(t_b) \equiv \hat{k}_b \cdot \underline{r}_n, \tag{8.1}$$

$$\dot{R}_n(t_b) \equiv \dot{\hat{k}}_b \cdot \underline{r}_n, \tag{8.2}$$

where $\hat{k}_b$ and $\dot{\hat{k}}_b$ are defined by equations (2.4) and (2.5) and are functions of the motion parameters $\theta(t_b)$, $\dot{\theta}(t_b)$, $\varphi(t_b)$, $\dot{\varphi}(t_b)$. For a scattering center located at $\underline{r}_n$, referenced to the target-centered coordinate system, changes in the target motion are nonlinearly coupled to the temporal evolution of the range, range-rate observable sequences.

Equations (8.1) and (8.2) form a nonlinear set of equations characterizing the behavior of $\theta(t)$ and $\varphi(t)$ over the observation time interval. The focus of this chapter is developing solution techniques for which $\theta(t)$ and $\varphi(t)$ can be estimated from the time evolution of the range, range-rate observable sequences, assuming some a priori knowledge about the target topology. Because, for torque-free Euler motion, $\theta(t)$ and $\varphi(t)$ are completely characterized by the transformation of equations (2.23) and (2.24), estimating the motion is equivalent to estimating κ and the five Euler parameters characterizing the transformation.

It is instructive to develop a simple example to illustrate the utility in estimating target motion using some a priori target information. Assume there is a known slipping scattering center located on the target, for example at the ring of the base edge of the canonical cylinder illustrated in figure 2.7, and it can be correlated to a measured range observable sequence. Referring to section 2.5, the location, $r_s(\varphi)$, referenced to target space is given by the expression $\underline{r}_s(\varphi) = \rho_0 \hat{\rho} + z_0 \hat{z}$ where ρ_0 is the radius of the slipping ring and z_0 is its z-axis location. $\hat{\rho}$ is the radial unit vector $\hat{\rho} = \cos\varphi \hat{x} + \sin\varphi \hat{y}$. Using equation (2.4) for $\hat{k}_b$ and $\underline{r}_n = \underline{r}_s(\varphi)$, equation (8.1) can be written as

$$\sin\theta(t_b)\rho_0 + \cos\theta(t_b)z_0 = R_s(t_b), \tag{8.3}$$

where $R_s(t_b)$ is the range observable extracted from the wideband data at block time $t = t_b$. After some manipulation, equation (8.3) can be solved for θ in the form

$$\theta(t_b) = \tan^{-1}(z_0/\rho_0) + \cos^{-1}(R_s(t_b)/\rho_0). \tag{8.4}$$

Assuming ρ_0 and z_0 are known, the measurement sequence $R_s(t_b)$, $b = 1, \ldots, B$ combined with this a priori knowledge is sufficient to yield solutions for $\theta(t_b)$ directly.

Simplistic as this example may be, it demonstrates the utility of having some knowledge of the target topology for estimating the motion. In subsequent sections, three techniques based on these assumptions are developed: the first is based on using the range observable sequence similar to the simplistic example shown previously; the second is based on using the range-rate sequence; and the third is based on incorporating a combination of the range, range-rate sequences that relaxes the criteria on the degree of a priori target information required.

Before proceeding to the development of the three techniques it is useful to digress and develop some special case properties characteristic of torque-free motion useful in developing the techniques.

8.3   Torque-Free Motion Special Cases

8.3.1   Precession and Aspect Angle

Equation (2.23) characterizes $\theta(t)$ as a function of $(\kappa, \theta_p, \phi_p)$. Although κ changes with time over the overall motion of the target trajectory, for the cases considered in this text the changes in κ are assumed constant over the processing interval (condition B as defined in section 5.1). Examination of equations (2.23) and (2.24) also indicates a basic decoupling of the motion relative to spin and precession: the aspect angle θ is dependent only on κ and the set of precession parameters (θ_p, ϕ_p), whereas the roll angle φ is dependent on all the motion parameters. This decoupling can be exploited if slipping scattering centers are present on the target, typically characteristic of base edges and rings. It is shown in section 8.4 that if one can isolate certain regions on the range sequence where $\dot\theta = 0$, these regions can be exploited to extract (κ, θ_p) independent of the other motion parameters.

To illustrate the importance of the motion in the region defined about $\dot\theta = 0$, consider the variation of the observable sequence $R_s(t)$ using equation (8.3) for an elongated target assuming $z_0 \gg \rho_0$. In this case, the range sequence time variation can be approximated by $R_s(t) = z_0 \cos(\theta)$, and $\cos(\theta)$ is given by equation (2.23). The range observable sequence passes through a region where periodically $\dot\theta = 0$, which can be exploited to extract (κ, θ_p). Using equation (2.23), $\dot\theta$ is characterized by

$$-\dot\theta \sin\theta = \dot\phi_p \sin\kappa \sin\theta_p \cos\phi_p \tag{8.5}$$

When $\dot{\theta} = 0$, $\theta(t)$ is either maximum or minimum, which, using equation (8.5), occurs when $\phi_p = \pi/2$ or $3\pi/2$, respectively. Using these two values of ϕ_p in equation (2.23) implies either $\theta = \kappa - \theta_p$ or $\theta = \kappa + \theta_p$ in the region $\dot{\theta} = 0$. This property is exploited in section 8.4 to develop solutions for κ and θ_p assuming two slipping scattering centers separated by a known length can be isolated on the target.

8.3.2   Spin Motion For $f_s \gg f_p$

A second special case that often occurs in measurement (field) data is a spinning-precessing target corresponding to the physical case I_x, $I_y \gg I_z$, where I_x and I_z are the respective x- and z-axis moments of inertia of the target (see appendix A). Such a case is indicative of an elongated target having its long axis along the z-direction of target space. In this case, the observable sequence $(R_n, \dot{R}_n)_b$, $b = 1, \ldots, B$ passes through a region where periodically $\dot{\theta} = 0$, and relative to the radar line of sight (LoS), the target appears locally in time as a spinning target at nearly constant aspect angle. In this region there is an effective roll angle rotation rate given by $\dot{\varphi} = 2\pi f_\varphi$, influenced by both the spin and precession frequencies. As noted previously, the aspect angle θ in this region is given by $\theta = \kappa - \theta_p$ or $\theta = \kappa + \theta_p$.

To develop the expression characterizing f_φ for this special case, examine the roll behavior $\varphi(t)$ defined by equation (2.24) in the region $\dot{\theta} = 0$. Consider the denominator, $\sin\kappa \cos\phi_p$, of the arctangent term in equation (2.24). Using equation (8.5), in the region $\dot{\theta} = 0$, $\sin\kappa \cos\phi_p$ can be expressed as

$$\sin\kappa \cos\phi_p = -\dot{\theta} \sin\theta / (\dot{\phi}_p \sin\theta_p). \tag{8.6}$$

Because $\dot{\theta} \to 0$, by assumption it follows that $\sin\kappa \cos\phi_p \to 0$ in this region, and ϕ_p must also approach $\pi/2$ or $3\pi/2$ in these regions. For $\phi_p \approx (\pi/2)$, the arctangent term in equation (2.24) can be approximated using the expansion

$$\tan^{-1}\left(\frac{a}{x}\right) \to \frac{\pi}{2} - \frac{x}{a}, \tag{8.7}$$

as $x \to 0$. Noting that in this region $\cos\phi_p = \sin(\pi/2 - \phi_p) \approx (\pi/2 - \phi_p)$, equation (2.24) reduces to

$$\varphi \to -\psi - \frac{\pi}{2} + \frac{(\phi_p - \pi/2)\sin\kappa}{\sin(\theta_p - \kappa)}, \tag{8.8}$$

so that

$$|\dot{\varphi}| \to \dot{\psi} + \dot{\phi}_p \sin\kappa / \sin(\kappa - \theta_p).$$ (8.9)

Thus, in the region $\dot{\theta} \approx 0$, φ has a linear time variation. Noting that the sign of $\dot{\varphi}$ is ambigious, depending on the direction of rotation in the target based measurement system, the rotation frequency $f_\varphi \equiv |\dot{\varphi}|/2\pi$ is given by

$$f_\varphi = f_s + f_p \frac{\sin\kappa}{\sin(\kappa - \theta_p)}$$ (8.10)

The spin and precession frequencies are additive when $\kappa > \theta_p$. Because f_φ is a radar observable that can be directly extracted from the range-rate sequence, equation (8.10) is useful in estimating the spin frequency f_s once f_p, κ, and θ_p are determined. (Note that in some analysis, f_φ is referred to as the roll frequency of the target, often denoted as f_R).

The degree to which various known characteristics of the target relative to their utility in estimating κ and the five Euler parameters characterizing the motion is illustrated in table 8.1.

The following sections consider each of the various target attributes in relationship to the degree to which the motion and component diffraction coefficients on the target can be estimated.

8.4 Motion Extracted from a Wideband Range Observable Sequence for a Target Having Known Separation between Two Slipping Scattering Centers

The first wideband motion estimation technique considered uses the range observable sequence. The technique is based on the results of section 8.3.1, and assumes that two slipping scattering centers, separated by known length L, are present on the target. Denote the range sequences of each slipping scattering center as $R_1(t)$ and $R_2(t)$, and the range difference $\Delta R(t) = R_2(t) - R_1(t)$. Note that ΔR is independent of the range bias offset R_{bias}, discussed in appendix F. Using equation (8.3) applied to each slipping scattering center, if each of the slipping scattering centers has the same radius (e.g., the cylinder example treated in section 2.5), $\Delta R = L \cos\theta$ is precisely the apparent range difference along the radar LoS between two slipping scattering centers separated by length L. For slipping scattering centers having unequal radii, the motion solution determined as

Table 8.1   Progression of a priori target knowledge

Data Observables	A Priori Target Knowledge Required	Extracted Euler Parameters and Kappa	Extracted Target Parameters
Wideband RTI $\{R_n, D_n\}_b$, $b = 1, \ldots, B$	Length (L) between two slipping scattering centers	T_p, α_p, K, θ_p	—
Narrowband/Wideband DTI $\{\dot{R}_n, D_n\}_b$, $b = 1, \ldots, B$	Radius (a) of fixed-point scattering center	T_p, α_p, K, θ_p, T_s	—
Wideband range, range-rate observable sequence $\{R, \dot{R}_n, D_n\}_b$, $b = 1, \ldots, B$	Target topology $S(x_n, y_n, z_n) = 0$	Complete motion solution	Detailed characterization of target components
Narrowband RCS—specular timing	Angle of specular returns referenced to target coordinates	T_p, α_p, θ_p, K	Refine topology estimate

RTI denotes range-time intensity plot...plotting range observables vs time correlated to amplitude.
DTI denotes Doppler-time intensity plot...plotting Doppler (range-rate) observables vs time.
Examples are illustrated in figure E.2

follows is only approximate, but is a close approximation for elongated targets having $I_x, I_y \gg I_z$. The range observable sequence associated with each slipping scattering center can typically be identified because the range observable sequence variation viewed as a function of block time for a slipping scattering center is typically different from the range-rate variation of the observables associated with fixed-point scattering centers, which exhibit both precession and roll variation (e.g., figures 8.1 and 8.2). Because the technique assumes only a priori knowledge of slipping scattering centers, it provides only an estimate of κ, θ_p and $\dot{\phi}_p = 2\pi f_p = 2\pi / T_p$, where T_p denotes the precession period.

Assume that regions of the range observable sequence can be isolated about $\dot{\theta} \approx 0$. When $\dot{\theta} \approx 0$ then ΔR is either maximum or minimum. Using the results of section 8.3.1, θ must take either of the values $\kappa - \theta_p$ or $\kappa + \theta_p$ values in these regions. Denote the spread in ΔR as ΔR_1, and ΔR_2 in these respective regions. Imposing the constraint $\Delta R = L \cos\theta$ results in two equations characterizing κ and θ_p

$$\Delta R_1 = L \cos(\kappa - \theta_p), \tag{8.11a}$$

$$\Delta R_2 = L \cos(\kappa + \theta_p), \tag{8.11b}$$

from which, because L is assumed known, yield solutions for κ and θ_p. The value of T_p is determined by visual observation of the periodic variation of the range sequence associated with either slipping scattering center.

The next section presents an example illustrating the use of equation (8.11).

Example 1: Canonical cylindrical target, small precession motion

Consider the canonical cylindrical target introduced in section 2.5, illustrated in figure 2.7. Use the same radar parameters as for the simulation in section 2.5, but now assume a motion having precession cone angle $\theta_p = 5°$, $\kappa = 35°$ and spin and precession periods $T_p = 1$ s, $T_s = 0.1$ s, such that $f_s \gg f_p$. The range sequence extracted from the simulation data is illustrated in figure 8.1. For this example, the extracted range sequence is illustrated over 3.5 precession periods over the time interval (0–3.5 s). Observe the relatively slow precession variation of the range sequence for the slipping scattering centers compared to the fast variation of the fixed-point scattering centers. Also note that because the fixed-point scattering centers for this target are attached to the cylinder ends, the precession variation would be evident even if the slipping scattering centers were not present. The extended time variation in the range observable sequence illustrates how one can isolate the regions of the range sequence where $\dot{\theta} = 0$ by visual identification of the inflection points associated with the slipping scattering center range sequence trajectories. Because the example target has two slipping scattering centers, one at each end, and eight fixed-point scattering centers, four at each end, examination of figure 8.1 also illustrates the inherent decoupling of the motion of the slipping scattering centers (precession) versus the fixed-point scattering centers (spin and precession). The precession period of the slipping scattering centers is identified as 1 s, and the coupling of the spin frequency into the roll angle motion of the fixed-point scattering centers is clearly evident.

As indicated on the figure, the presence of the two slipping scatterers defines inflection regions as a function of time where the range sequence variation versus time exhibits periodic minima and maxima, which correspond to the desired regions for which $\dot{\theta} \approx 0$. It is straightforward to extract the envelope separation distances $\Delta R_1 = 1.5$ m and $\Delta R_2 = 1.725$ m in the regions $\dot{\theta} \approx 0$. Using the known length $L = 2$ m between the ends of the cylinder, equations (8.11a) and (8.11b) are solved for the solutions $\kappa \sim 35.1°$ and $\theta_p \sim 5°$, which correspond well to the simulation values $\kappa = 35°$ and $\theta_p = 5°$.

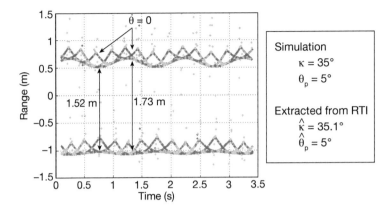

Figure 8.1   Range observable sequence versus block time for the small precession example applied to the cylindrical target illustrated in figure 2.7. Simulation parameters are $\theta_p = 5°$, $\kappa = 35°$, $T_p = 1$ s, and $T_s = 0.1$ s.

8.5   Motion Extracted from a Wideband Range-Rate Observable Sequence for a Target Having Known Radius

Motion extraction using the range-rate observable sequence is also based on isolating regions for which $\dot{\theta} \approx 0$, but using a priori target knowledge based on a known radius of a fixed-point scattering center. The technique development parallels that in section 8.4, except that it uses the range-rate observable sequence. However, because the motion of a fixed-point scattering center is dependent on both the spin and precession frequencies, a more robust set of motion parameters can be determined, namely the set $(\kappa, \theta_p, f_s, f_p)$. The range-rate sequence is characterized by equation (8.2), where $\hat{k}$ is defined by equation (2.5):

$$\dot{R}_n = \dot{\theta}\hat{\theta}\cdot\underline{r}_n + \dot{\varphi}\sin\theta\,\hat{\varphi}\cdot\underline{r}_n \tag{8.12}$$

where $\underline{r}_n$ denotes the position of the known fixed-point scattering center. Isolate the range-rate sequence about the region $\dot{\theta} \approx 0$ so that in this region only the second term in equation (8.12) contributes to $\dot{R}_n$. Observe from (8.12) that in the regions about $\dot{\theta} \approx 0$, since $\hat{\varphi}\cdot\hat{z} \equiv 0$, only the radial dimension of the rotating scattering center contributes to the range-rate observable sequence. In these regions the maximum of $\dot{R}_n$ is given by $\max(\dot{\varphi}\sin\theta\,\hat{\varphi}\cdot\underline{r}_n) = a\dot{\varphi}\sin(\theta)$, where a denotes the radius of the fixed-point scattering center. Thus in a manner similar to the wideband range observable sequence, in the regions about $\dot{\theta} \approx 0$, two equations are determined relating the range-rate observables $(\dot{R}_n)_+$ and $(\dot{R}_n)_-$; these correspond to $\theta = \kappa + \theta_p$ and $\theta = \kappa - \theta_p$, respectively. They are

$$(\dot{R}_n)_+ = a \, \dot{\varphi} \sin(\kappa + \theta_p)$$
$$(\dot{R}_n)_- = a \, \dot{\varphi} \sin(\kappa - \theta_p). \tag{8.13}$$

In equation (8.13), the roll angle φ is characterized by the roll frequency $f_\varphi = 1/T_\varphi$, so that $\dot{\varphi} = 2\pi/T_\varphi$. Thus if a is known and T_φ can be measured from the range-rate observable sequence, one can solve equation (8.13) for κ and θ_p. Now consider a large precession example illustrating the use of equation (8.13).

Example 2: Large precession and $f_s \gg f_p$

Assume the same cylindrical target geometry used in the example of section 8.4, except now enhance the precession by choosing $\theta_p = 30°$ and a look angle to the angular momentum vector $\kappa = 60°$. For this example, the minimum and maximum aspect angles as referred to the z-axis of the target, that is, $\kappa - \theta_p$ and $\kappa + \theta_p$, are given by 30° and 90°, respectively, indicating that for one region where $\dot{\theta} = 0$, the radar LoS to the target is broadside to the cylinder. The observable sequence $\{R_n, \dot{R}_n\}_b, b = 1, \ldots, B$ for this example, as well as the motion, are illustrated in figure 8.2 over the time interval $(0, 1.5 \text{ s})$. Because $f_s \gg f_p$, the motion of the slipping scattering centers versus the fixed-point scattering centers is clearly evident.

Comparing figure 8.2(a) to figure 8.2(b) for those regions where the envelope of range observable sequence undergoes maximum separation, the envelope of the range-rate observable sequence undergoes minimum separation, and conversely, when the envelope of the range observable sequence has minimum spread, the envelope of the range-rate observable sequence has maximum spread. This complimentary behavior can be used to identify the regions where $\dot{\theta} = 0$. Note also that for this example, when $\theta = \kappa + \theta_p = 90°$ aspect angle (i.e., broadside to the cylinder), the range observable sequence minimum separation for the slipping scattering centers at the edge of the cylinder is zero, and the spread in the range-rate observable sequence is maximum. Thus, when looking at the cylinder broadside, the range separation between the two cylinder ends (having equal radii) is zero and the Doppler spread is maximum.

Examine the regions of the range-rate observable sequence corresponding to $\dot{\theta} \approx 0$ illustrated in figure 8.3.

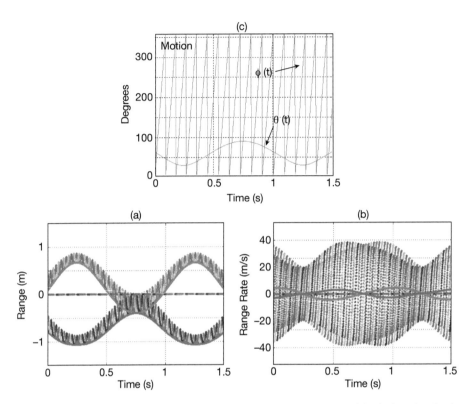

Figure 8.2   Range and range-rate observable sequences versus block time for the large precession example applied to the cylindrical target illustrated in figure 2.7. Simulation parameters are $\theta_p = 30°$, $\kappa = 60°$, $T_p = 1$ s, and $T_s = 0.1$ s. (a) Range observable sequence. (b) Range-rate observable sequence. (c) Motion $\theta(t)$, $\varphi(t)$ versus time

In most cases it is more accurate to measure the two-sided envelope spread of the range-rate sequence rather than a one-sided offset. To estimate the motion parameters, isolate the regions $\dot{\theta} = 0$ and determine the maximum and minimum range-rate envelope separation, that is $\Delta \dot{R}_1 = 29.5$ m/s and $\Delta \dot{R}_2 = 59$ m/s as indicated in figure 8.3. Multiply each side of equation (8.13) by a factor of 2, to obtain

$$\Delta \dot{R}_1 = 2\dot{R}_{n+} = (2a\dot{\varphi})\sin(\kappa + \theta_p)$$
$$\Delta \dot{R}_2 = 2\dot{R}_{n-} = (2a\dot{\varphi})\sin(\kappa - \theta_p). \tag{8.14}$$

The roll period, T_φ, can be determined by expanding the range-rate sequence about the region $\dot{\theta} = 0$, as depicted in figure 8.4. Because the target geometry is assumed

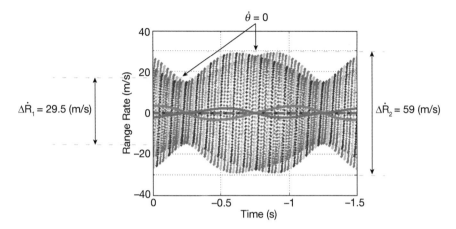

Figure 8.3 Expanded view of the range-rate observable sequence for example 2. Isolate and expand the region about $\dot{\theta} = 0$.

known, the four fixed-point scattering centers at each end of the cylinder result in a four-fold variation in the time evolution of the range-rate sequence as each point scattering center on the cylinder edges rotates and becomes visible to the sensor. Using this four-fold symmetry, the estimate $T_\varphi = 0.085$ s corresponding to $\dot{\varphi} = 2\pi / T_\varphi = 73.9$ Hz is obtained and illustrated in figure 8.4.

Substituting in the measured values, for $\Delta \dot{R}_1$ and $\Delta \dot{R}_2$, as well the estimated value of $\dot{\varphi}$ and the known radius, $a = 0.4$ m, equation (8.14) can be solved for κ and θ_p to be $\theta_p = 29.3°$, $\kappa = 60.7°$; these estimates are very close to the simulation parameters.

Spin and precession parameters

It remains to estimate the spin and precession frequencies f_s and f_p. However, now the procedure is much simpler due to the reduced dimensionality of the problem. Using equation (8.10) the expression for the spin period T_s can be expressed in the form

$$T_s = \frac{1}{1/T_\varphi - \dfrac{\sin \kappa}{\sin(\kappa - \theta_p)} \dfrac{1}{T_p}}. \tag{8.15}$$

The tumble period T_p is estimated from the range observable sequence associated with the slipping scattering centers to be $T_p \approx 1$ s. Substituting in the estimated values

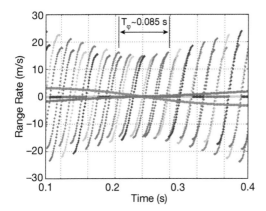

Figure 8.4   Range-rate observable sequence expanded about the region $\dot{\theta} = 0$. Estimate $T_\varphi \approx 0.085$ s.

for T_p and T_φ into equation (8.15) we obtain the estimated value $T_s = .099$ s, which is very close to the simulation value of $T_s = 0.1$ s.

Physical constraints on the spin and precession parameters

Solutions to Euler's dynamical equations pose specific constraints on the behavior of f_s and f_p. These are fully developed in [88]. To summarize these here, assume the body is nearly symmetrical so that the x and y moments of inertia I_x and I_y, respectively, are nearly equal. Then the motion of the target is constrained to satisfy

$$f_s = f_p \left(\frac{I_x}{I_z} - 1 \right) \cos\theta_p, \tag{8.16}$$

where I_z is the z-axis moment of inertia. Note that if $\theta_p < 90°$ and $I_x > I_z$ (typically the case for elongated targets), then f_s and f_p must have the same sign. Using this result in equation (8.10) for $\kappa > \theta_p$ means that f_s and f_p are additive. Equation (8.16) is often useful in eliminating ambiguities that can occur in solutions for $\theta(t)$, $\varphi(t)$. Note also that because f_s, f_p, and θ_p can be determined from the radar observables, the target moment of inertial ratio I_x/I_z can be indirectly inferred using equation (8.16).

8.6   Joint Solutions for *L*, *a*, *κ*, and θ_p

One of the value-added features of using the two-dimensional sequential spectral estimation technique applied to time sequences of data blocks is that the technique results in simultaneous range and range-rate observable estimates, because the observables are extracted as range, range-rate pairs. Using these paired observables, the motion estimation techniques described previously using either the range observable sequence or the range-rate observable sequence can be combined to relax the requirement of a known length and radius. The joint estimation scheme is illustrated in figure 8.5. As previously noted, the corresponding maxima and minima of the range observable sequence and the range-rate observation sequence for each region where $\dot{\theta} \cong 0$ are complementary—that is, the range sequence maximum separation corresponds to the range-rate minimum separation for one region, and conversely for the subsequent region in time where $\dot{\theta} \cong 0$ occurs. Given this fact, equations (8.11) and (8.14) can be combined in the coupled form

$$\Delta R_1 = L \cos(\kappa - \theta_p) \tag{8.17a}$$

$$\Delta R_2 = L \cos(\kappa + \theta_p) \tag{8.17b}$$

$$\Delta \dot{R}_1 = W_D \sin(\kappa - \theta_p) \tag{8.17c}$$

$$\Delta \dot{R}_2 = W_D \sin(\kappa + \theta_p) \tag{8.17d}$$

where the normalized width W_D is defined as $W_D \equiv 2a\dot{\varphi}$, and the notation ΔR_1, ΔR_2, $\Delta \dot{R}_1$, $\Delta \dot{R}_2$ denotes the two-sided maximum/minimum separation of the envelope of the respective range and range-rate observable sequences as indicated in figure 8.5. Equations (8.17a)–(8.17d) present four independent equations in the four unknowns L, W_D, κ, and θ_p. Solutions can be determined in the two-step process:

1. Eliminate κ and θ_p to obtain two equations for L and W_D:

$$\left(\frac{\Delta R_1}{L} \right)^2 + \left(\frac{\Delta \dot{R}_1}{W_D} \right)^2 = 1, \tag{8.18}$$

$$\left(\frac{\Delta R_2}{L} \right)^2 + \left(\frac{\Delta \dot{R}_2}{W_D} \right)^2 = 1, \tag{8.19}$$

which yields solutions for L and W_D in the form

$$\begin{pmatrix} \dfrac{1}{L^2} \\ \dfrac{1}{W_D^2} \end{pmatrix} = \begin{pmatrix} \Delta R_1^2 & \Delta \dot{R}_1^2 \\ \Delta R_2^2 & \Delta \dot{R}_2^2 \end{pmatrix}^{-1} \cdot \begin{pmatrix} 1 \\ 1 \end{pmatrix}. \tag{8.20}$$

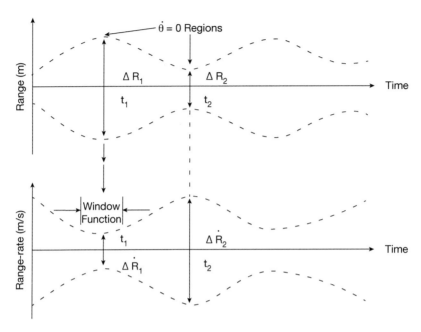

Figure 8.5   Depiction of the envelope structure of the range and range-rate observable sequences.

2.   Given L and W_D, now solve the following two linear equation for κ and θ_p:

$$\kappa + \theta_p = \cos^{-1}\left(\frac{\Delta R_2}{L}\right), \tag{8.21}$$

$$\kappa - \theta_p = \cos^{-1}\left(\frac{\Delta R_1}{L}\right). \tag{8.22}$$

Applying this joint solution technique to the joint range, range-rate observable sequences illustrated in figure 8.5 for the large precession case results in the estimates

$$W_D = 59 \text{ m/s, } L = 2 \text{ m, } \theta_p = 29.1°, \text{ and } \kappa = 60.9°$$

Using $W_D \equiv 2a\dot{\varphi}$ and $\dot{\varphi} = 73.9$ radians/s as determined earlier, the solution for the radius is

$$2a = 59/73.9 \Rightarrow a = 0.399 \text{ m}$$

consistent with the simulation value of $a = 0.4$ m.

9

Joint Motion and 2D/3D Characterization of Tumbling Targets

9.1   Introduction

In chapter 8 the development of a methodology for estimating the target's motion parameters when some target characteristics such as its length and width are known was addressed in detail. In this and the following chapter, techniques for jointly estimating the target's motion and its scattering center locations using wideband data blocks are developed in several steps of increasing complexity. This chapter focuses on two closely related joint target-motion solution techniques for the case of a target in torque-free tumble motion, but for which radar line of sight (LoS) does not lie in the tumble plane. Such a sensor-target scenario is considered in [96], leading to the development of a technique using two multistatic sensor observations (analogous to the formulation in chapter 7) on the target to develop three-dimensional (3D) estimates of the target topology for the case where the motion is assumed known. Using the techniques developed in this chapter, joint solutions for the motion and 3D scattering center location estimates become possible using a single sensor. Most importantly, the range, range-rate and range-acceleration observable sequences need not be correlated in time to apply the techniques.

The problem of estimating Euler motion from a time sequence of observation data has been discussed by numerous authors. The formulation developed in [97] is most representative of the approach used in this chapter. In this reference, the time sequence of data obtained from an instrument mounted on the target, a magnetometer, is extracted over an extended time duration, long enough that the harmonic nature of the motion of the object becomes readily manifested in the magnetometer data. A parameter-based model of the motion is developed, and a nonlinear optimization of the fit of the model to the data is used to estimate the Euler parameters. This chapter presents a generalization of this approach, which incorporates an estimate of the target scattering center location parameters and the motion variables, but is restricted to pure tumble motion. Recall from the discussion in chapter 8, for torque-free Euler motion, the joint target-motion solution framework is completely characterized by the five Euler variables, the look angle κ to the target angular momentum vector, and the scattering center locations on the target. For this special case, solutions to the joint target-motion estimation problem defined by equation (5.7) can be reduced to a $6 + 3(N) + 1$ variable search space, where the additional variable (+1) corresponds to a range bias estimate. This search space is considerably reduced for the case when the target exhibits a pure rotational motion about the target angular momentum vector. This tumble motion condition requires the estimation of three parameters, $(T_p, \alpha_p, R_{bias})$; these are the tumble period, tumble phase, and range bias, respectively. For the case where the radar LoS to the target lies outside the tumble plane, the value of κ can be estimated from the composite target space mapping of the target [scaled to $\sin(\kappa)$] if some characteristic of the target, such as its length, is known. Once $(T_p, \alpha_p, R_{bias})$ are determined, the time evolution of the range, range-rate and range-acceleration observables associated with each scattering center are used to extract the 3D location coordinates of the scattering centers referenced to target space.

For the development presented, the radar data must be aligned to the rotational component of the motion, registered to center of rotation (CoR) of the target as described in appendix F. As discussed there, various factors can contribute to errors in the estimate of the translational motion $R_T(t)$ used to register the range observable sequence to the rotational motion, and must be addressed. For long track times and closely spaced data blocks, the estimate becomes increasingly more accurate. When applied to the observables sequence over the shorter time frame of constant κ, the range alignment in equation (F.10) reduces to a constant range offset, denoted as R_{bias}, which must be estimated. Two solution techniques for estimating the parameters $(T_p, \alpha_p, R_{bias})$ are developed. In the first case, the motion parameters are estimated independently: Tp is extracted from the observables using a unique property that relates the scattering center range observables to the corresponding range-acceleration observables and is developed in section 9.3; the range bias offset, R_{bias}, is estimated using the wideband range autocorrelation contour described in chapter 4, and is illustrated in section 9.5; finally, the phase α_p is set to orient the target in target space. The second technique developed in section 9.6 jointly estimates these motion parameters by applying the motion optimization dispersion metric defined in equation (9.28) over a composite search space that covers the three parameters. The dispersion metric introduced in section 9.6 is applicable to a wider class of Euler motion other than pure tumble, but is not addressed in this text. Once the motion and range bias parameters are determined, the 3D scattering center location estimates are obtained using the appropriate mapping of the range, range-rate and range-acceleration sequences $(R_n, \dot{R}_n, \ddot{R}_n)_b, n = 1, \ldots, N, b = 1, \ldots, B$ to target space as developed in section 9.2. In chapter 10 a joint solution technique is developed extending these techniques to more general torque-free Euler motion. However, to apply the techniques developed there, one must first correlate the range, range-rate observable sequences in time. It is also assumed that the range offset, R_{bias}, is known and/or independently estimated. Solutions for this more general problem, assuming a known range bias, appear in the literature [98, 99] but a novel approach developed by Dr. M. L. Burrows (private communication, see list of contributors) is presented in chapter 10.

9.2   Development

As discussed earlier in section 2.2 of chapter 2, when integrating measurement data into a computational measurements-based signature model, it is often necessary to introduce two separate coordinate system orientations: one for processing the measurement data, the other for characterizing the computational model. Both generally use a spherical coordinate system. However, the CoR and axis orientation of the computational system and measurement system are typically different. Typically, the orientation of the (x, y, z)-axes for the computational system is chosen so the z-axis is colocated with the long axis of the target and the CoR lies on the z-axis, as illustrated in figure 9.1(a). For targets in pure tumble, it is convenient to process the field measurement data by defining the (x', y', z') measurement coordinate system so that the (x', y') plane corresponds to the tumble plane and the z'-axis is perpendicular to the tumble plane aligned with the angular momentum vector. The LoS of the radar to the angular momentum vector is given by $\theta' = \kappa$. Appendix F addresses the relationship between the computational and measurement coordinate systems and develops the techniques required to integrate the field and static range measurements on the target into the computational signature model. The measurement system oriented in this manner is illustrated in figure 9.1(b).

Targets in pure tumble for which the radar LoS lies in the tumble plane, $\kappa = 90°$, can be characterized using the two-dimensional (2D) composite mapping techniques based on the wideband target space mapping developed in section 6.3.2. The resulting target model is registered in a 2D target-centered coordinate system, and the range, range-rate observable sequence $(R_n, \dot{R}_n)_b, b = 1, \ldots, B$ is mapped into this fixed frame using the 2D, wideband target space mapping equations defined by equation (6.14).

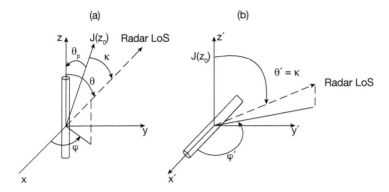

Figure 9.1 Two target orientations in target space coordinates for use in modeling the motion and geometry of tumbling targets. (a) The (x, y, z) computational coordinate system and its relationship to the Euler motion parameters. (b) The (x', y', z') measurement system introduced to process the measurement data for a target in pure tumble motion.

The mapping requires the tumble period of the target be known. If the radar LoS lies in the tumble plane, the resulting 2D mapping provides an accurate projection of the target scattering center locations onto the (x', y')-plane. However, if the radar LoS lies out of the tumble plane ($\kappa < 90°$), the (x', y') estimates obtained in this manner are distorted, and the composite mapping distorts further as the look angle moves out of the (x', y') plane. Note that this distorted mapping does provide useful information (see for example figure F.3 of appendix F), but the 3D mapping developed in section 9.3 must be employed to obtain accurate location estimates.

In this chapter, the target space mapping equations characterized by equation (5.22), defined relative to the measurement coordinate system illustrated in figure 9.1(b), that incorporate the scattering center acceleration observable, are used to obtain 3D estimates of the target scattering center locations. The technique provides true coordinate location estimates in the (x', y') plane, scaled to the look angle, $\sin(\kappa)$, and develops an estimate of the out-of-plane z'-axis location of each scattering center, scaled to the look angle $\cos(\kappa)$. This more general technique offers several advantages:

The assumption of a known tumble period is relaxed. The parameters $(T_p, \alpha_p, R_{bias})$ are estimated from a minimal number of sequential data blocks versus the requirement of viewing the target over a full tumble period.

The location estimates of the target's scattering centers (x'_n, y'_n) can be carried out independent of the out-of-plane distance, z'_n. The location estimates in the (x', y')

plane correspond to the true scattering center (x'_n, y'_n) locations, are uncoupled from the z'_n coordinate location estimates, and are independent of the range estimate. As a consequence, the (x'_n, y'_n) plane scattering center location coordinates are insensitive to range alignment errors caused by R_{bias} and any wideband delay mechanisms that might occur.

The z'_n location estimates (which are dependent on the factors listed in step 2) are directly correlated to the (x'_n, y'_n) location estimates, and thus offer the potential for 3D scattering center location estimates.

For applications utilizing static range measurements, raising the LoS of the transmitter out of the plane of rotation offers a way of collecting data that allows for 3D target characterization.

The technique is also applicable to targets undergoing pure spin motion. This is because a spinning target can be viewed as tumbling about the angular momentum vector that is colinear with the target's z'-axis, and proper orientation of the target space coordinate axes allows one to apply this approach.

The following sections present the formulation that validates these statements, followed by simulation examples to illustrate some potential scattering center location interpretation anomalies that arise using wideband 2D processing based on equation (6.14) for $\kappa < \pi/2$. In section 9.5, the effects of the CoR range offset error on the location estimation accuracy are considered, and techniques using the wideband autocorrelation filter developed in chapter 4 to estimate the range offset correction are presented.

9.2.1   Target Coordinate Systems and Motion Models

The choice of coordinate system to be used for the 3D target-centered frame of reference can make a significant difference in terms of the ease of analysis going forward. Figure 9.1(a) illustrates the typical target-fixed coordinate system used to analyze targets undergoing general spinning, precessing Euler motion and also used to characterize the computational signature model.

In figure 9.1(a) an elongated cylindrical target is oriented in the target (x, y, z) reference system such that the longitudinal axis of the target is colinear with the z-axis of the target reference coordinate system. Assume the tumble motion is constrained to the (x, z) plane. First consider the case where the radar LoS is restricted to lie in the (x, z) tumble plane. For this coordinate system orientation, when the target is in pure tumble, the angular momentum vector J is colinear with the $\hat{y}$-axis and $\theta_p = \pi/2$. Let κ

define the angle between the radar LoS and the angular momentum vector J. Because the radar LoS is coplanar with the (x, z) tumble plane, $\kappa = \pi/2$. The time evolution of the motion angles $\theta(t)$, $\varphi(t)$ are characterized by equations (2.23) and (2.24) of chapter 2:

$$\cos\theta = \cos\kappa \cos\theta_p + \sin\kappa \sin\theta_p \sin\phi_p \qquad (2.23)$$

$$\varphi = -\psi + \tan^{-1}\left(\frac{\cos\kappa \sin\theta_p - \sin\kappa \sin\phi_p \cos\theta_p}{\sin\kappa \cos\phi_p}\right) \qquad (2.24)$$

Using these values for θ_p and κ in equations (2.23) and (2.24), the time evolution of (θ, φ) is illustrated in figure 9.2(a). The standard spherical coordinates used in defining the look angle to the sensor in the target reference coordinate system, (θ, φ), are constrained to the limiting values

$$0 \le \theta \le \pi$$
$$0 \le \varphi \le 2\pi. \qquad (9.1)$$

Referring to figure 9.1(a), because θ is bounded by the interval $(0, \pi)$, the value of φ must flip $180°$ as θ approaches either 0 or π radians in order to cover the full $360°$ rotation cycle as defined in the (x, z) plane. As a consequence of this discontinuity,

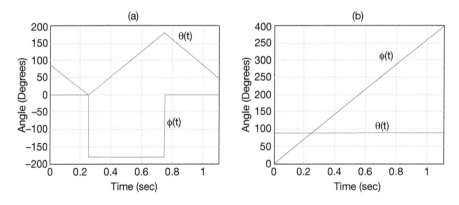

Figure 9.2　Motion variables $\theta(t)$, $\varphi(t)$ versus time for two differing target geometries ($\kappa = 90°$) having tumble period $T_p = 1$ s. (a) Target cylindrical axis oriented on z-axis, $\theta_p = \pi/2$, $\kappa = \pi/2$. (b) Target cylindrical axis oriented on x-axis, $\theta_p = 0$, $\kappa = \pi/2$.

both $\dot{\theta}$ and $\dot{\varphi}$ are not continuous functions, and the rotational motion for a target in pure tumble is not conveniently modeled using this target orientation in target space.

This is easily remedied by defining a target-fixed measurement coordinate system that constrains the tumble plane of the target to the (x', y') plane. In this case, the target-fixed coordinate system is oriented so the z'-axis aligns with the angular momentum vector J as shown in figure 9.1(b). As a result, for $\kappa = \pi/2$, the radar LoS lies in the (x', y') tumble plane such that $0 < \varphi' \leq 2\pi$.

For the target orientation in figure 9.1(b), the elongated axis of the target is positioned along the x'-axis of the target coordinate system. The radar LoS is elevated to the tumble plane, and the tumble motion is modeled as the radar viewing angle to the target rotating about the z'-axis in target space such that $\theta' = \kappa \leq \pi/2$. The target coordinate system z'-axis and inertial system axis (i.e., J) are coincident, and the motion model has zero precession, $\theta_p = 0$. The tumble motion of the target is constrained to the (x', y') plane so that when the radar LoS is in the tumble plane, then $\kappa = \pi/2$. For this target orientation and pure tumble motion, the motion angle rates, $\dot{\theta}$ and $\dot{\varphi}$ are continuous for all values of κ. Applying equations (2.23) and (2.24) for this target orientation, $(\theta'(t), \varphi'(t))$ are given by

$$\theta'(t) = \kappa$$
$$\varphi'(t) = (2\pi/T_p)t + \alpha_p,$$

(9.2)

where T_p denotes the tumble (rotation) period. Because the target coordinate system z'-axis and J are coincident, both κ and the radar aspect angle θ' are identical and the motion defined by $\varphi'(t)$ corresponds to pure tumble. Furthermore, the motion angles $\theta'(t)$ $\varphi'(t)$ are characterized by the continuous functions illustrated in figure 9.2(b), and correspond to the more intuitive linear variation $\varphi'(t)$ covering the interval $0 < \varphi'(t) \leq 2\pi$ commonly used in most 2D composite image processing techniques.

Once the target scattering center locations and diffraction coefficients have been estimated in this measurement system, a coordinate system mapping is required to transform the measurement system angles to the computational system angles. Appendix F addresses the association between these two systems and develops the techniques required to integrate both field and static range measurements into the computational model. For the remainder of this chapter, because it is strictly the measurement system in figure 9.1(b) that is the focus of interest, it is convenient to drop the primed notation and denote the origin of the target-centered field measurement system as having

CoR located at O_p, with axes denoted by (x, y, z) and spherical coordinate motion angles denoted by (θ, φ).

9.3 Target Space Mapping Equations

Consider the target-centered coordinate system depicted in figure 9.1(b) having the elongated axis of the target located on the x-axis. Assume for this section that the range offset correction, R_{bias}, has been applied to the range observables, so that the range observable sequence $(R_n)_b$, $n = 1, \ldots, N$, $b = 1, \ldots, B$ is registered to the target CoR. Techniques for independently estimating R_{bias} are considered in section 9.5.2, and jointly in section 9.6.

Assume the target is in the rotational motion defined by equation (9.2). Because the target space system is fixed, the look angle, $\hat{k}$, to the sensor, using standard spherical coordinates, changes with time and is given by

$$\hat{k} = \sin\kappa \cos\varphi\,\hat{x} + \sin\kappa \sin\varphi\,\hat{y} + \cos\kappa\,\hat{z}. \tag{9.3}$$

where $\theta = \kappa$ from equation (9.2). In equation (9.3), $\varphi(t)$ is linear as described by equation (9.2) so that $\dot{\varphi} = 2\pi/T_p$ and $\ddot{\varphi} = 0$. Thus $\dot{\hat{k}}$ and $\ddot{\hat{k}}$ are determined to be

$$\dot{\hat{k}} = -\dot{\varphi}\sin\kappa \sin\varphi\,\hat{x} + \dot{\varphi}\sin\kappa \cos\varphi\,\hat{y}, \tag{9.4}$$

$$\ddot{\hat{k}} = -(\dot{\varphi})^2 \sin\kappa \cos\varphi\,\hat{x} - (\dot{\varphi})^2 \sin\kappa \sin\varphi\,\hat{y}. \tag{9.5}$$

Consider a *fixed-point* scattering center having Cartesian coordinates (x_n, y_n, z_n). Using equation (5.22), the three target space mapping equations, applied at each block time $t = t_b$ are given by

$$\hat{k}_b \cdot \underline{r}_n = (R_n)_b, \tag{9.6}$$

$$\dot{\hat{k}}_b \cdot \underline{r}_n = (\dot{R}_n)_b, \tag{9.7}$$

$$\ddot{\hat{k}}_b \cdot \underline{r}_n = (\ddot{R}_n)_b, \tag{9.8}$$

where the range, range-rate and acceleration observable sequence $(R_n, \dot{R}_n, \ddot{R}_n)_b$, $b = 1, \ldots, B$ is extracted from the data as described in chapter 3, using 2D blocks of wideband data each centered about $t = t_b$. Because each triplet $(R_n, \dot{R}_n, \ddot{R}_n)_b$ at time $t = t_b$ provides an independent estimate for (x_n, y_n, z_n), it is notationally convenient to drop the subscript b from the following discussion. Substituting equations (9.3)–(9.5) into equations (9.6)–(9.8), the target space mapping equations are given by

$$\sin\kappa\cos\varphi\, x_n + \sin\kappa\sin\varphi\, y_n + \cos\kappa z_n = R_n, \tag{9.9}$$

$$-\dot{\varphi}\sin\kappa\sin\varphi\, x_n + \dot{\varphi}\sin\kappa\cos\varphi\, y_n = \dot{R}_n, \tag{9.10}$$

$$-\left(\dot{\varphi}\right)^2\sin\kappa\cos\varphi\, x_n - \left(\dot{\varphi}\right)^2\sin\kappa\sin\varphi\, y_n = \ddot{R}_n. \tag{9.11}$$

Define the scaled variables $(\tilde{x}_n, \tilde{y}_n, \tilde{z}_n)$ according to

$$\begin{aligned}
\tilde{x}_n &= x_n\sin\kappa \\
\tilde{y}_n &= y_n\sin\kappa \\
\tilde{z}_n &= z_n\cos\kappa.
\end{aligned} \tag{9.12}$$

Using equation (9.12) in equations (9.9)–(9.11), solutions for $(\tilde{x}_n, \tilde{y}_n, \tilde{z}_n)$ are determined from the equation set

$$\tilde{x}_n\cos\varphi + \tilde{y}_n\sin\varphi = R_n - \tilde{z}_n, \tag{9.13}$$

$$\tilde{x}_n\sin\varphi - \tilde{y}_n\cos\varphi = -\dot{R}_n/\dot{\varphi}, \tag{9.14}$$

$$\tilde{x}_n\cos\varphi + \tilde{y}_n\sin\varphi = -\ddot{R}_n/(\dot{\varphi})^2. \tag{9.15}$$

Observe that the effect of the term $\tilde{z}_n$ in equation (9.13) is similar to a range bias applied to the range observable R_n. If the 2D wideband mapping equations (9.13) and (9.14) omitting this term are used to obtain a 2D composite target space mapping in the (x, y) plane when $\tilde{z}_n \neq 0$, the mapping will be distorted similar to a range bias as discussed in section 9.5.1 (e.g. note analogy to equation (9.19)).

Examination of equations (9.13)–(9.15) leads to the following observations:

Equations (9.13)–(9.15) present three independent equations characterizing the solutions for $(\tilde{x}_n, \tilde{y}_n, \tilde{z}_n)$, given the observables $(R_n, \dot{R}_n, \ddot{R}_n)$ and motion $\dot{\varphi}$. They can be applied independently to the observables extracted from each data block in the sequence. Most importantly, note that solutions for $\tilde{x}_n, \tilde{y}_n$ using equations (9.14) and (9.15) are uncoupled from solutions for $\tilde{z}_n$ in equation (9.15).

Equations (9.13) and (9.14) are directly applicable to characterizing a target in pure tumble for the case where the radar LoS lies in the tumble plane ($\kappa = 90°$, $z_n' = 0$). For this case, the range, range-rate observables (R_n, $\dot{R}_n$) are used in the mapping to target space. They are directly analogous to the wideband target space mapping using equation (6.14) in chapter 6.

Equations (9.14) and (9.15), treated separately, form a set of two equations for solutions for $\tilde{x}_n$, $\tilde{y}_n$. They are independent of the range observable R_n, which requires proper range alignment referenced to the CoR of the target, and is subject to range time delay artifacts discussed in chapter 6. For $\kappa = 90°$ they become identical to the wideband Doppler-only scattering center location mapping equations (6.23) developed in chapter 6 for tumbling targets.

Once $\tilde{x}_n$ and $\tilde{y}_n$ are determined, the $\tilde{z}_n$ out-of-plane component is obtained using equation (9.13).

Examination of equations (9.13) and (9.15) suggests a novel technique for estimating the tumble rate $\dot{\varphi} = 2\pi / T_p$. Comparing the left sides of equations (9.13) and (9.15), it follows that $\ddot{R}_n$ and R_n satisfy the linear relationship

$$\ddot{R}_n + (\dot{\varphi})^2 R_n = \dot{\varphi}^2 \tilde{z}_n. \tag{9.16}$$

Equation (9.16) can be used to determine the tumble rate as follows: for B data intervals of the sequence $(R_n, \ddot{R}_n)_b$, equation (9.16) is used to plot the observable $(\ddot{R}_n)_b$ as a function of $(R_n)_b$. Because the relationship is linear, the slope of the resulting graph is $-(\dot{\varphi})^2$, from which $\dot{\varphi}$ and T_p are determined. Furthermore, when $\kappa = \pi/2$ and the sensor is located in the tumble plane, $\tilde{z}_n = 0$ so that equation (9.16) yields

$$(\dot{\varphi})^2 = -\ddot{R}_n / R_n, \tag{9.17}$$

that is, the tumble period can be deduced from observables extracted from a single expanded data block. Conversely, once $\dot{\varphi}$ is estimated, the $\tilde{z}_n$ coordinate is given by

$$\tilde{z}_n = R_n + \ddot{R}_n / \dot{\varphi}^2. \tag{9.18}$$

Once $(\dot{\varphi})^2$ is determined, equation (9.16) can also be used to graphically determine the $\tilde{z}_n$ coordinate by estimating the vertical displacement of the linear graphs applied to each of the N scattering centers used for estimating the slope.

Equation (9.17) is a remarkable result. Typically, the range sequence must be observed for a large fraction of the tumble period to estimate the period. Using equation (9.17) the estimate is obtained from a single expanded data block used to estimate $\ddot{R}_n$. A simulation is used in the following section to illustrate the attributes discussed previously.

9.4 Simulation: Tumbling Target

Assume the fixed-point target geometry illustrated in figure 9.3. A set of five scattering centers are located on the x-axis, having $z_n = 0$, as well as a set of five scattering centers displaced 0.125 m out of the (x, y) plane in the $\hat{z}$-direction, all constrained to the $y = 0$ plane. The overall length on the x-axis is 2 m. The simulation data for this example is developed using the geometrical theory of diffraction model defined in equation 2.6. The diffraction coefficient amplitudes are set to unity, The sensor and motion parameters used in the simulation are summarized in table 9.1. Each scattering center has amplitude $D_n = 1$.

Figure 9.4 illustrates the temporal evolution of the observable sequence $(R_n, \dot{R}_n, \ddot{R}_n)_b$ for $\kappa = 60°$ in block increments of Δt s for $b = 1, \ldots, B$, where each value of b represents a data block of 64 pulses centered at about $t = t_b$ and t_b covers the time interval $(0, 1\text{ s})$. The tumble period is easily estimated from the data when viewed over the complete 1 s tumble period.

Now examine the estimate of the tumble period using sparse segments of observable data using equation (9.16). Consider, for example, processing four data blocks,

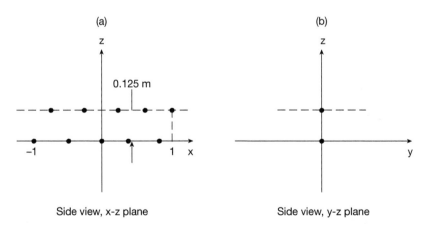

(a)   (b)

Side view, x-z plane   Side view, y-z plane

Figure 9.3 Target geometry used for simulation. Target constrained to x, z-plane having five scattering centers on x-axis for $z = 0$, and five scattering centers displaced vertically by 0.125 m. (a) Side view, x, z-plane. (b) Side view, y, z-plane.

Table 9.1   Sensor and motion parameters used for simulation

$f0$	10 GHz
BW	2 GHz
PRF	1,333 Hz
Tp	1 s
Kappa	60°, 90°

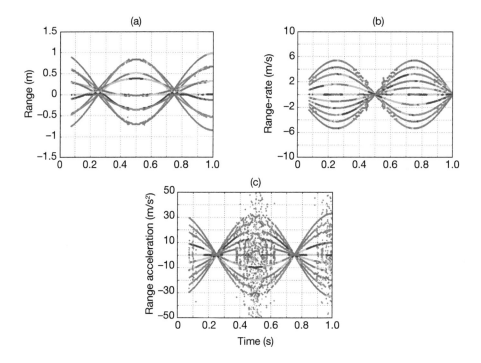

Figure 9.4   Observable sequence $(R_n, \dot{R}_n, \ddot{R}_n)_b$ extracted from the simulation data for the example illustrated in figure 9.3 and table 9.1. Kappa $= 60°$. (a) Range observable sequence. (b) Range-rate observable sequence. (c) Range-acceleration observable sequence.

each of 64 pulses in length, spaced 50 pulses (0.0375 s) between block centers. Figure 9.5 illustrates the extracted $(R_n, \dot{R}_n, \ddot{R}_n)_b$, $b = 1, \ldots, 4$ estimates obtained from these blocks, for $\kappa = 60°$. Clearly it would be difficult to estimate the tumble period from this sparse observable set. Plotting the R_n, $\ddot{R}_n$ pair data of figure 9.5 yields figure 9.6, illustrating the linear relationship between the two $\ddot{R}_n$ and R_n observables from which the slope is clearly seen. The true slope $-(\dot{\phi})^2 = -(2\pi)^2$ is indicated on the figure, illustrating that the estimate of $\dot{\phi}$ obtained from the slope of the linear relationship is quite good. The vertical displacement of the two linear graphs in figure 9.6 corresponds to the slope offset $\dot{\phi}^2 \tilde{z}$ on the right side of equation (9.16).

Using the motion estimate obtained from figure 9.6, the observables extracted from the same four data blocks can now be mapped into target space using equations (9.13)–(9.15). Figure 9.7 illustrates the 3D scattering center locations estimates (dots)

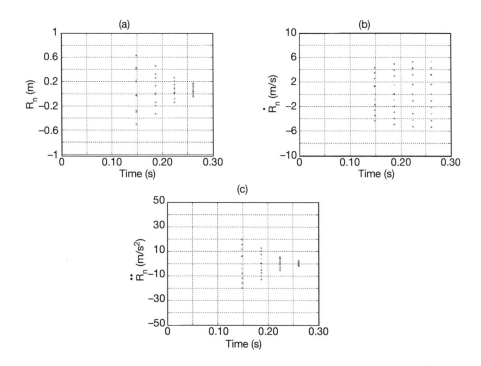

Figure 9.5 Extracted $(R_n, \dot{R}_n, \ddot{R}_n)_b$ sequence from four sequential data blocks spaced 0.0375 s, $N = 10$ scattering centers. Kappa = 60°. (a) Range sequence for each of four data blocks. (b) Range-rate sequence. (c) Range-acceleration sequence.

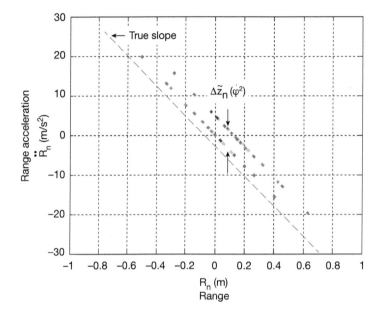

Figure 9.6   Graph of $\ddot{R}_n$ versus $(R_n)_b$, $n = 1, \ldots, N$, with slope of linear graphs equal to $-\dot{\phi}^2$. Kappa $= 60°$. True slope indicated by dashed red line. Vertical displacement corresponds to $\dot{\phi}^2 \Delta \tilde{z}$, i.e. the vertical displacement between the in-tumble-plane and out-of-tumble-plane scattering center locations.

compared to the actual scattering center locations used in the simulation (circles). The agreement is quite good.

9.4.1   Simulation: Canonical Spinning Cylinder Example

In order to illustrate the robustness of using equation (9.16) for motion estimation, it is instructive to revisit the canonical spinning cylinder example treated earlier in chapters 3 and 5, and in particular, the example in section 5.6. There, the scattering center location estimates were obtained assuming the motion (roll frequency) was known, as well as a priori knowledge that each cylinder edge had four fixed-point scattering centers located around the edge. The sequence $(R_n, \dot{R}_n, \ddot{R}_n)_b$, $b = 1, \ldots, B$ for this example is illustrated in figure 3.8 of chapter 3. Perusal of the range-rate observable sequence exhibits this four-fold periodicity for the rotating fixed-point scattering centers. However, without a priori knowledge of the target geometry, there is an ambiguity in roll frequency estimate ranging from the true frequency, denoted as f_ϕ, and the apparent

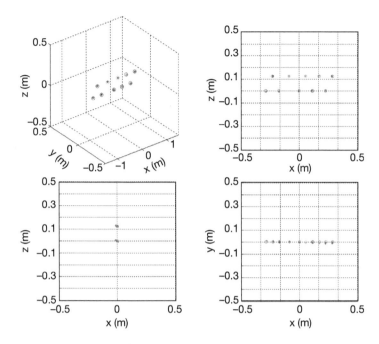

Figure 9.7   The 3D scattering center location estimates using four widely spaced data blocks incorporating the acceleration estimation observable. Three-dimensional view and three orthogonal projections.

frequency of $4 \cdot f_\varphi$ observed in the data. There is no way of resolving this ambiguity by simple observation of the range-rate observable sequence. However, by applying equation (9.16) to the range, range-acceleration sequence, this ambiguity is resolved, because equation (9.16) characterizes the motion of each individual scattering center leaving no ambiguity in the motion estimate.

Figure 9.8 illustrates the relationship of $\ddot{R}_n$ versus R_n collectively for all the observables in the sequence illustrated in figure 3.8, superimposed on the graph illustrated in figure 9.8 for this example. Note the linear relationship separates into two parts: for the slipping scattering centers, because there is no precession, R_n does not change, leading to the vertical spread (caused by spurious estimates for $\ddot{R}_n$) indicated in the figure; for the fixed-point rotating scattering centers, the slope of the linear relationship between the range and range-acceleration observables characterizing $(\dot{\varphi})^2$ is clearly evident for both the upper and lower portions of the cylinder. Enlargement of this area of the graph shows the slope estimate to be 39.1, which is close to the simulation value of $(2 \cdot \pi)^2$.

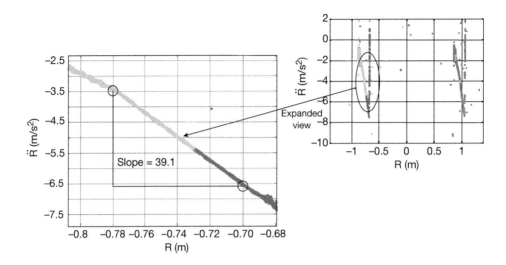

Figure 9.8   The $\ddot{R}_n$ versus R_n motion estimation for the spinning cylinder example described in section 2.6, using the range, range-acceleration observables illustrated in figure 3.8.

Once the motion is determined, the 3D location estimates are obtained as in chapter 5, section 5.6.

Most importantly, observe that the ambiguity in extracting the true roll frequency of the target from the range-rate observable sequence is resolved. The slope correctly characterizes the actual roll motion of the target for each individual scattering center independent of the number of scattering centers rotating about the z-axis.

9.5   Applications of the Wideband Autocorrelation Filter to Range Offset Estimation

9.5.1   Target CoR and Range Offset

As discussed previously, when processing multiple blocks of data corresponding to differing look angle views of the target, one of the most important corrections required is accurate estimation of the range offset, R_{bias}. To gain insight into the effect of this range offset, and the role it plays in multiple block processing, it is useful to contrast two measurement situations: data collected from the target on a precisely calibrated radar cross-section range, and field data collected on the same target as it moves along a specified dynamic trajectory. For example purposes, choose the target to be that of example 1 illustrated in figure 9.3 undergoing pure tumble (rotational) motion. First assume the target is placed on a rotating pylon on a static radar cross-section range, and rotated $360°$ in angle during the data collection. Assume the center of the target located at $(x, y) = (0, 0)$ is positioned on the pylon axis of rotation, which also corresponds to the target CoR. Figure 9.9(a) illustrates the range observable sequence $(R_n)_b$, $b = 1, \ldots, B$, extracted from the data for this measurement situation ($\kappa = 90°$).

Observe from figure 9.9(a) that, for this example, the scattering center located at the CoR of the target (as positioned on the pylon), appears at zero-range in the time evolution of the range observable sequence. Now assume the target is displaced 0.4 m from the pylon CoR, creating a data set emulating a field measurement scenario where the range bias R_{bias} encountered in correcting for the translational motion as discussed in appendix F contains an unknown error of 0.4 m. Figure 9.9(b) then emulates the measurement situation for data collected in the field, where the measured wideband range across the target over any particular block time is relative range, that is, there is an inherent uncertainty on where to place the target CoR range mark.

To illustrate the effects of not correctly aligning the measurement data to the target CoR, consider the wideband 2D target space mapping equations defined by equations

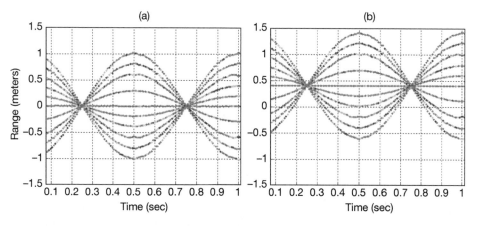

Figure 9.9   Range sequence versus time for target illustrated in figure 9.3 as placed on a static range pylon for two cases. (a) Target centered on pylon. (b) Target displaced 0.4 m from original pylon zero-range reference.

(9.13) and (9.14) for $z'_n = 0$. Assume the scattering center location estimates are obtained by processing the sequence of observables $(R_n, \dot{R}_n)_b, b = 1, \ldots, B$ and applying the target space mapping equations over a complete tumble cycle, assuming the target is properly registered relative to the CoR $R_{bias} = 0$. This sequence represents truth for the target space mapping. To demonstrate the effect of an error in the CoR location consider what happens when a range offset $R_{bias} = 0.4$m is embedded in the range data sequence illustrated in figure 9.9(a). Equations (9.13) and (9.14) take the form

$$\tilde{x}_n \cos\varphi + \tilde{y}_n \sin\varphi = R_n + R_{bias}, \tag{9.19}$$

$$\tilde{x}_n \sin\varphi - \tilde{y}_n \cos\varphi = -\dot{R}_n/\dot{\varphi}, \tag{9.20}$$

where $R_{bias} = 0$ when range observable data are properly aligned, and $R_{bias} = 0.4$ m for the field measurement simulation. It can be shown that solutions for the scattering center locations using equations (9.19) and (9.20) are in error relative to the true locations by

$$\Delta\tilde{x}_n = R_{bias} \cos(\varphi), \quad \Delta\tilde{y}_n = R_{bias} \sin(\varphi). \tag{9.21}$$

When processed over a complete 360° rotation period to form the composite target space map, the error forms a circle of radius R_{bias} about the true scattering center location. This follows from equation (9.21) by noting that

$$(\Delta\tilde{x}_n)^2 + (\Delta\tilde{y}_n)^2 = R_{bias}^2. \tag{9.22}$$

Figure 9.10 illustrates the resultant composite target scattering center location estimates for the example target illustrated in figure 9.3 using the target space mapping equations for the range and range-rate sequences with and without range offset correction.

The presence of the 0.4-m error estimate encircling the true location of each scattering center is clearly evident in the resultant composite set of scattering center locations in figure 9.10(b). It is useful to point out here that, as discussed previously for the case when Kappa is elevated from the tumble plane such that $\kappa < 90°$, $z'_n \neq 0$, even if $R_{bias} = 0$ and the data is aligned to the CoR, the effect of not including z'_n in the wideband mapping equations acts as an effective range bias applied to the composite mapping. The distortion of the resulting 2D composite image caused by $z'_n \neq 0$ would be similar to that shown in figure 9.10b.)

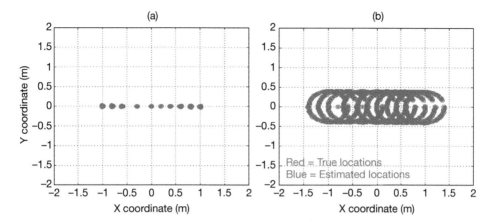

Figure 9.10   Composite target scattering center location estimates for two cases: (a) $R_{bias} = 0$ m; (b) $R_{bias} = 0.4$. When processed over a complete 360° rotation period to form the composite target space map, the range bias error forms a circle of radius R_{bias} about the true scattering center location.

   Determination of a range bias present in measurement data is necessary for accurately estimating the target scattering locations, and is essential in obtaining a complete joint target-motion solution. Typically, the range bias is estimated on examination of the range observable sequence extending over a complete rotation cycle of the motion. Clearly, if a data set containing an unknown range offset were processed, the resulting spread in scattering center location estimates could be employed in a focusing scheme to minimize the spread. In the next section, the concept of the wideband autocorrelation contour introduced in chapter 4 is applied to estimating the range offset using only a single 2D data block. Section 9.6 examines an optimization technique to solve for the motion and the target parameters, including the range bias.

9.5.2   The Wideband Autocorrelation Contour Applied to Estimate the Range Offset

The general characteristics of the wideband autocorrelation contour were discussed in some detail in chapter 4. The contour is generated by processing a block of Q_0 pulse data samples using the methodology described in figure 4.12. For the application considered in chapter 4, the primary interest was to contrast the estimate of the tumble motion of two targets where each target is contained in a single range resolution cell. In this section, the advantages of employing a wideband waveform, and applying the autocorrelation contour concept over the set of range resolution cells that encompass all the target scattering components is examined with the objective of estimating the range offset.

A simple intuitive understanding of the connection between the target wideband autocorrelation contour and the target CoR can be developed by examining the coherence of the time samples associated with each wideband range gate of each 2D data block. Recall that the autocorrelation function is a measure of the coherence time extent of the data samples in each range gate, and the coherence time is dominated by the phase changes in the data sequence for each range gate. Consider data samples in a data block associated with the example target in figure 9.3 having CoR referenced to zero-range on the pylon. For scattering centers located precisely on the pylon CoR, there is minimal phase change as the target rotates (due only to the slowly varying scattering center diffraction coefficient phase variation that can be viewed as constant over the length of the data block; see the discussion in chapter 2, section 2.2) because the field scattered from this scattering center exhibits no phase change in range as the pylon rotates. Thus, the autocorrelation function applied to data samples in this gate results in perfect coherence. For scattering centers located progressively further from the CoR, the phase of the data samples changes more rapidly with turntable motion as the location distance increases away from the CoR. Thus, one would expect the wideband autocorrelation contour applied across all the gates to be most coherent over the range gate containing the target CoR. If the pylon is displaced, as in the second example considered in figure 9.9, the additional phase change induced in each range gate due to the displacement is the same constant for each gate, independent of rotation angle, and does not affect the autocorrelation contour. Thus, one would expect the autocorrelation contour as described in chapter 4 to show maximum coherence for the range gate closest to the CoR of the target. Because this intuitive example applies independent of the look angle to the target, this coherence is ideally present for every data block as the rotation angle changes. This implies that the target CoR can ideally be estimated from a single data block, versus requiring processing of the range sequence over a complete 360° rotation of the target.

Figure 9.11 illustrates an application of the autocorrelation contour to a given data block of 96 pulses obtained using the simulation data developed in the examples in section 9.4 for $\kappa = 90°$.

Figure 9.11(a) illustrates a partition of the data within the data block using a shorter block of 32 pulses for developing a range, range-rate image of the target component

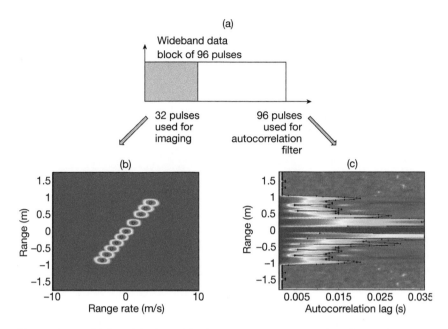

Figure 9.11   (a) Data block partitioning used for range-rate imaging and the auto-correlation contour. (b) Range, range-rate image formed using the smaller data block. (c) Autocorrelation contour formed using the larger data block.

locations, and a longer data block for forming the autocorrelation contour for the target. As described in chapter 4, for a Q_0 pulse data block, the length of the autocorrelation contour formed from the data is $Q_0/2$ pulses. Thus, it is necessary to use a longer data block to form the autocorrelation contour than for forming a range, range-rate image of the target. Figures 9.11(b) and 9.11(c) contrast the information extracted from the data block: the wideband range, range-rate image provides an estimate of the instantaneous target scattering center locations in range, range-rate space as viewed from the radar LoS; the autocorrelation contour provides a measure of the projected length of the target, and the coherence properties for each of the scattering centers present in the data.

Examine the autocorrelation contour illustrated in figure 9.11(c) in light of the previous discussion. When viewed as a function of range on the vertical axis, the scattering center located directly on the CoR exhibits maximum coherence, and the coherence time for each scattering center generally decays as a function of displacement away from the CoR. Figure 9.12 contrasts the two autocorrelation contours when the wideband

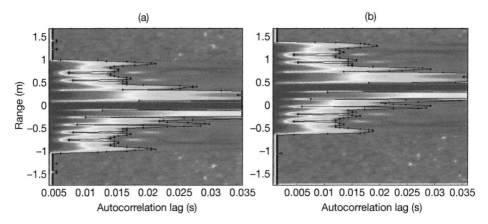

Figure 9.12   Autocorrelation filter contours for the static range simulation example corresponding to figure 9.9 for a zero-range offset and a 0.4-m range offset. (a) No range offset. (b) The 0.4-m range offset. The two contours are identical in form, except the one is displaced 0.4 m in range relative to the other.

pulse data are used to form the autocorrelation contour with and without the 0.4 m range offset. The two contours are identical in form, except one is displaced 0.4 m in range relative to the other.

The range gate exhibiting maximum coherence is displaced at a distance equal to the range offset. The zero-range reference associated with the target CoR can be estimated from the autocorrelation contour as the range exhibiting maximum coherence which occurs at the 0.4m gate in the autocorrelation contour in figure 9.12b. For the simplistic example depicted in figure 9.12, the range offset can be estimated from a single data block. In practice, for field data on complex targets, a number of blocks may need to be processed. This happens for at least two reasons: first, and most importantly, we cannot expect a scattering center to occur exactly at the target CoR; and second, scattering centers on complex targets often exhibit scintillation and occlusion, changing the coherence properties. The concept of the *coherency envelope* is now introduced to mitigate these factors.

To form the coherency envelope, search each range gate for the value of τ_0 for which $R_0(\tau_0) < R_{\min}$, where $R_{\min}$ is a prescribed threshold value. Extract each value of τ_0 for each range gate and define the composite of these values as a function of range gate as the coherency envelope. Figure 9.13 illustrates a typical coherency envelope for the

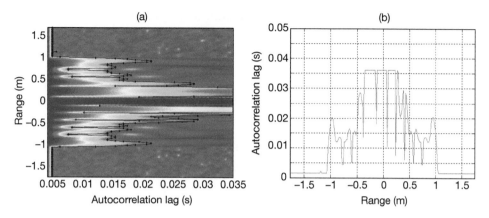

Figure 9.13   (a) Autocorrelation contour for a typical data block. Coherency envelope depicted as black dots at autocorrelation lag τ_0 as determined by the condition $R_0(\tau_0) < R_{min}$ for each range gate. (b) Single block coherence envelope extracted from autocorrelation contour of (a).

static range simulation for a typical data block having zero-range offset. To mitigate the single block artifacts discussed previously, is useful to superimpose the coherence envelopes extracted from a number of blocks and fit a Gaussian envelope to the resultant set. The peak of the Gaussian fit best approximates the range gate corresponding to the range offset.

It should be mentioned that using the coherency envelope contour to estimate the range offset is dependent on the ability to fully compensate for the translational motion. For example, assume the correction for translational motion, $\hat{R}_T(t)$, defined in equation (F.10) in appendix F does not result in a constant range bias error, but in a linear error. This leads to an error in range bias defined by $R_{bias} + \dot{R}_{bias}t$ over the region over which κ is constant. This would result in a linear shift in phase of each pulse over the duration of the data collection. For this case, the autocorrelation peak would shift in range as the block of pulses increases in time, resulting in a range bias estimate that changes linearly with time. Such an error would also manifest itself in the resulting composite image formed by processing the sequence of data blocks over the entire rotational motion. This emphasizes the importance of proper data alignment and compensation for the translational motion over the duration of the entire data collection used for processing.

9.6   Joint Solutions Using Scattering Center Dispersion

9.6.1   Introduction

The results of the previous sections provide joint solutions for the target scattering center locations and rotational motion using independent estimates for the parameters T_p and R_{bias}. The parameter T_p is first determined using a unique property defined by equation (9.16) relating the scattering center range observables to their corresponding range-acceleration observables. The range offset R_{bias} is estimated using the wideband range autocorrelation contour. The phase parameter α_p is then chosen to set the orientation of the scattering center locations present in the composite image in target space. The technique uses the motion estimate and range bias estimate to map the range, range-rate and range-acceleration observables into target space to form an estimate for the scattering center locations.

In this section a technique is introduced to jointly estimate the motion parameters $(T_p, \alpha_p, R_{bias})$ using a 3D optimization technique applied over a composite search space formed covering the three parameters. The optimization metric is based on the persistence of each scattering center location estimate in target space obtained from sequential applications of the target space mapping equations when the correct motion and range offset are applied to the mapping. Using the results developed in section 5.6, recall that by including the acceleration observable, $\ddot{R}_n$, along with the range, range-rate observables extracted from a given data block [i.e., $(R_n, \dot{R}_n, \ddot{R}_n)_b$], the set of three independent equations defined by equation (5.22) provide for a direct inversion of the target-motion coupling equations. Each choice of the triplet parameters $(T_p, \alpha_p, R_{bias})$ in the composite search space results in N scattering center location estimates for each data block. For B data blocks, B estimates for the location of each fixed-point scattering center are obtained. For correct motion and range bias, these estimates ideally result in B identical estimates for each of the N scattering center locations. For incorrect motion or range bias, the set of B sequential location estimates will be widely dispersed over target space. The optimization metric and technique are based on the persistence of the correct estimates and are described in the following section.

9.6.2   Development

Assume the radar collects wideband, high pulse repetition frequency data over a given time span. Further assume the target can be described by N fixed-point scattering centers and extract the $N \cdot B$ sequential range, range-rate and range-acceleration observables over B data blocks:

$$(R_n, \dot{R}_n, \ddot{R}_n)_b, n = 1, \ldots, N, b = 1, \ldots, B. \tag{9.23}$$

As discussed in section 9.1.1, determining a joint solution for a target in tumble motion condition requires the estimation of three parameters $(T_p, \alpha_p, R_{bias})$: the tumble period, tumble phase, and range bias, respectively. The value of κ can be estimated from the composite map of the target [scaled to $\sin(\kappa)$] if some characteristic of the target, such as its length, is known. The following introduces the parameter-based motion function defined by

$$\varphi(t) = (2\pi/T_p)\mathsf{t} + \alpha_p, \tag{9.24}$$

where α_p and T_p are to be estimated. Assume the range observable sequence $(R_n)_b$, $n = 1, \ldots, N$, $b = 1, \ldots, B$ corresponds to the range observable sequence extracted directly from the field measurement data and is offset from the target CoR. Denote the range sequence properly aligned to the CoR as $(R_n)_{b, CoR}$ and introduce the search parameter R_{bias}, having optimum value aligning the range observable data relative to the target CoR:

$$(R_n)_{b, CoR} = (R_n)_b + R_{bias}. \tag{9.25}$$

$(R_n)_{b, CoR}$ represents the range observable sequence defined relative to the CoR. Define the modified observable set for incorporation into the target space mapping equations:

$$(R_n + R_{bias}, \dot{R}_n, \ddot{R}_n)_b, n = 1, \ldots N, b = 1, \ldots, B. \tag{9.26}$$

Form a search space for each of the three parameters $(T_p, \alpha_p, R_{bias})$. For each set of possible parameters within the composite three parameter search space, use the motion and target space mapping equations defined by equations (9.13)–(9.15) applied to each of the B sequential range, range-rate and range-acceleration observables defined in equation (9.26) to estimate $N \cdot B$ scattering center locations. Note that each set of $N \cdot B$ location estimates will be widely dispersed over target space for incorrect motion and range offset, and strongly localized for the correct motion and range offset. Denote the accumulated $N \cdot B$ location estimates for each set of possible search parameters as

$$(\tilde{x}_p, \tilde{y}_p, \tilde{z}_p), p = 1 \ldots NB. \tag{9.27}$$

Define the motion optimization metric by choosing the set of motion and range offset parameters that minimize the mean square dispersion measure, M,

$$M = \frac{Min}{(\alpha_p, T_p, R_{bias})} \left(\sum_p^{NB} \sum_{p'}^{NB} (\tilde{x}_p - \tilde{x}_{p'})^2 + (\tilde{y}_p - \tilde{y}_{p'})^2 + (\tilde{z}_p - \tilde{z}_{p'})^2 \right), \quad (9.28)$$

where the minimum in equation (9.28) is carried out over the 3D composite search space. Once the optimal motion parameters are determined, apply the target space mapping equations to estimate the target scattering center locations. Incorporate a target space filter to eliminate nonpersisting location estimates. The following comments apply:

  The resolution of each scattering center location estimate is determined by the grid size chosen for the target space filter, as discussed in section 5.6.2. Multiple location estimates contained within a grid interval can be combined as a single location on the target.

  The target space mapping used is in normalized coordinates, independent of the look angle κ of the radar LoS to the tumble plane.

  The metric defined by M is a dispersion metric, and strictly speaking not a location estimate. For example, if the scattering center locations were known, as determined from mechanical drawings of the target, then equation (9.28) could be modified to a single summation over the index p, and the locations $(\tilde{x}_{p'}, \tilde{y}_{p'}, \tilde{z}_{p'})$, $p' = 1 \ldots N$ used as the known locations. In this case, the minimum for M results in a motion that most closely maps the observables into target space as a best fit to the assumed known locations. The dispersion metric differs in that it minimizes the very wide dispersion of the mapping for incorrect motion, but if the assumed motion parameters cover too wide a range, a grossly inaccurate solution can result that maps the observables into a volume much smaller than the target dimensions. Thus, it is most useful in fine-tuning the initial motion estimates obtained independently as discussed previously. Alternately, it is possible to constrain the solutions to the dispersion metric by discarding those values of M contained within a small fraction of the bounding volume of the target, but such a constraint is not considered in the text.

As discussed in section 9.4.1, the range, range-acceleration technique described by equation (9.16) for estimating the motion of a spinning target is dual to that for a tumbling target. Thus, the dispersion metric developed previously is applicable to a target in pure spin by introducing the three motion parameters

$(T_p, \alpha_s, R_{bias})$. It is easily shown that in this case, the motion optimization is in fact independent of R_{bias}.

9.6.3   Example: The Slotted Cylindrical Target

As an example in applying the dispersion metric to obtain a joint solution, consider the slotted cylindrical target introduced in chapter 6. This target was used there to simulate synthetic static range data and used to illustrate the applications of the techniques introduced in the text to narrowband processing for a target in rotational motion. Wideband data simulated using this target were also used in appendix E to illustrate the utility of the autocorrelation filter introduced in chapter 4 for choosing the parameters used for estimating the range-acceleration observable using the nonlinear signal model scattering center acceleration estimation technique developed in chapter 3. An example range, range-rate and range-acceleration observable sequence assuming a 2 GHz bandwidth at X-band for this target is illustrated in figure E.2. In this section an example using the dispersion metric to jointly estimate target scattering center and motion parameters using this wideband synthetic static range data set is developed. A 1 GHz subset of the 2 GHz wideband synthetic data is used in this example to model field data collected on a target in pure tumble motion having radar LoS in the tumble plane. Thus, it inherently presents a joint 2D motion and scattering center location estimation problem.

In order to simulate a target in rotational motion, the synthetic static range data simulated versus rotation angle must be converted to radar pulses versus time. This requires choosing an appropriate rotational motion for the simulation. For the example in appendix E, a rotation period of $T_p = 2\pi$ s was used to map the data versus angle to time. For the example considered here, a rotation period $T_p = 1$ s is assumed. Otherwise, the same block size parameters for extracting the range, range-rate and range-acceleration observables used in appendix E are incorporated. Note that the range observable sequence extracted from static range data is inherently registered to the target CoR. In order to simulate field conditions for this example, a range offset of 0.2 m is added to the range observable sequence extracted from the synthetic static range data. Thus, the motion and range offset parameters $T_p = 1$ and $R_{bias} = -0.2$ m represent the optimal solutions for motion and range offset. The phase α_p is included as an optimization parameter and represents the typical measurement situation where the (x, y)-axes aligned to the target would not necessarily line up with the (x, y) target space axes, corresponding to scattering center location estimates oriented in target space characterized by orientation angle α_p.

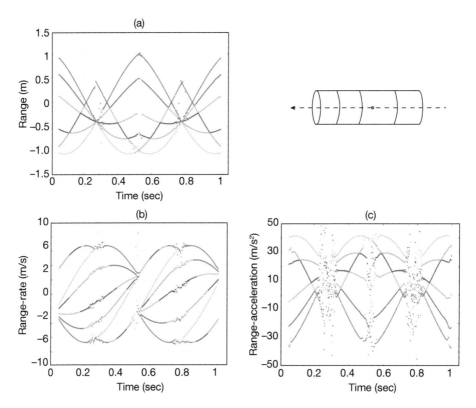

Figure 9.14 Range, range-rate, and range-acceleration observable sequences versus time obtained using synthetic static range data for the slotted cylindrical target at X-band, 1 GHz bandwidth, one second rotation period. Data block parameters chosen are identical to example in appendix E. (a) Range observable sequence. (b) Range-rate observable sequence. (c) Range-acceleration observable sequence.

Figure 9.14 illustrates the range, range-rate and range-acceleration observable sequences extracted from the synthetic static range data using block size parameters corresponding to those used in appendix E applied to each data block. The range off-set, $R_{bias} = 0.2$ m, is added to the extracted range sequence illustrated in figure 9.14(a) to simulate a field data environment.

Form the three-parameter search space defined by

$$T_p = 0.5: 0.1: +1.5\text{s}$$
$$\alpha_p = 0°: 45°: 315°$$
$$R_{bias} = -0.3: 0.1: +0.3 \text{ m}$$

(9.29)

The composite search space results in 616 possible triplets for the assumed parameters $(T_p, \alpha_p, R_{bias})$. Apply the motion to target space mapping using the modified observable sequence indicated in equation (9.26) for each set of parameters $(T_p, \alpha_p, R_{bias})$. Because the radar LoS lies in the plane of rotation, the target mapping geometry is 2D and either the range, range-rate sequence $(R_n, \dot{R}_n)_b$ or range-rate, range-acceleration observable sequence $(\dot{R}_n, \ddot{R}_n)_b$ can be used for the target space mapping. To illustrate the optimization technique, choose the 2D range, range-rate target mapping equations defined by equations (9.13) and (9.14) with $\tilde{z}_n = 0$ to obtain each set of $N \cdot B$ location estimates. For the case where the radar LoS lies out of the tumble plane, the range-rate, range-acceleration mapping equations defined by equations (9.14) and (9.15) would be used. Figure 9.15 illustrates the inverse of the optimization metric, $1/M$, as a function of the 616 composite search parameters covering the search space defined by equation (9.29).

Observe that there are four optimal values extracted from the optimization, each corresponding to $T_p = 1$ s and $R_{bias} = -0.2$ m, and four possible values of the phase search space defined by α_p. Each value of α_p corresponds to a different possible orientation of the target in target space, all of which result in a nearly optimum solution.

Figure 9.16 compares the motion $\dot{\varphi} = 2\pi/T_p$ estimated from the data using the joint target-motion solution for the technique described in section 9.6.2 to that developed in section 9.3, that is, the range, range-acceleration relationship defined by equation (9.17).

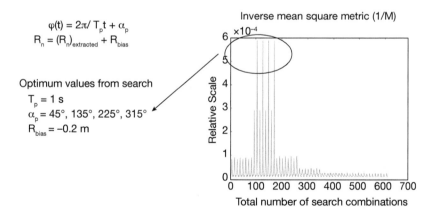

Figure 9.15   The inverse of the optimization metric, $1/M$, as a function of the 616 composite search parameters covering the search space for the parameters (T_p, a_p, R_{bias}): $T_p = 0.5$: 0.1: 1.5s; $a_p = 0°$: 45°: 315°; $R_{bias} = -0.3$: 0.1: 0.3 m.

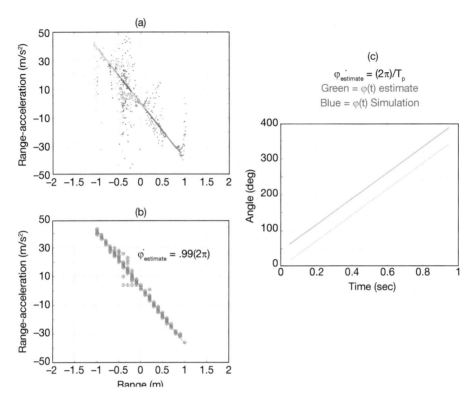

Figure 9.16 Estimate of the angular rotation rate $\dot{\varphi} = 2\pi/T_p$. (a) Range versus range-acceleration observables for all N scattering centers over B data blocks. (b) Estimate $\dot{\varphi} = (.99)2\pi$ using the range versus range-acceleration slope as defined by equation (9.17), after applying a threshold to the data in (a). (c) Estimate $\dot{\varphi} = 2\pi$ using the joint solution over the composite search space resulting in $T_p = 1$ s.

First consider the independent estimate of T_p using equation (9.17) with $z_n' = 0$. Figure 9.16(a) illustrates the superposition of each of the range versus range-acceleration estimates for each of the N observables extracted from each sequential data block. Figure 9.16(b) illustrates a linear fit to the data after a threshold is applied to eliminate the spurious estimates in figure 9.16(a). The independent estimate of $\dot{\varphi}$ is determined to be $\dot{\varphi} = 0.99(2\pi)$ corresponding to $T_p = 1.01$ s. Figure 9.16(c) illustrates the estimate $T_p = 1$ s obtained from the joint three parameter search space, as well as the motion $\varphi(t)$ compared to the simulation motion. The joint estimate $T_p = 1$ s is identical to the simulation value.

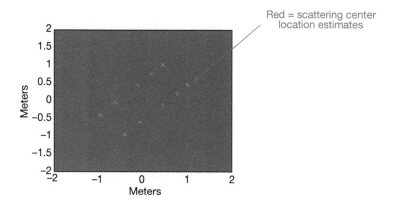

Figure 9.17   Scattering center location estimates for the slotted cylindrical target introduced in chapter 6 using the motion estimated from the dispersion metric applied to the composite target space mapping. Red indicates the estimated scattering center locations overlayed on the target space mapping contour .

Figure 9.17 illustrates the estimate of the target scattering center locations using the estimated motion and range-offset parameters applied to the observable to target space mapping. A grid size of 0.05×0.05 m is used to characterize the resolution in target space, covering a target space region having limits ± 2 m for the (x, y) axes. Note the target orientation angle of $45°$ relative to (x, y) target space coordinates, corresponding to the value $\alpha_p = 45°$. For some cases, more than a single scattering center is present within the grid size resolution in target space.

10

Joint Target-Motion Solution from Range-Only Data

10.1 Introduction

In this chapter a second technique for obtaining joint motion and target solutions from wideband radar data is developed. For this solution technique the observable sequence $(R_n, \dot{R}_n)_b, b = 1, \ldots, B$ is assumed to be correlated as time progresses. The joint solution technique presented is, in some sense, the most general as it relaxes the constraint of Euler motion and allows for a more general motion solution. The range-rate estimates are not directly used in the scattering center location estimation techniques in the development presented in this chapter, but because they largely prevent track-jumping at range crossovers during processing of the radar returns, they are important to the process of extracting accurate correlated range estimates.

Two techniques are presented. One technique factors the full correlated range observable matrix to extract the target location and motion estimates and includes a treatment of the situation in which the range matrix has gaps in the sequence but can be subdivided into a number of full submatrices. The other technique fills in the missing elements of the observable sequence using a steepest descent method.

The essence of the solution technique can be seen by careful observation of equation (5.7). The observable matrix decomposes into the product of a motion matrix times a target matrix[1]. By applying a singular-value decomposition (SVD) to the correlated range observable sequence versus time, the SVD factorization procedure in effect decomposes the observable matrix into a target model and a motion model. The formulation

---

1. The analysis in this chapter was carried out independently by Dr. M.L. Burrows and adapted for incorporation into this text. A slight change in notation is introduced. The notation change for the motion matrix $M \rightarrow -C$ and target matrix $[r_1^T, \ldots, r_N^T] \rightarrow X$ are used throughout this chapter, where $r_n = [x_{n1}, x_{n2}, x_{n3}]$. The notational block time t_b is replaced by the block time label t_m and there are $M = B$ observables.

presented here uses only the sequence of wideband range observables correlated in time for each scattering center. However, range-rate estimates are used in the development to prevent incorrect correlation at multiple scattering center range observable intersections.

Using the tabulation of the range observables—denoted as the range observable matrix—of the target's scattering centers taken at a number of different unknown observation angles, a matrix factorization procedure (referred to as the target-motion CX decomposition) provides three-dimensional (3D) estimates of the targets scattering center locations, and if a time tag t_m is associated with each observation angle measurement, its motion model as well. The range lists can be incomplete—missing entries due to dropped or shadowed tracks can be accommodated. The range origin used for each continuous listing resulting in the motion solution for this interval can be arbitrary and unknown.

The minimum number of observables needed is six, not all in one plane, and the minimum number of scattering centers is four, also not all in one plane. A crucial requirement is that the identification of each scattering center be preserved from one observable to the next, that is, that the scattering centers be correlated (e.g., as described in chapter 5, section 5.5.1). Then, if the scattering center locations do not move in a coordinate system fixed with respect to the target (a requirement met only approximately, e.g., by a slipping scattering center) and the range estimates are accurate, the solutions are exact, except for a right-side/left-side ambiguity in the coordinate system in which the scattering center coordinates and the motions are defined. In other words, the inferred target image may be the mirror image of the actual target.

Although the approach presented here was developed independently, the basic concept of using a singular-value decomposition to separate out target and motion is not new. Tomasi and Kanade [100] presented a similar approach in 1992 that extracts both a 3D image and the camera motion from correlated features in a series of two-dimensional video images. And those authors acknowledged, in a 1998 paper, an even earlier formulation by [101] in 1987. However, certain specifics associated with the implementation of the technique have been incorporated. See also the paper by [102] for an alternate derivation.

10.2 Target-Motion (*CX*) Decomposition

The technique is an application of the mathematical technique of singular-value decomposition applied to the sequential range observable matrix for a number of scattering centers at a number of different (unknown) observation angles. It depends on the fact that the range observable matrix R can be represented as the product of a motion matrix C and a target coordinate matrix X

$$R = -CX^T. \tag{10.1}$$

where the minus sign indicates that range is measured away from the radar. Using the notation introduced in chapter 5, equation (5.7), $-C$ is identical to the motion matrix M, referenced to direction cosine space. In direction cosine space, the look angle to the sensor, $\hat{k}$, is expressed in the form $\hat{k} = [C_x, C_y, C_z]$, where $C_x = \sin(\theta)\cos(\varphi)$, $C_y = \sin(\theta)\sin(\varphi)$, $C_z = \cos(\theta)$, and X is identical to the scattering center location matrix $[\underline{r}_1^T, \dots, \underline{r}_N^T]$, where the superscript T indicates *transpose*.

The problem solution reduces to carrying out the factorization of R. For that it is enough to know that C and X are both of rank 3 and that each row of C has a norm of 1.

First, we will provide some definitions. Let element r_{mn} of the $M \times N$ matrix R be the range² at motion time t_m of the scattering center located at $\underline{r}_n = [x_{n1}, x_{n2}, x_{n3}]$. The notational subscript m is introduced to refer to the observable time sequence, versus the index b used elsewhere in the text to denote the block time corresponding to the range, range-rate observables $(R_n, \dot{R}_n)_b$. The motion solution (in direction cosine space, the time sequence of direction cosines) is a direct output of the technique. Let row m of the motion matrix C list the direction cosines to the radar at observation time t_m in a target-fixed coordinate system, and let row n of the coordinate matrix X list the Cartesian coordinates of scattering center n in the same coordinate system. Equation (10.1) can be expanded in the form

$$R = \begin{bmatrix} r_{11} & r_{12} & r_{13} & \cdots & r_{1N} \\ r_{21} & r_{22} & r_{23} & \cdots & r_{2N} \\ r_{31} & r_{32} & r_{33} & \cdots & r_{3N} \\ \vdots & \vdots & \vdots & \ddots & \vdots \\ r_{M1} & r_{M2} & r_{M3} & \cdots & r_{MN} \end{bmatrix} = -CX^T$$

$$= -\begin{bmatrix} c_{11} & c_{12} & c_{13} \\ c_{21} & c_{22} & c_{23} \\ c_{31} & c_{32} & c_{33} \\ \vdots & \vdots & \vdots \\ c_{M1} & c_{M2} & c_{M3} \end{bmatrix} \begin{bmatrix} x_{11} & x_{21} & x_{31} & \cdots & x_{N1} \\ x_{12} & x_{22} & x_{32} & \cdots & x_{N2} \\ x_{13} & x_{23} & x_{33} & \cdots & x_{N3} \end{bmatrix}, \tag{10.2}$$

---

2. At each motion, the ranges are expressed relative to a chosen reference scattering center, common to all motions, eliminating the need for a range bias correction.

where $c_{m1} = \cos\varphi_m \sin\theta_m$, $c_{m2} = \sin\varphi_m \sin\theta_m$ and $c_{m3} = \cos\theta_m$ are the direction cosines to the target at observation time t_m expressed in terms of the standard spherical coordinate aspect and azimuth angles (θ, φ).

Thus C and X are both of rank 3, and so R, ideally, is of rank 3. But noise, and the fact that the scattering center locations may vary with motion, introduce errors. So R in general will be of full rank.

First the best rank 3 approximation R_3 of R is evaluated by retaining, from its singular-value decomposition, just the first three columns of U and V, and the first three columns and rows of Σ. That is,

$$R = U\Sigma V^T \Rightarrow R_3 = U_3\Sigma_3 V_3^T = -CX^T.$$

Then $U_3\Sigma_3 V_3^T = -CX^T$ is rewritten as

$$(U_3\alpha)(\alpha^{-1}\Sigma_3 V_3^T) = -CX^T,$$

where the as-yet unknown 3×3 normalizing matrix α has been inserted to make the norm of each row of $U_3\alpha$ equal, in the least-squares sense, to 1. The motion matrix C and the coordinate matrix X can be identified with the two groups of factors. That is,

$$C = -U_3\alpha,$$
$$X = (\alpha^{-1}\Sigma_3 V_3^T)^T.$$

The first step toward determining α is evaluating the 3×3 symmetric matrix $\alpha\alpha^T$ by requiring its six independent elements to minimize, in the least-squares sense, the deviation of the diagonal elements of $U_3\alpha\alpha^T U_3^T$ from 1. This satisfies the constraint that the norm of each row of the motion matrix is equal to 1 in the least-squares sense. That is, the equation

$$\begin{bmatrix} \vdots & \vdots & \vdots & \vdots & \vdots & \vdots \\ u_{m1}^2 & u_{m2}^2 & u_{m3}^2 & 2u_{m1}u_{m2} & 2u_{m2}u_{m3} & 2u_{m3}u_{m1} \\ \vdots & \vdots & \vdots & \vdots & \vdots & \vdots \end{bmatrix} \begin{bmatrix} a_{11} \\ a_{22} \\ a_{33} \\ a_{12} \\ a_{23} \\ a_{31} \end{bmatrix} \approx \begin{bmatrix} \vdots \\ 1 \\ \vdots \end{bmatrix}, \quad (10.3)$$

must be satisfied, also in the least-square sense, for all m, where the u_{mn} are the elements of U_3 and the a_{ij} are the elements of the symmetric matrix $A = \alpha\alpha^T$. Denote Λ to represent the $M \times 6$ matrix whose mth row is delineated previously, and J the M-element column of ones on the right side of equation (10.3). Then the six independent elements of $\alpha\alpha^T$ are determined by $(\Lambda^T\Lambda)^{-1}\Lambda^T J$. Apply a singular-value decomposition to A, $A = \alpha\alpha^T = u\sigma u^T$, from which the normalizing matrix α is given by[3] $\alpha = \pm u\sigma^{1/2}$. The sign ambiguity implies two solutions, mirror images of one another.

The presence of six unknowns in the previous equation makes it clear that the solution requires at least six different range observables be used.

10.3   Shadowing and Dropped Tracks

Shadowing of scattering centers, and dropped tracks as the target-motion angle changes, give rise to blanks in the range matrix. Simply filling in the blanks with zeros would be introducing random errors in the matrix. There are two better approaches.

The first takes advantage of the redundancy implied by the small rank of the range matrix to estimate its missing elements and so fill in the blanks. This works if the matrix is mostly filled.

The second is to avoid the missing elements by subdividing the range matrix into filled submatrices, apply the CX decomposition to each of the submatrices separately, and then, by coordinate transformations, align the solutions of successive submatrices. This alignment is necessary because the coordinate system in which the CX decomposition expresses its solution is undefined. Ultimately, this produces estimates in a common coordinate system of the locations of all the scattering centers. For the CX decomposition to be successful each submatrix must include at least four scattering centers, and for the coordinate alignment to be successful adjacent submatrices in a submatrix pair must have at least three scattering centers in common.

Consider the following approaches.

10.3.1   Estimate Missing Elements

Data redundancy allows factorization of the range matrix even if it has missing elements. Now the goal of the initial rank-reducing singular-value decomposition operation is the

---

3. In general, the real decomposition of a real symmetric matrix can be written as $A = u\sigma u^T$ only if σ can have negative elements. However, in this case, because A has the form $\alpha\alpha^T$ and α is real, the elements of σ are nonnegative.

evaluation of the matrix pair U_3 and S_3 that minimize the sum of the squares of the elements of the Hadamard product[4]

$$W \cdot (R' - U_3 S_3^T), \tag{10.4}$$

where R' is the incomplete range matrix and W is a weighting matrix having zeros where R' has missing elements. (The non-zero entries of W can be used to weight the different range estimates of R' to enhance, e.g., those with greater signal-to-noise ratio). Because the technique uses only the product $\Sigma_3 V^T$, there is no need to evaluate Σ_3 and V_3 separately, so only their product $S_3 = V_3 \Sigma_3$ is evaluated here. In the language of the calculus of variations, the solution requires that

$$\delta \sum_{m,\,n} w_{mn}^2 (\sum_l u_{ml} s_{nl} - r_{mn})^2 = 0.$$

Varying u_{ml} and s_{nk} yields the following $3M + 3N$ equations:

$$\sum_l u_{ml} \sum_n w_{mn}^2 s_{nl} s_{nk} - \sum_n w_{mn}^2 r_{mn} s_{nk} = 0, \quad m = 1, 2, \ldots, M; k = 1, 2, 3$$

$$\sum_l s_{nl} \sum_m w_{mn}^2 u_{ml} u_{mk} - \sum_m w_{mn}^2 r_{mn} u_{mk} = 0, \quad n = 1, 2, \ldots, N; k = 1, 2, 3. \tag{10.5}$$

These are solved iteratively starting from the approximate values of u_{ml} and s_{nk} given by the singular-value decomposition of the range matrix in which the data gaps could be filled, for example, by interpolation or by local averaging, or even simply by zeros.

Each iteration proceeds in the following manner. First, from the $3M$ equations of the first set, update the u_{mk} for each value of m using

$$\begin{bmatrix} u_{m1} & u_{m2} & u_{m3} \end{bmatrix} = \begin{bmatrix} \alpha_{m1} & \alpha_{m2} & \alpha_{m3} \end{bmatrix} \begin{bmatrix} \beta_{m11} & \beta_{m12} & \beta_{m13} \\ \beta_{m21} & \beta_{m22} & \beta_{m23} \\ \beta_{m31} & \beta_{m32} & \beta_{m33} \end{bmatrix}^{-1}, \tag{10.6}$$

where the α_{mk} and β_{mlk} are the column sums $\sum_n w_{mn}^2 r_{mn} s_{nk}$ and $\sum_n w_{mn}^2 s_{nl} s_{nk}$. Similarly, from the $3N$ equations of the second set, update the s_{nk} using row sums.

---

4. Each element of the product matrix is the product of the corresponding elements of the two multiplicand matrices. All three have the same dimensions.

At each iteration, the columns of U_3, but not those of S_3, are re-orthonormalized. This keeps the columns of U_3 spatially separated, does the same for the columns of S_3, and keeps the solution properly bounded. The end result is the desired decomposition of the reconstituted range matrix as $R = U_3 S_3$.

All that remains now to complete the solution is to introduce the normalizing matrix α as $R = (U_3\alpha)(\alpha^{-1}S_3)$, evaluating it as described in section 10.2.

Figure 10.1 shows an occupancy sketch of a simulated range matrix whose six columns list the ranges at different observation angles of six scattering centers located at the vertices of a lantern-shaped object. The blanks in the matrix denote shadowed or dropped scattering centers. The figure also shows two different views of the object, constructed from the true and the estimated scattering center locations.

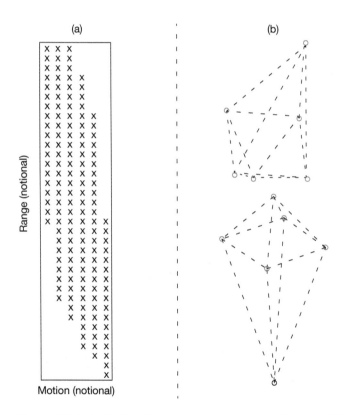

Figure 10.1 (a) Notional occupancy of an incomplete simulated range matrix for six scattering centers. (b) Two 3D views of the scattering center locations ("o" markers) estimated from it. The dashed lines connect the true scattering center locations.

10.3.2   Avoid Missing Elements: The Divide and Conquer Method

If the target has a sufficient number of scattering centers, the problem of shadowed and dropped observations can be handled by subdividing the range matrix along its motion dimension into a sequence of adjacent submatrices, which by excluding specific rows and columns, are all complete—each one has no missing elements. Figure 10.2 illustrates the procedure.

The *CX* technique applied to each submatrix separately produces a list of the locations of the scattering centers that are included in that submatrix together with a list of the motions it includes, where both lists are expressed in a coordinate system specific to that particular submatrix. For this first step to be successful, at least four scattering centers and six motions must be included in the submatrix.

The next step is to align the coordinate systems of all the separate solutions sequentially by taking the solutions of adjacent pairs of submatrices and aligning the scattering centers common to both by coordinate transformation. For this to succeed the separate solutions must have at least three scattering centers in common.

One method of accomplishing the transformation is the Kabsch technique[5]. Specifically, if X_{m-1} and X_m are the coordinates of the same set of scattering centers expressed in the coordinate system of the $(m-1)$th and mth submatrix, respectively, then the required alignment is the rotation of X_m by the rotation matrix Q to align it with X_{m-1}, and Q is given by $Q = VDU^T$. Here V and U are obtained by the singular-value decomposition $USV^T = X_{m-1}^T SX_m$ and D is a modified form 3×3 identity matrix—instead of a one in its $(3, 3)$ element, it has $\text{sgn}(\det(VU^T))$.

Once Q is evaluated, it is used to update all the scattering centers in the mth submatrix, whether they are common to the $(m-1)$th one or not. It is also used to update the observation angles. Thus, the updated X_m and C_m are evaluated by post-multiplying the original ones by Q.

The result, when this procedure is applied to all pairs of adjacent solutions, is a complete listing of the coordinates of all scattering centers, and a listing of all the observation angles specified by their corresponding direction cosines, both lists expressed in a common coordinate system. This coordinate system is still in a sense unmoored, in that if a mildly truncated set of motions were used, there is no assurance that the resulting coordinate system would be aligned with the one emerging from processing the full set of motions.

---

5. http://en.wikipedia.org/wiki/Kabsch_technique.

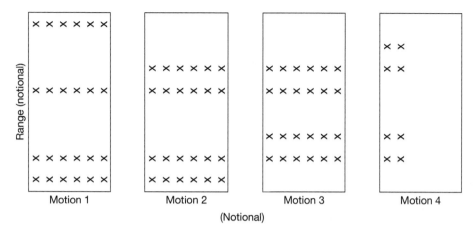

Figure 10.2 Occupancy sketch of the range matrix, with shadowed and dropped scattering centers indicated by blanks. By subdividing it as shown into submatrices with complete motion entries, and then dropping the empty columns from each one, a sequence of complete submatrices can be realized.

10.3.3 Example: The Canonical Cylinder

In this section, the application of the CX decomposition technique is applied to the simulated range, range-rate observable sequence for the spinning cylinder example considered in chapters 2 and 5. The simulated range, range-rate observable sequence versus time, with shadowing accounted for, is illustrated in figures 3.8(a) and 3.8(b). Figure 10.3 shows the result for extracting the target location estimates applying the CX decomposition directly to the complete unshadowed version of the synthetic range matrix (the blue markers) and compares that to the result of applying the divide and conquer technique to a shadowed version of the same range matrix. The Kabsch rotation technique was used to align the two solution sets. The slipping scattering centers in the shadowed range matrix were omitted because, not unexpectedly, they caused the divide and conquer technique to fail. No noise was included in the simulation.

Figure 10.4 shows the motion solution (in direction cosine space) and the corresponding comparison, after the same least-squares alignment rotation, between the motions implied by the complete unshadowed version of the synthetic range matrix (blue) and the motions implied by applying the divide and conquer technique to its

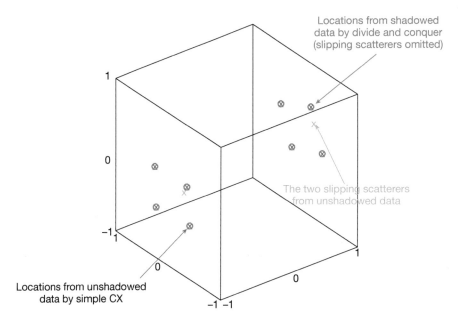

Figure 10.3   3D target location estimates. The locations of all 10 scattering centers estimated by applying *CX* decomposition to the unshadowed range matrix compared with the locations of all eight fixed scattering centers estimated by the divide and conquer technique to the shadowed range matrix.

shadowed version (red). The curves shown are the loci of the tips of the unit motion vectors implied by the two data sets.

The motion estimates indicate that the synthetic target is essentially in a state of pure spin[6] with the spin axis inclined at an angle of approximately 30° to the radar's LoS. A comparison of figures 10.3 and 10.4 indicates that the centers of the two separated groups of four scattering centers lie on this axis. The angular velocity is readily computed from the fact that over the whole length of the simulated data collection, to which time stamps would in practice be attached, the target undergoes about one and one-quarter revolutions.

---

6. The only other interpretation, given that the motion locus is so smoothly circular, is that two of the three principal moments of inertia of the target are equal to one another and the radar motion lies along the angular momentum axis.

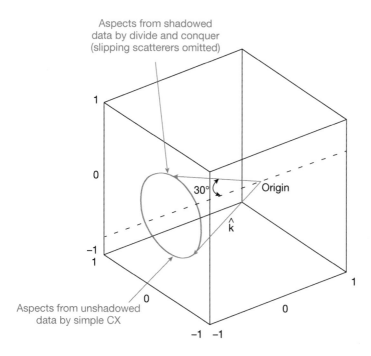

Figure 10.4   Motion solution estimate. (a) The loci of the tips of the unit vector $\hat{k}$ representing all 1,199 motions estimated by direct *CX* decomposition of the unshadowed range matrix (blue). (b) The loci of the tips of the unit vector $\hat{k}$ representing all 1,099 motions estimated by applying the divide and conquer technique to the shadowed range matrix (red). The two solutions lie so close to one another that the red markers all but mask the blue ones.

Note that although the results of figure 10.4 would be the same no matter what the time ordering or specific timing of the motions, attention to ordering would typically be necessary during the subdivision of the range matrix. That is because the track of each scattering center, and each dropped track, tend to persist in time over a continuous sequence of many samples. The range samples in each submatrix will therefore be associated in time. However, within the submatrix, the time ordering can be arbitrary and the timing of the samples can be irregular.

III

Data Extrapolation and Sensor Fusion Processing

11

Multisensor Fusion and Mutual Coherence

11.1 Introduction

Section 1.4.2 briefly introduced the concept of estimating a target's ultrawideband frequency response by fusing data received by separate but closely located sensors that operate over different frequency bands. The bandwidth fusion problem as posed there is actually a special case of the more general problem of estimating a target signature model from multisensor data that is sparse in both angle and frequency. The techniques discussed in chapters 12–15 can be used to develop a target signature model capable of generating simulation data over extended regions in both frequency and angle space to complement the measurement data set, and provide a comprehensive signature model that is predictive for regions outside the measurement space. Such a model provides an important capability for generating a continuous data set for use in comparing model generated data, measured data, and electromagnetic code computations.

The data fusion techniques developed in this text are based on the following comments:

The radar data recorded at each sensor are collectively processed to develop a comprehensive signature model of the target.

The recorded data are compensated for the amplitude, delay, and phase errors introduced by each sensor's receive channel as discussed in section 1.3.1 and later in this chapter to remove sensor artifacts that could corrupt the target's radar signature.

To fuse the radar data into a common signature model, the effects of any differences in viewing angle, and range to the target center of rotation (CoR), embedded in the individual radar observations must also be accounted for. Details of these techniques are discussed in the following sections.

The key to fusing signature data received by any number of sensors is to register each sensor to a common fixed target-centered coordinate system with origin located at the target CoR, as discussed in appendix F. In this text the term mutual coherence between separate sensors is used to indicate that this adjustment has been made. Mutual coherence of separate independently operating sensors, to be precise, means that once the recorded data for each sensor is properly adjusted, the phase in each recorded pulse, referenced to the observed target's CoR, is constant and invariant over time and from sensor to sensor. When fusing data between independent sensors, adjusting the data of each sensor to achieve mutual coherence is referred to in this text as *cohering the sensors*.

The concept of mutual coherence can be viewed as compensating the data recorded at each sensor such that it appears to have been taken by one sensor at a fixed location. For example, assume a static range measurement where the target is viewed over two nonoverlapping angular observation regions, Ω_1 and Ω_2, and denote the measured amplitude and phase over each angle region as $\sigma_c(\Omega_1)$ and $\sigma_c(\Omega_2)$, respectively. The measurements of $\sigma_c(\Omega_1)$ and $\sigma_c(\Omega_2)$ are inherently registered to the CoR location on the target and are phase coherent. Now assume a field experiment such that the frequency-time data block over Ω_1 is received by sensor 1 at location 1, and the frequency-time data block over Ω_2 is received by sensor 2 at location 2. Mutual coherence corrections are required for each sensor measurement so that, to the signature analyst, $\sigma_c(\Omega_1)$ and $\sigma_c(\Omega_2)$ appear as if viewed from a common sensor, as described in the static range experiment.

It is convenient to handle the adjustment required to fuse the recorded data in two steps, first by addressing the differences in range to the target CoR for each sensor, followed by compensating for differences in the relative baseline look angles to the target for each sensor. Assume the target is in the far field of the two nearly colocated sensors such that the *far-field* criteria $R_0 \gg 2L^2 / \lambda$ is met, where L is the sensor separation distance measured perpendicular to the line of sight (LoS) to the target, and R_0 the range to the target. Under this assumption one may visualize the two sensors as being separate sub-apertures of a large aperture sparsely filled array radar where the target is in the far field defined relative to the large array of aperture L. In this case the sensors have the same look angle to the target, but a range adjustment is required to account for the difference in range delay from the target CoR to the individual apertures. If the *far field* criteria is not met, but each sensor satisfies its individual far-field criteria, sparse-angle data compensation techniques will also have to be employed, and each sensor's look angle to the target must be taken into account. The sensor fusion techniques considered in this text consist of combining the data blocks obtained by each sensor using a single global fused signature model defined over the composite of each sensor observation space. The fused signature model used for both the sparse band and sparse angle problem is based on the geometrical theory of diffraction (GTD) scattering model defined by equation (2.6). Using this model, the field scattered from the target is dependent on the frequency and angle of arrival of the incident wave registered in the target-centered coordinate system, and is a function of the frequency and angle variation of the diffraction coefficients $D_n(f, \theta, \varphi)$ that characterize each scattering center on the target.

The coherent combination of sparse-band data blocks is the simplest of the fusion problems. When sparse band sensors are colocated, the viewing angles to the target are identical and the sensor fusion techniques need only account for the frequency behavior of the diffraction coefficients for each scattering center, denoted as $D_n(f)$. The frequency-time data blocks from each sensor are collected simultaneously, and the fusion techniques are independent of target motion. This permits fusion techniques of the type developed in chapter 13 where for simplicity the fusion is demonstrated on a single pulse basis. This is done by fitting a single global all-pole model to all the data covering the composite sparse frequency bandwidth.

For the sparse-angle fusion problem, the global signature model must account for the angular variation of the diffraction coefficient denoted as $D_n(\theta, \varphi)$. The data samples collected by each sensor are labeled as a function of time, and a *joint* multisensor motion solution is required to map the time samples received by each sensor into the target-centered polar angle reference space. Additionally, because the diffraction coefficient $D_n(\theta, \varphi)$ is typically unknown, a parameter-based model characterizing $D_n(\theta, \varphi)$ as a function of angle for each scattering center must be defined. For small angular separations between sensors, the variation of $D_n(\theta, \varphi)$ with angle can be approximated by an exponential growth or decay, and a global all-pole model having complex poles may be used to model the multisensor data. However, as the angular separation increases, the variation of $D_n(\theta, \varphi)$ with angle typically prevents the applicability of a global all-pole model over the fused data region. In this case other models for the variation of $D_n(\theta, \varphi)$ must be used. The challenge for the sparse-angle sensor fusion problem is to develop a suitable diffraction coefficient model that accounts for the angular variation of the response.

In general, the parametric representation of the component diffraction coefficient can be defined as $D_n(\theta, \varphi, c_1, c_2, \ldots, c_m)$, characterized by the parameter set $(c_1, c_2, \ldots, c_m)$. $D_n(\theta, \varphi, c_1, c_2, \ldots, c_m)$ approximates the behavior of $D_n(\theta, \varphi)$ over the entire fused data region, and the parameters $(c_1, c_2, \ldots, c_m)$ are determined using a best metric fit to the measured diffraction coefficients extracted from the composite set of measurement data.

In a broader context, diffraction coefficient modeling represents a key element in creating a signature model for the target using the GTD scattering model. The choice of diffraction coefficient basis functions becomes particularly important when integrating the diffraction coefficients extracted from measurement data into a comprehensive target model characterized by the diffraction coefficient basis sets. This problem is discussed in greater detail in chapter 14, where the angular behavior of $D_n(\theta, \varphi)$ is modeled using a set of parametric basis functions.

The two signature models used to characterize the field scattered from the target for the sparse-band and sparse-angle data fusion scenarios are summarized as follows:

$$E(f) \sim \sum_n D_n(f)e^{-j2\pi f\tau_n}, \tag{11.1a}$$

$$E(\theta, \varphi) \sim \sum_n D_n(\theta, \varphi, c_1, c_2, \ldots c_m)e^{-j\frac{4\pi f}{c}\hat{k}\cdot\underline{r}_n}, \tag{11.1b}$$

where $\tau_n = \dfrac{2}{c}\hat{k}\cdot\underline{r}_n$ represents the propagation delay along the radar line of sight to the target, referenced to the nth scattering center. The signature model for sparse band fusion defined by equation (11.1a) is a function of the one-dimensional (1D) variation of $D_n(f)$ for each scattering center and the propagation delays τ_n. However, the dual problem, fusing sparse-angle sensor data is more complex. The global signature model defined by equation (11.1b) generally requires the existence of a two-dimensional (2D) representation of $D_n(\theta, \varphi)$ for the composite angular observation space, as well as an estimate of the three-dimensional (3D) scattering center locations on the target. Thus, the sparse-angle fusion problem requires each sensor to obtain a 3D location estimate for each of the target's scattering centers, as well as the parameters characterizing each scattering center diffraction coefficient. Multisensor fusion, properly formulated, allows the noncoherent combination of these location estimates into a common 3D target-centered coordinate system. Special case sensor location scenarios relax this general requirement somewhat and are considered later in this chapter. Range and system delay mismatches between sensors, as well as errors in the mapping from time to angle will produce errors in the fused global signature model.

The essence of the sensor data fusion problem is to determine the parameters of the global signature model that provides the minimum mean square error (MMSE) between the model predictions and the mutually coherent data blocks over the two separate observation regions Ω_1 and Ω_2, that are properly registered in angle space. The resultant global signature model provides synthetic data estimates over those regions of the angle observation space not observed by each sensor. One example of extending the observation space using all-pole signature models is illustrated using the frequency-angle extrapolation techniques developed in chapter 12 to generate synthetic data extending the observation region. Another corresponds to the 1D ultrawideband sparse-band fusion problem discussed in chapter 13.

Assuming the sensors are mutually coherent as defined previously, the sparse-band and sparse-angle fusion problems can be summarized as follows:

Sparse-band data fusion: The look angles to the target are such that the target is in the far field relative to the baseline separation of both sensors. In this case, the sensors have the same look angle to the target referenced to the target-centered coordinate system. Frequency-time data blocks recorded by each sensor can be coherently fused using a global all-pole signature model applied over the composite observation space. The fusion can be carried out on either a single pulse or data block basis, independent of the motion of the target.

Sparse-angle data fusion: This is much more complex than the sparse band problem. A more in-depth discussion that addresses the additional complexity is provided in section 11.2.1. A joint, multisensor motion solution is required to map the time samples received by each sensor into a common target-centered polar angle reference space. Additionally, a parameter-based functional behavior characterizing the diffraction coefficient as a function of angle for each scattering center must be defined. For cases where the angular separation gap between each sensor's observation space is small, and the angular variation of the component diffraction coefficients can be modeled by an exponential variation, a global all-pole model having complex poles can be used as the data fusion mechanism; otherwise, the more general signature model defined by equation (11.1b) must be used.

The remainder of this chapter discusses several multisensor fusion situations that address the differences in sensor viewing geometry and operating frequency band. The signal processing flow introduced in section 1.3.1, figure 1.5, is generalized to the case of two independent sensors in order to identify the parameters that must be adjusted to achieve multisensor mutual coherence. In chapter 12, 1D and 2D data extrapolation techniques appropriate for a single sensor are developed. Data extrapolation outside each measurement region is an essential step to achieving coherence between independently operating sensors. It is used in the sparse-band data fusion problem as a critical step in adjusting the measurement data in each frequency band to allow the bands to be processed coherently. In chapter 13 the problem of coherently fusing the data of pulses received by two wideband sensors operating independently over separate frequency bands, but having a common look angle to the target is considered. Chapter 14 examines the dual problem of fusing sparse angle data from two sensors viewing the target simultaneously at different look angles. Finally, chapter 15 discusses the fusion of processed measurement data into a signature extrapolation and prediction model.

11.1.1   Frequency-Angle Data Fusion Scenarios

A general overview of the frequency-angle sensor fusion problem is gained by examining the composition of the GTD scattering model discussed in chapter 2, equation (2.6). The field scattered from the nth scattering center located on the target is given by

$$E_n(f,t) = E_n(f, \theta, \varphi) = D_n(f, \theta, \varphi)e^{-j\frac{4\pi f}{c}\hat{k}\cdot r_n}, \qquad (11.2)$$

where the relationship between time and angle is governed by target motion. Using the direction cosine angle characterization (C_x, C_y, C_z) of the motion variables introduced in section 10.2, the field scattered from the nth scattering center can be expressed in the form

$$E_n(f, \theta, \varphi) = D_n(f, \theta, \varphi)e^{-ju x_n}e^{-jv y_n}e^{-jw z_n}, \qquad (11.3)$$

and the polar variables (u, v, w), commonly referred to in the literature as *polar formatting*, are defined by

$$u = \frac{4\pi f}{c}C_x$$

$$v = \frac{4\pi f}{c}C_y \qquad (11.4)$$

$$w = \frac{4\pi f}{c}C_z.$$

Using the transformation defined by equation (11.4), the frequency angle variables (f, θ, φ) are replaced by three new variables (u, v, w). The resulting expression is still exponential in form so that if the diffraction coefficient $D_n(f, \theta, \varphi)$ were also exponential in form over significant regions of (u, v, w) space, then the expression for the scattered field could be represented by an all-pole signature model over these regions. However, for most components, the angular behavior of $D_n(f, \theta, \varphi)$ over extended regions of observation space deviates considerably from an exponential variation, in contrast to the variation of $D_n(f, \theta, \varphi)$ with frequency, which is typically well modeled by an exponential behavior over ultrawide bandwidths. Thus, data fusion using all-pole signature models is generally possible for data that is obtained in separate frequency bands. Fusion of data recorded over extended angular sectors is a more difficult problem. The implications of the contrasting characterization of $D_n(f, \theta, \varphi)$ versus frequency and angle form the basis for, and limitations of, the data fusion techniques developed in chapters 13 and 14.

Using (u, v, w) space notation, the data extrapolation and fusion scenarios considered in this text are notionally illustrated in figure 11.1 with a two-dimensional presentation for the special case $w = 0$, $(\theta = \pi/2)$. The mapping from (f, φ) space is now characterized by the two-dimensional (u, v) space where $u \propto f \cos(\varphi)$ and $v \propto f \sin(\varphi)$, so that in the (u, v) coordinate system the frequency f corresponds to the radial distance emanating from the origin, and the angle φ corresponds to arcs of constant radius.

Figure 11.1 illustrates four possible scenarios [103] relative to data extrapolation and sensor fusion, three of which are considered in chapters 12, 13, and 14, respectively. Figure 11.1(a) shows 2D bandwidth and angle extrapolation for a single sensor referred to as enhanced bandwidth extrapolation (EBWE); figure 11.1(b) shows sparse-band fusion between two separate nonoverlapping frequency bands, but with common look angles to the target; and figure 11.1(c) shows sparse-angle fusion for sensors with common operating frequency bands, but having separate viewing angles to the target. The more general case of simultaneous sparse-band and sparse-angle data fusion depicted in figure 11.1(d) is not considered in this text but can be treated by extension of the three special cases (e.g., see the treatment in [103]; however, note that in reference [103] the target components consist of ideal point scattering centers having constant amplitudes over global frequency-time space for which a global all-pole model is valid over the complete 4π steradian sphere.) Although the sparse-band and sparse-angle fusion problems are dual in nature, the fusion techniques differ and are dependent on the specific functional models for the diffraction coefficients appropriate to each situation.

11.2   Mutual Coherence Examples

11.2.1   Fusion of Multisensor Angle Data

The sparse angle problem, illustrated in figure 11.1(c), is more complex than the sparse frequency problem and is discussed in greater detail in this section. Consider the

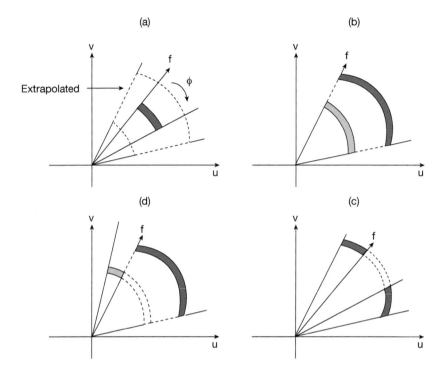

Figure 11.1   Different sensor fusion situations. (a) Two-dimensional bandwidth and angle extrapolation for a single sensor band (EBWE). (b) Sparse band fusion between two separate sparse frequency bands having common look angles to the target. (c) Sparse angle fusion from sensors having a common operational frequency band, but not colocated. (d) The general case of sparse band, sparse angle sensors.

situation where the sensors have separate viewing angles to the target but operate with common frequency bands. It is useful to compare the dimensional complexity of the multisensor angle data fusion problem with the sparse band single pulse fusion problem with a simple example, namely two sensors operating over a common frequency band but viewing a rotating target from two different locations. For notational simplicity assume the data are collected simultaneously but independently by each sensor so that the time sample labels are the same for each sensor. To coherently integrate each sensor's data samples into a common signature model, they must be properly registered in a target-centered polar coordinate system. Assume the case where data are collected simultaneously by sensor 1 over angular sector Ω_1 centered at look angle ϕ_{10} and by sensor 2 over angular sector Ω_2 centered at look angle ϕ_{20}. The proper association of Ω_1 and

Ω_2 within the composite observation space defined in a common target-fixed coordinate system is critical to the success of the angle fusion problem.

Assume a global signature model characterized by equation (11.1b) is used to fuse the data over the composite observation space of both sensors. To properly associate each sensor's time samples to a composite sampling grid in angle space, the following steps are required:

1. The effects of translational motion are removed from each sensor's data independently using only its own measurements as discussed in appendix F. In the ideal case the resultant data set for each sensor is dependent only on the rotational motion of the target and unknown range biases, R_{10} and R_{20} for each sensor resulting from the removal of the translational motion. The rotational motion, registered to the CoR of the target, is used to characterize the viewing angle time history to the target for each sensor.

2. Determine a target-motion solution independently for each sensor using the adjusted data sets determined in step 1. The result of each sensor's independent target-motion solution provides an estimate of the range bias, R_{10} and R_{20}, for each sensor [e.g. , as used in the optimization defined in equation (9.28)], and registers the observation regions Ω_1 and Ω_2 each in a target-centered polar angle coordinate system. The estimates of R_{10} and R_{20} are used to compensate the data for each sensor so that it is referenced to the target CoR as described in appendix F. Assume the look angle κ to the angular momentum vector, J, is represented by the values κ_1 and κ_2, respectively, for each sensor. As discussed in chapter 5, section 5.1, κ_1 and κ_2 must both remain constant over each observation interval for the data processing to be valid. The objective is to solve for the motion parameters $\theta_p, f_p, \alpha_p, f_s, \alpha_s$ and κ introduced in appendix A for each sensor that define the target's rotational motion in the measurement system used to process the data observed by each sensor over this time interval. The Euler parameters θ_p, f_p, f_s characterizing the rotational motion of the target relative to the angular momentum vector J are common to each sensor; only the angles κ_1 and κ_2, and phase references α_p, α_s depend on sensor position. Using the techniques developed in chapters 5–10, each sensor can form a separate estimate of the target motion as viewed from its location dependent on $\theta_p, f_p, f_s, \kappa_1, \alpha_{p1}, \alpha_{s1}$ and $\theta_p, f_p, f_s, \kappa_2, \alpha_{p2}, \alpha_{s2}$ according to the transformation developed in appendix A.

$$F\{\theta_p, f_p, f_s, \kappa_1, \alpha_{p1}, \alpha_{s1}, t\} \rightarrow [\theta_1'(t), \varphi_1'(t)]$$
$$F\{\theta_p, f_p, f_s, \kappa_2, \alpha_{p2}, \alpha_{s2}, t\} \rightarrow [\theta_2'(t), \varphi_2'(t)]. \tag{11.5}$$

Done independently, each motion solution introduces a relative phase (e.g., $\alpha_{p1}=0$, $\alpha_{p2}=0$) defining the motion state at time $t=0$. For the single sensor case, this relative phase determines the orientation of the resultant target scattering center location estimates in the measurement target-centered coordinate system used to process the data.

The motion solution for $\theta_1'(t), \theta_1'(t)$ defines the angular sector Ω_1, and the motion solution for $\theta_2'(t), \theta_2'(t)$ defines the angular sector Ω_2, each defined in a fixed target-centered polar framework. The motion to target space mapping equations can be used to estimate the 3D scattering center locations in the target-centered measurement coordinate system that was defined to process the data for each sensor. Because the motion estimates are done independently, the axes of the two measurement coordinate systems, as well as the orientation of the target location estimates, although referenced to the target CoR, are not necessarily aligned. Association of the composite set of 3D scattering center location estimates obtained from each sensor into a common fixed target-centered coordinate system depends on the baseline look angle of sensor 1 and sensor 2 to the target, and requires accurate knowledge of each sensor's location defined in a common target-fixed coordinate system. If the two sensors collect data simultaneously, determining the baseline look angles to the target, registered to a target-fixed coordinate system, is facilitated by the triangle formed connecting the measured range to each sensor to the target CoR connected by the baseline distance, L, between the two sensors.

3. For the two-sensor problem, the motion solutions are coupled and only one relative phase defining the motion at $t=0$ can be chosen independently. For fusion of the data into a global signature model one must determine the baseline look angles for each sector, ϕ_{10} and ϕ_{20}, referenced to the common target-centered fixed coordinate system, and properly associate them with Ω_1 and Ω_2. An error in these estimates, denoted as $\Delta\phi_{B1}$ and $\Delta\phi_{B2}$, results in an error when fusing the data from sensor 1 with that of sensor 2. Knowledge of the baseline look angles defines a coupling between the phases α_{p1}, α_{s1} and α_{p2}, α_{s2} obtained for each independent motion solution.

To illustrate this coupling, consider the motion of a target in pure tumble examined in chapter 9. Assume for simplicity that two narrowband sensors operate at the same pulse repetition frequency (PRF) and view the target simultaneously. Denote the time samples $E_1(t_q)$, $q=1,\dots Q$ for sensor 1 to correspond to $E_1(\Omega_1)$ and $E_2(t_q)$, $q=1,\dots Q$ to correspond to $E_2(\Omega_2)$ for sensor 2. Because the data are collected simultaneously, the time sample labels are the same for each sensor. To coherently fuse $E_1(t_q)$ and $E_2(t_q)$ into a global signature model each sensor's data samples must be properly associated in angle space. This requires a joint motion solution and a knowledge of ϕ_{10} and ϕ_{20}. For this example, in the measurements target-fixed system defined in chapter 9, the angles $\theta_1'(t_q)$, $\varphi_1'(t_q)$ for sensor 1 are given by equation (9.2), repeated here as

$$\theta_1'(t_q) = \kappa_1$$
$$\varphi_1'(t_q) = (2\pi f_p)t_q + \alpha_{p1}.$$

(11.6)

For sensor 2, referenced to the same time interval defined relative to $t_q = 0$, the angles $\theta_2'(t_q)$, $\varphi_2'(t_q)$ are given by

$$\theta_2'(t_q) = \kappa_2$$
$$\varphi_2'(t_q) = (2\pi f_p)t_q + \alpha_{p2}.$$

(11.7)

Assume $\kappa_1 = \kappa_2 = \pi/2$, so that both sensors are located in the tumble plane of the target. Define the baseline (azimuthal) look angle from each sensor to target at $t_q = 0$ as φ_{10} and φ_{20}. Using each sensor motion solution, it follows that $\alpha_{p1} = \varphi_{10}$ and $\alpha_{p2} = \varphi_{20}$ so the phases α_{p1} and α_{p2} must be related by

$$\alpha_{p2} - \alpha_{p1} = \varphi_{20} - \varphi_{10}.$$

(11.8)

and are not independent, but define the orientation of the target and sensor locations along the φ-axis in a common target-fixed measurements system. The phases $\alpha_{p1} = \varphi_{10}$ and $\alpha_{p2} = \varphi_{20}$ define a joint motion solution and a mapping associating the time samples $E_1(t_q)$ with $E_2(t_q)$ within a composite sampling grid connecting each sensor observation space. Errors in the estimate of φ_{10} and φ_{20} result in errors in associating the data samples from each sensor onto the composite grid connecting the angular sectors. Also note that for $\kappa_1 \neq \kappa_2$ forming a composite sampling grid poses a more complex fusion problem requiring a 2D composite angular sampling grid. Further discussion of the dependence of the baseline look angle difference $\varphi_{20} - \varphi_{10}$ on the poles extracted from the frequency-time data of sensor 2 relative to those extracted from the frequency-time data of sensor 1 for this special case example are examined later in section 11.3.2.

11.2.2   General Formulation versus Range, Angle Bias

The general development tracing the signal flow through the transmit/receive chain for independently operating sensors is presented in this section and used to identify the parameters required to achieve sensor coherence.

Figure 11.2 illustrates the signal paths for a two-sensor configuration as a general-ization of the single sensor signal flow developed in section 1.3.1. Two sensors operate independently and interrogate the target at notional angles ϕ_1 and ϕ_2, respectively. Sensor 1 has bandwidth BW_1 centered at operational frequency f_1 and sensor 2 has bandwidth BW_2 centered at operational frequency f_2.

Denote the signal received by each sensor using the subscripts 1 and 2, respectively. The signal received at each sensor is processed using the analysis leading to equation (1.2) for a single sensor, with appropriate subscript labeling. The baseband received waveforms, before pulse compression, are given by

$$E_1(\omega, \phi_1) = |T_1(\omega)|^2 \, F_1^2(\omega + \omega_1) e^{-j2\left(\frac{\omega + \omega_1}{c}\right)R_1} \times \sigma_c(\omega + \omega_1, \phi_1) e^{j\psi_1} \, e^{-j\omega\tau_1},$$
$$-\frac{BW_1}{2} \le f \le \frac{BW_1}{2}, \tag{11.9}$$

$$E_2(\omega, \phi_2) = |T_2(\omega)|^2 \, F_2^2(\omega + \omega_2) e^{-j2\left(\frac{\omega + \omega_2}{c}\right)R_2} \times \sigma_c(\omega + \omega_2, \phi_2) e^{j\psi_2} \, e^{-j\omega\tau_2},$$
$$-\frac{BW_2}{2} \le f \le \frac{BW_2}{2}, \tag{11.10}$$

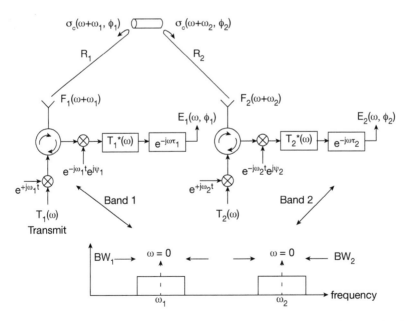

Figure 11.2   Signal model for two-sensor mutual coherence characterization.

where R_1 and R_2 denote the ranges to the target CoR for sensors 1 and 2, respectively. $T_1(\omega)$ and $T_2(\omega)$ are the baseband transmitted waveforms and $\omega = 2\pi f$. $F_1(\omega + \omega_1)$ and $F_2(\omega + \omega_2)$ are the antenna frequency response functions. $\sigma_c(\omega + \omega_1, \phi_1)$ and $\sigma_c(\omega + \omega_2, \phi_2)$ denote the target's complex radar cross section (introduced in section 1.3.1) and are functions of both viewing angle ϕ and frequency. The phase and delay offsets (ψ_1, τ_1) and (ψ_2, τ_2) account for hardware differences in the two sensor's receiver chains.

Consider fusing multisensor signature data over a given time interval. Assume for simplicity the sensors operate at the same PRF and a number of pulses are processed. Furthermore, because the sensors are not colocated the rotational motion of the target observed by each sensor corresponds to different values of κ, and the viewing angle history of the target observed at each sensor is not the same. Denote the angle change (notional) as viewed by each sensor over time interval Δt as $\Delta\phi_1$ and $\Delta\phi_2$, each referenced to baseline look angles ϕ_{10} and ϕ_{20}. Changes in ϕ_1 and ϕ_2 over time interval $Q\Delta t$, for Q pulses having pulse repetition interval $(\Delta t = PRI)$ are given by:

$$\left.\begin{array}{l} \phi_1 = \phi_{10} + q\Delta\phi_1 \\ \phi_2 = \phi_{20} + q\Delta\phi_2 \end{array}\right\} \quad q = 0, 1, \cdots Q. \tag{11.11}$$

Based on the previous mutual coherence discussion, sensor 1 covers angular sector $\Omega_1 = Q\Delta\phi_1$ referenced to ϕ_{10} and sensor 2 covers the angular sector $\Omega_2 = Q\Delta\phi_2$ referenced to ϕ_{20}. For each sensor, the response functions $|T_1(\omega)|^2 F_1^2(\omega + \omega_1)$ and $|T_2(\omega)|^2 F_2^2(\omega + \omega_2)$ are generally known and equalized via calibration as discussed in chapter 1, section 1.3.1. Thus, this factor can be omitted from each response. Express the remaining factors in $E_1(\omega, \phi_1)$ and $E_2(\omega, \phi_2)$ in the form

$$E_1(\omega, \phi_1) = \sigma_c(\omega + \omega_1, \phi_1) \times \left\{ e^{j\psi_1} e^{-j\omega\tau_1} e^{-j\frac{2\omega_1}{c}R_{10}} e^{-j\frac{2\omega}{c}R_{10}} \right\}$$

$$E_2(\omega, \phi_2) = \sigma_c(\omega + \omega_2, \phi_2) \times \left\{ e^{j\psi_2} e^{-j\omega\tau_2} e^{-j\frac{2\omega_2}{c}R_{20}} e^{-j\frac{2\omega}{c}R_{20}} \right\}. \tag{11.12}$$

The signal received by each sensor, $E_1(\omega, \phi_1)$ and $E_2(\omega, \phi_2)$, is divided into two parts: to the left of the multiplication symbol are the two components of the target response, $\sigma_c(\omega + \omega_1, \phi_1)$ and $\sigma_c(\omega + \omega_2, \phi_2)$, which characterize the complex signature of the target, phase referenced to the CoR of the target; the terms to the right of the multiplication symbol indicate corrections required to cohere the multisensor signature data. These terms must be removed from the data before sensor fusion can take place. Consider each of these factors separately.

11.2.3   Hardware Phase Shift Offset and Time Delay

The phase shift offset $e^{j\psi}$ and time delay $e^{-j\omega\tau}$ factors for each sensor are similar to that considered in chapter 1. They represent a combination of hardware propagation phase and time delay mechanisms. These factors are removed by each sensor before data recording, as discussed in section 1.3.1.

11.2.4   Frequency-Dependent Range and Phase Offsets

The remaining factors on the right side of equation (11.12) represent a range and phase offset resulting from removing the translational motion in each respective sensor measurement data set as discussed in appendix F. For a given pulse, the effect of these terms is to offset the range marking and phase of the compressed wideband received pulse of each sensor relative to the location of the target CoR. This results in range offsets R_{10} and R_{20}, and in phase modifications $\Psi_1 = 2\dfrac{\omega_1}{c} R_{10}$ and $\Psi_2 = 2\dfrac{\omega_2}{c} R_{20}$ for the range alignment of the compressed wideband pulses for each sensor relative to the target CoR.

11.2.5   Sensor Location Estimate Errors

For each sensor, the rotational motion is registered to the target CoR and is determined from the motion solution. Errors in estimating each sensor's baseline location registered in a common target-fixed coordinate system are significant in how they affect data fusion as described in the simple example of a target in pure tumble in section 11.2.1. For multisensor data fusion, ϕ_{10} and ϕ_{20} must be properly associated to viewing angle sectors Ω_1 and Ω_2. This association is facilitated by the knowledge of the location of each sensor. An error in estimating ϕ_{10} and ϕ_{20} will result in misplacing the sectors Ω_1 and Ω_2 relative to one another in the common target-fixed polar coordinate framework, and introduce errors in the sparse-angle data fusion result.

11.3   Mutual Coherence Compensation

The previous discussion summarizes the compensation coherence parameters that must be applied to multisensor data in order to fuse the signature data over differing look angles and frequency bands, and bring the data sets into mutual coherence. The net effect of the compensation required can be partitioned into the error sources delineated in table 11.1

Table 11.1   Error sources and mutual coherence effects

Coherence Parameter	Mutual Coherence Effect
Hardware Phase/Time Delay Offsets	Single Sensor Calibration
Range Dependent Phase Offsets	$e^{-j\Psi_1}, e^{-j\Psi_2}$
Range Delay Offsets	$e^{-j2\frac{\omega}{c}R_{10}}, e^{-j2\frac{\omega}{c}R_{20}}$
Sensor Location Estimation Errors	$\Delta\phi_{B1}, \Delta\phi_{B2}$

The effect of these terms is to cause an error in the estimation of the complex radar cross section $\sigma_c(f, \phi)$ of the target after fusing the multisensor data into a global signature model. It is useful to reference the net effect of the frequency-dependent range and phase offset error sources in table 11.1 as occurring in sensor 2 relative to sensor 1. Denote by $E_1(f, \phi_{1,q})$ and $E_2(f, \phi_{2,q})$ the estimate of σ_c and Ω, the angular sector characterizing the target motion over the processing time interval.

$$E_1(f, \phi_{1,q}) \approx \sigma_c(f + f_1, \phi_{10} + \Delta\phi_{B1} + q\Delta\phi_1), \quad \begin{cases} |f| \le BW_1 \\ |q\Delta\phi_1| \le \Omega_1 \end{cases},$$

$$E_2(f, \phi_{2,q}) \approx \sigma_c(f + f_2, \phi_{20} + \Delta\phi_{B2} + q\Delta\phi_2) e^{j(\Delta\Psi)} e^{-j\omega\tau_0}, \quad \begin{cases} |f| \le BW_2 \\ |q\Delta\phi_2| \le \Omega_2 \end{cases}, \quad (11.13)$$

where the delay and phase errors have been combined into the common terms

$$\tau_0 \equiv 2(R_{20} - R_{10})/c$$
$$\Delta\Psi = \Psi_2 - \Psi_1. \quad (11.14)$$

Two cases that deserve special consideration are now considered. The technique for estimating the coherence parameters τ_0 and $\Delta\Psi$ is summarized in this chapter in order to illustrate the duality of the sparse-band fusion technique to that used to fuse sparse angle data. A brief overview of the technique is described in the next section and illustrates the application of the data extrapolation techniques developed in chapter 12.

11.3.1   Common Look Angle, Differing Frequency Bands

The sparse-band fusion problem depicted in figure 11.1(b) is the least complicated of the multisensor fusion problems. As discussed earlier, data fusion can be performed on either a single pulse or data block basis, and is independent of target motion. For simplicity the following discussion uses single pulse fusion to describe the process. In this case the sensors share a common look angle to the target, so the scattering centers projected to the radar line of sight that give rise to the propagation delays are identical for each sensor. An overview of various techniques that could be used to estimate the coherence parameters τ_0 and $\Delta\Psi$ is presented in this section. The problem to be solved is to bring the data from each sensor into mutual coherence. The key steps described as follows are applied to a single pulse data set. Because each sensor observes the same target motion, the process can be readily generalized and applied to all the pulses in the data blocks received by each sensor to increase the signal-to-noise ratio. Data are processed in each sub-band independently to determine a separate all-pole model appropriate for each sub-band. Denote the all-pole models representing the data in each sub-band as $M_1(f)$ and $M_2(f)$, respectively. If the sub-bands were mutually coherent, the locations of the poles characterizing $M_1(f)$ and $M_2(f)$ common to each sensor's sub-band data would be the same. For the example treated in chapter 13, the two sub-bands are not mutually coherent, and the pole estimates for the two sub-band data sets are repeated here and illustrated in figure 11.3. The poles shown in blue and red are determined to be the dominant poles for the lower and upper sub-bands, respectively. Notice that the poles in the lower sub-band do not align with the poles in the upper sub-band, but they share a common pattern. They are misaligned by a pole rotation angle $\Delta\theta = 2\pi\Delta f\tau_0$ and phase offset $\Delta\Psi$, due to lack of mutual coherence.

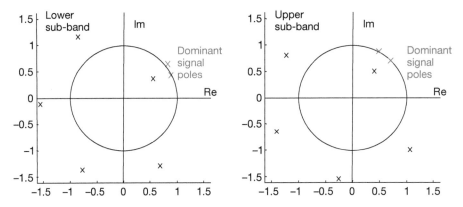

Figure 11.3 Pole estimates for the two sub-band data set examples treated in chapter 13. The dominant poles in the lower and upper sub-bands are shown in blue and red, respectively.

Several ad hoc approaches to estimate $\Delta\Psi$ and τ_0 are possible and could be used to bring the poles into alignment. A distance metric could be applied to the separation of the two pole patterns in the complex plane and minimized with respect to the rotation. A second approach introduced in this section brings the measured data directly into play, and is shown in equation (11.15).

$$C(A, \Delta\theta) = \underset{(A, \Delta\theta)}{Min} \left(\sum_{band\,1} |M_2(f_k) - AE_{d1}(f_k)e^{jk\Delta\theta}|^2 \right.$$

$$+ \left. \sum_{band\,2} |AM_1(f_k)e^{jk\Delta\theta} - E_{d2}(f_k)|^2 \right). \qquad (11.15)$$

In this case a normalization constant, $A = A_0 e^{j\Delta\Psi}$, and pole rotation parameter $\Delta\theta = 2\pi\Delta f\tau_0$ are used to bring the pole patterns into alignment. The phase constant and rotation are applied to both the data and model of sub-band 1. Each of the sub-band models are used to generate model data in the opposing sub-band.

A third approach applies a metric similar to equation (11.15) to minimize the distance between the two models $M_1(f)$ and $M_2(f)$, suitably modified by $A = A_0 e^{j\Delta\Psi}$ and pole rotation parameter $\Delta\theta = 2\pi\Delta f\tau_0$. In this case the MMSE between models is computed over the entire fusion bandwidth. The coherence parameters are estimated using the cost function given by equation (13.3), chapter 13, and [104]. The cost function provides the optimum match between the two sub-band models, each extrapolated over the ultrawide bandwidth. This is contrasted to the sub-band model alignment process depicted in equation (11.15) which determines the best adjustment between the data and the opposing sub-band models for mutual coherence between the two sub-bands. Once the two data sets are adjusted for mutual coherence, the adjusted data sets are used to estimate the parameters of a single global all-pole signature model that minimizes the mean square error (MSE) over the composite sampling grid covering the two sub-bands as described in chapter 13.

The degree to which each of these approaches provides mutual coherency has not been evaluated. However, in practice the third approach used in chapter 13 has yielded satisfactory results.

11.3.2   Sparse Angle, Common Frequency Bands

The dual problem to sparse band fusion is sparse angle fusion, depicted in figure 11.1(c), for sensors with common operating frequency bands, but having separate viewing angles to the target. In this case errors in estimating baseline reference look angles ϕ_{10} and ϕ_{20} dominate the association of Ω_1 and Ω_2 into a common target-fixed coordinate system. As discussed in section 11.2, for the wideband case, the result of each sensor's independent target-motion solution provides an estimate of the range bias, R_{10} and R_{20}, for each sensor, for example as used in the optimization defined in equation (9.28). These values are used to compensate the data from each sensor so that it is referenced to the target CoR.

Now consider generalizing the approach used in the previous section to estimate the errors $\Delta\phi_{B1}$, $\Delta\phi_{B2}$. For simplicity, consider the one dimensional (in angle space) example treated in section 11.2.1 of two narrowband sensors having baseline location look angles to the target, φ_{10}, φ_{20}, located in the plane of rotation of a target in pure tumble operating at the same PRF. Denote the data samples collected by sensor 1 over time interval $[t_1, \ldots t_Q]$ characterizing angular sector Ω_1 as $E_{d1}(t_q)$. Similarily, for sensor 2, the data samples denoted as $E_{d2}(t_q)$ cover angular sector Ω_2 over the same time interval $[t_1, \ldots t_Q]$. Assume the data length Q is such that the data for each sensor can each be represented by an all-pole signature model. Denote the all-pole signature models representing the data samples $E_{d1}(t_q)$ and $E_{d2}(t_q)$ over the time interval $[t_1, \ldots t_Q]$ as $M_1(t_q)$ and $M_2(t_q)$, respectively. Using equation (2.13) with $s_n^k = 1$ the narrowband signature models $M_1(t_q)$ and $M_2(t_q)$ can be expressed in the form

$$M_1(t_q) = \sum_n D_{1n} p_{1n}^q \qquad M_2(t_q) = \sum_n D_{2n} p_{2n}^q. \qquad (11.16)$$

The joint motion solutions for each sensor are defined by equations (11.6) and (11.7). The motion solution for this scenario is one-dimensional where the angle variation is characterized by the azimuthal angle $\varphi(t)$. For this special case sensor-target-motion scenario, it will be shown later in this section that the true baseline locations φ_{10}, φ_{20} are directly related to a cross-range pole rotation that aligns the poles of $M_1(t_q)$ to those of $M_2(t_q)$, and this property can be used to develop a relationship between φ_{20} and φ_{10}. Knowledge of each sensor's baseline look angle allows proper association of each sparse-angle sector data sets Ω_1 and Ω_2 into a common angular coordinate system relating the motion $\varphi(t)$ as defined by equations (11.6) and (11.7) with $\alpha_{p2} - \alpha_{p1} = \varphi_{20} - \varphi_{10}$. The estimate of φ_{10} and φ_{20} is equivalent to coupling each independent motion solution as determined by the phase relationship connecting the two motion solutions, that is, equation (11.8).

Define the pole rotation aligning the two fusion model poles as $\Delta\theta$. To estimate $\Delta\Psi$ and $\Delta\theta$, a normalization constant $A = A_0 e^{j\Delta\theta}$ and the two parameters A and $\Delta\theta$ are adjusted until the error between the separate models is minimized over the measurement space. Using the dual cost function defined in chapter 13, equation (13.3), these values are determined by minimizing the cost function

$$C(A, \Delta\theta) = \underset{(A, \Delta\theta)}{Min} \left(\sum_{[t_1 \ldots t_Q]} | AM_1(t_q) e^{jq\Delta\theta} - M_2(t_q) |^2 \right). \qquad (11.17)$$

with respect to the pole rotation angle $\Delta\theta$ and complex amplitude coefficient A. The optimization $C(A, \Delta\theta)$ in equation (11.17) defines a cross-range pole rotation in angle space, the dual of the time delay (range) pole rotation defined by equation (13.3) for the sparse band problem.

The relationship of φ_{10}, φ_{20} to the pole rotation $\Delta\theta$ extracted from the data samples in each observation sector for this special case scenario can be illustrated using the following example. Consider a two-point scattering center model referenced to the two-dimensional target-centered coordinate system illustrated in figure 11.4. Two scattering centers denoted as P_1 and P_2 are located on the x-axis separated by distance a. The angular variation of the scattered field $E(\varphi)$ at center frequency $f=f_1$ is given by

$$E(\varphi) = D_1 + D_2\, e^{-j\frac{4\pi f_1}{c}a\cos\varphi} \tag{11.18}$$

Using this two scattering center example, assume the data for each sensor collected over time interval $[t_1, \ldots t_Q]$ corresponds to angular regions Ω_1 and Ω_2 defined about the angles φ_{10}, φ_{20}, respectively. The fusion problem for this example is then one dimensional in angle space. Sensor 1 has baseline look angle φ_{10} to the target and sensor 2

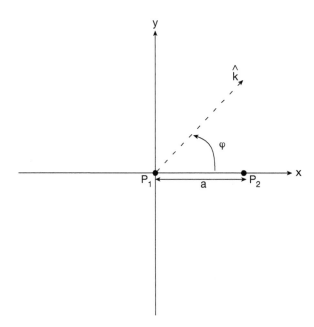

Figure 11.4 Two-point scattering center example.

has baseline look angle φ_{20}. Assume the angular region Ω_1 about φ_{10} is characterized by the expansion $\varphi = \varphi_{10} + q\Delta\varphi$ and similarly about φ_{20} by $\varphi = \varphi_{20} + q\Delta\varphi$ where $\Delta\varphi$ denotes the angular increment corresponding to the sensor pulse repetition interval time Δt, and $(q = 0,1,\ldots,Q) \cdot \Delta\varphi$ covers Ω_1 and Ω_2. Define the signature model associated with $M_1(t_q)$ expressed in the angle domain as $M_1(\varphi_q)$, obtained using the motion solution time to angle mapping. Using the angle sum formula for $\cos(\varphi_{10} + q\Delta\varphi)$ in equation (11.18) and assuming $q\Delta\varphi \ll 1$, $M_1(\varphi_q) \equiv E(\varphi_{10} + q\Delta\varphi)$ can be written in the form

$$M_1(\varphi_q) = D_1 + D_2\, e^{-j\frac{4\pi f_1}{c}a\cos\varphi_{10}}\, e^{j\frac{4\pi f_1}{c}a\sin\varphi_{10}(q\Delta\varphi)}. \qquad (11.19)$$

Express the sequence $M_1(\varphi_q)$ in the form

$$M_1(\varphi_q) = D_1(1)^q + D_2\, e^{-j\frac{4\pi f_1}{c}a\cos\varphi_{10}}\left[e^{j\frac{4\pi f_1}{c}a\Delta\varphi\sin\varphi_{10}}\right]^q. \qquad (11.20)$$

Two poles for $M_1(\varphi_q)$ corresponding to each of the two scattering centers are identified:

$$(p_{11}, p_{12}) = \left(1,\ e^{j\frac{4\pi f_1}{c}a\Delta\varphi\sin\varphi_{10}}\right).$$

The poles for sensor 2, denoted as (p_{21}, p_{22}) centered about $\varphi = \varphi_{20}$, are identical to those of sensor 1 with φ_{10} replaced by φ_{20}.

The poles associated with look angles φ_{10} and φ_{20} are located on the unit circle. The pole locations associated with the scattering center P_1 located at the origin of the target coordinate system do not change with look angle. The poles associated with scattering center P_2 are positioned in angle by α_1 and α_2 for sensor1 and sensor 2, respectively, and are illustrated in figure 11.5:

$$\alpha_1 = \frac{4\pi f_1}{c}a\,\Delta\varphi\sin\varphi_{10} = \frac{4\pi f_1}{c}\Delta\varphi\rho_{CR_1}$$

$$ \qquad (11.21)$$

$$\alpha_2 = \frac{4\pi f_1}{c}a\,\Delta\varphi\sin\varphi_{20} = \frac{4\pi f_1}{c}\Delta\varphi\rho_{CR_2},$$

where the cross range ρ_{CR} is defined by $\rho_{CR_1} = a\sin\varphi_{10}$ and $\rho_{CR_2} = a\sin\varphi_{20}$ for each sensor look angle.

In order to examine the effects of errors in φ_{10} and φ_{20} on the fusion of multisector data, consider the estimate of the fused target space scattering center locations obtained

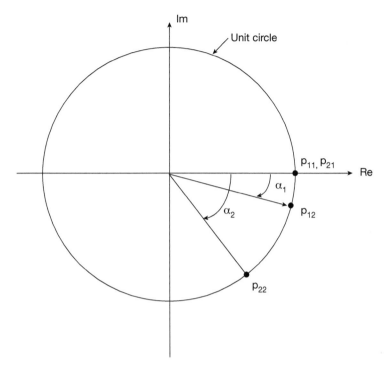

Figure 11.5   Cross-range poles on unit circle for two-point scattering center example. The poles for each sensor associated with scattering center P_2, (p_{12}, p_{22}) are positioned in angle by a_1 for sensor 1 and a_2 sensor 2.

by applying the narrowband mapping illustrated in figure 6.10, technique 2, for the two-sensor case. Fusion is accomplished using the cross-range poles obtained by each sensor, and noncoherently integrating the amplitude build up at intersecting lines spaced by the cross-range poles. Figure 11.6 compares the fused scattering center location estimates for each of the sensor look angles. Two possible mappings are illustrated, namely those with and without sensor location errors.

With perfect knowledge of the baseline look angles to the target, the cross-range mappings using $\varphi = \varphi_{10}$ and $\varphi = \varphi_{20}$ obtained separately for each sensor, but mapped into a common target-based coordinate system, intersect precisely at the true scattering center locations, so that the fused cross-range mappings reinforce at the true scattering center locations. However, when applying the mapping from sensor 2 with baseline sensor location error $\Delta\phi_B$, the error results in a displacement of the cross-range

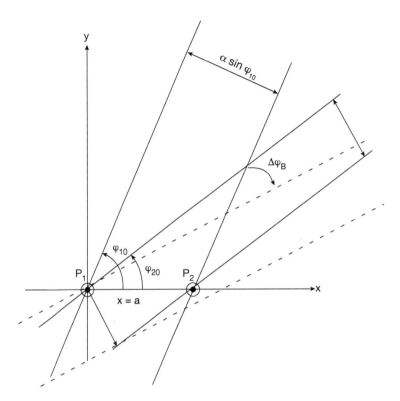

Figure 11.6   Cross-range target space mapping using the cross-range poles of sensor 1 fused with those of sensor 2. Two possible mappings are illustrated, namely, those with and without sensor location errors.

lines of sensor 2 away from the true scattering center location defined at P_2, and the intersecting lines no longer build up at the location of scattering center P_2, causing an error in the fused scattering center location estimate.

It is important to note that the signature model pole estimates determined for each sensor are obtained independent of the knowledge of the actual look angle to the target; they are directly extracted from the received data samples over the observation interval. It is the error in baseline look angles to the target when applied to the target space mapping that causes the scattering center location estimation error.

For this one-dimensional example, the relationship connecting φ_{10} and φ_{20} is determined by a simple pole rotation, the dual of sparse-band coherence compensation technique. This can be shown by examining the rotation of the poles obtained from sensor 2 into those of sensor 1 when plotted on a common reference system. Assume $\varphi_{20} > \varphi_{10}$, for which $\sin \varphi_{20} > \sin \varphi_{10}$. Referring to figure 11.5, the pole rotation angle $\Delta \theta = \alpha_2 - \alpha_1$ can be determined using the optimization $C(A, \Delta \theta)$ defined in equation (11.17). Using equation (11.21), it follows that

$$\Delta \theta = \frac{4 \pi f_1}{c} (a \, \Delta \varphi \sin \varphi_{20} - a \, \Delta \varphi \sin \varphi_{10}). \tag{11.22}$$

Equation (11.22) defines the relationship connecting φ_{10} and φ_{20}. Because the angle space connecting φ_{10} and φ_{20} is defined by the difference $\varphi_2 - \varphi_1$, one can arbitrarily choose $\varphi_{10} = 0$ so that the baseline look angle of sensor 2, relative to sensor 1 is directly measurable by the pole rotation $\Delta \theta$.

11.4   Range Bias and Sensor Location Mutual Coherence Duality

It is useful to contrast the two signature models and mutual coherence compensation techniques required for the cases of sparse band fusion versus sparse-angle fusion. First consider the signature model parameters characterized by equation (11.1) required for each case that are compared in figure 11.7. The all-pole model associated with sparse band fusion assumes a complex pole represented by parameters (α_n, τ_n) for each scattering center, as well as the coherence parameters (Ψ, τ_0) required to cohere the sensors, where $\tau_0 = 2 * \Delta R / c$ represents the LoS sensor delay difference to the target CoR between the sensors. For sparse-angle fusion, the signature model characterized by the 3D scattering center locations (x_n, y_n, z_n) must be augmented by parameter-based basis functions, characterized by parameters $(c_1, c_2, \ldots)$, which approximate the behavior of $D_n(\theta, \varphi)$ over the entire fused data region. These basis functions become particularly important when approximating components extracted from the data to develop a measurements-based signature model of the target considered in chapters 15 and 16.

Also represented in figure 11.7 is the contrast in the sparsity dimension of the data fusion regions for each of the techniques: interpolation in frequency is inherently a one-dimensional problem, as contrasted to angle interpolation where the observation space as a function of angle can cover a full 4π steradian sphere, and the observation data for a typical field data collection covers only a small angular sector defined over a specified $(\theta(t)\phi(t))$ trajectory. However, for angle-only sensor fusion over small intervals about the measurement regions for closely spaced sensors, an all-pole signature model with complex poles is usually valid.

The mutual coherence correction techniques applied for each case exhibit an interesting duality: both the range bias correction and the sensor location estimate errors are related by pole rotations, the first using the range poles, the second using the cross-range poles. This duality is summarized in figure 11.8 where the coherence compensation

Sparse-band	Sparse-angle

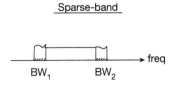 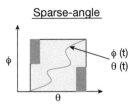

- All-pole model generally applicable
- Signal parameters (target space)
  $$\alpha_n, \tau_n$$
- Coherence parameters
  $$(\Psi, \tau_0)$$

- All-pole model locally valid
- Signal parameters (target space)
  - $x_n, y_n, z_n, D_n(c_1, c_2, ...)$
- Coherence parameters
  - Range/phase bias per sector
  - Angle biases
  - Motion estimation error compensation

Figure 11.7 Comparison of the sparse band, sparse angle signature models relative to the parameter characterization of each model.

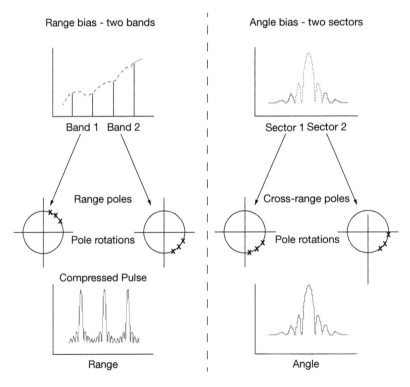

Figure 11.8 Coherence compensation duality between sparse-band frequency processing and sparse-sector angle processing. Both the range bias correction and the sensor location estimate are related by pole rotations, the first using the range poles, the second using the cross-range poles .

duality between sparse-band frequency processing and sparse-sector angle processing is illustrated.

It is not difficult to generalize the special case scenario of two sensors located in the plane of rotation considered previously to the general case of two wideband sensors operating in different frequency bands and located at different angular positions. Each sensor is able to form a composite set of scattering center location estimates with location estimate quality dependent on sub-bandwidth (range resolution) and angular sector size (cross-range resolution). Multisensor fusion, properly accomplished, allows the 3D noncoherent combination of these location estimates into a common target-based reference system using the extended coherent processing techniques developed in chapter 5. Range delay mismatches between sensors result in a range misalignment of the fused scattering center location estimates; errors in the estimate of the relative baseline look angles to the target for the two sensors result in a cross-range misalignment. It is only when the sensors are *mutually cohered* to eliminate these misalignments that signature fusion can be accomplished. In fact, for arbitrarily positioned sensors, the noncoherent combination of independent sensor data must necessarily be correlated to composite, three-dimensional scattering center location estimates referenced to a common target-based coordinate system.

12

Data Extrapolation and the Composite Target Space Mapping

12.1 Introduction

Chapter 11 introduced the concept of *synthetic data* and the various frequency-time synthetic data types that can be derived from measurement data, and which are applicable to either a single or two independently operating sensors. This chapter addresses the general problem of data extrapolation in frequency-time space for the single sensor case. Traditionally, the single pulse data extrapolation techniques described in [105] are applied to single pulse Fourier-based processing. A new Fourier-based compressed pulse is formed combining the observation and extrapolated data, providing enhanced resolution relative to pulse compressing the observation data directly. However, as was demonstrated in chapters 2 and 3, application of the one-dimensional (1D) or two-dimensional (2D) high resolution spectral estimation (HRSE) techniques to the observation data provides the enhanced resolution, and the range and range-rate sequence extracted from the observation data in this way circumvents the need for using Fourier-based processing. In fact, the signal models used for data extrapolation developed in this chapter incorporate the HRSE poles extracted from the observation data.

Nonetheless, data extrapolation can be useful for several other reasons:

1. Comparing the extrapolated data to truth data outside the data observation region provides a measure of the quality of the HRSE pole estimates.
2. The extrapolated data set allows comparison to signature predictions obtained from electromagnetic code computations for regions outside the measurement space using various estimates of the target model.
3. The enhanced data set improves the range-Doppler image quality (sharpness and resolution) obtained using Fourier-based processing.

4. Data extrapolation outside each measurement region is key to adjusting the data from multiple sensors for use in coherent sensor fusion.

The single sensor data extrapolation objective is illustrated in figure 12.1c. Measurement data are collected over a specified frequency-time (angle) observation space, and extrapolated outside the measurement space in both dimensions. The range, range-rate poles and complex amplitudes extracted from a data block contained within the measurement observation interval are used to form an all-pole signal model, which can be extrapolated outside the measurement observation region in both frequency and time. The resultant 1D and 2D extrapolation techniques based on using the HRSE extracted poles are referred to as enhanced bandwidth extrapolation (EBWE). Because the all-pole model is analytic in nature, the EBWE technique is particularly robust in data extrapolation over very wide bandwidths and angles. Coherent processing of the 2D observation data results in extrapolation techniques considerably more robust in performance relative to signal-to-noise limitations when compared to traditional single pulse bandwidth extrapolation (BWE) extrapolation techniques.

In chapter 13, the single pulse data extrapolation techniques introduced in this chapter are applied to the sparse band, colocated sensor fusion scenario illustrated in figure 11.1(b). Chapter 14 addresses the dual scenario illustrated in figure 11.1(c), contrasting the fusion techniques required for two sensors operating in the same frequency band, but having differing look angles to the target.

This chapter is divided into two major parts:

1. The development of the EBWE data extrapolation techniques using observables extracted from 1D and 2D frequency-time data blocks.

2. The application of the radar observable to target space motion mapping developed in chapter 5 to obtain composite, HRSE target space scattering center location estimates.

Two example targets described in figure 12.2 are used to illustrate the applications of the EBWE and HRSE scattering center location estimation techniques.

12.2   Data Extrapolation Techniques

Three different data extrapolation techniques are illustrated in figure 12.1. They are described as follows.

Method 1

Figure 12.1(a) illustrates the traditional single pulse BWE approach used to enhance range resolution in the compressed pulse. A good overview of the conventional single pulse BWE technique is provided in [105]. Conventional BWE recursive techniques do not adjust the in-band measured data, and they extrapolate the in-band data outside the observation region, constraining continuity at the boundaries of the data and extrapolation regions. These constraints limit the size of the extrapolation space before degrading the extrapolation.

Method 2

Figure 12.1(b) illustrates the single pulse EBWE technique using the one-dimensional all-pole signal model defined by equation (2.13) (choosing $p_n = 1$) applied to the measurement band. This model uses the HRSE 1D poles extracted from the observation data to enhance both the range resolution between closely spaced scattering centers as well as to characterize the frequency variation of each diffraction coefficient associated with each scattering center. The signal model coefficients are determined by a best root-mean-square fit to the data over the measurement data set, and the model data may differ slightly from the measurement data over the observation space.

Method 3

Figure 12.1(c) illustrates the extension of the one-dimensional signal model to the 2D all-pole signal model defined by equation (2.13), applied to a frequency-time data block, and used for data extrapolation in both frequency and angle (time) space.

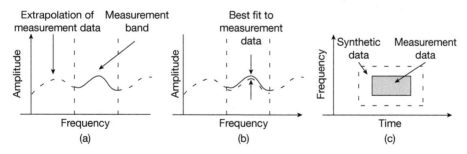

Figure 12.1 Progression in synthetic bandwidth techniques. (a) Traditional single pulse BWE described in [105]. (b) Single pulse BWE using the one-dimensional all-pole signal model. (c) EBWE using the two-dimensional all-pole signal model.

The single pulse EBWE technique described in method 2 is developed in section 12.3.2. The 2D EBWE technique is developed in section 12.3.3.

Because the all-pole signal model is analytic in nature, it is particularly robust in extrapolating data over very wide bandwidths and angles. The 1D and 2D HRSE data extrapolation techniques use the analytic all-pole signal model to extrapolate the measurement data outside the measurement band. The signal model amplitudes are determined using a best fit data match over the measurement space, compared to the conventional BWE technique that preserves the in-band data. Relaxing the constraint on the in-band data fit enhances the extent of the extrapolation space. Another benefit of using the all-pole signal model is that the synthetic data are noise free: that is, although the poles extracted from the measurement band are signal-to-noise ratio dependent, the signal model itself has no noise. The impact of the noise is manifested in the variance of the pole location estimate. This is compared to conventional BWE processing that extrapolates the noisy in-band data into the extrapolation space. The EBWE synthetic data are particularly useful in low signal-to-noise ratio environments where it becomes difficult to identify the signal components in a conventional BWE compressed pulse relative to peaks in the noise level.

Typically, the rationale for using a BWE technique is to achieve enhanced sharpness and resolution on the closely spaced scattering centers located on the target. As discussed in chapter 1, resolution is inversely proportional to bandwidth, and wider bandwidth leads to more precise location estimates. When using conventional Fourier-based pulse compression, the frequency domain data are weighted to reduce the compression side-lobes and this attenuates the measurement data at the band edges, reducing the effective resolution [see equation (5.1), where typically $\alpha \sim 1$ for very low side-lobe weighting]. Low side-lobe pulse compression is applied so that lower amplitude adjacent scattering centers are not masked by the range pulse compression side-lobes. However, the weighting function reduces the contribution of the data at the edge of the measurement band; thus low side-lobes are achieved at the cost of resolution. If the back-end processor uses conventional Fourier-based processing, extrapolating the data over a wider data band prior to applying the weighting function can improve resolution because the weighting function applied to the extrapolated data attenuates the data at the edges of a synthetic outer band, thus leaving the in-band data essentially unweighted. However, note that, as discussed earlier, there are fundamental limits on achievable resolution for a given measurement bandwidth and signal-to-noise ratio, so that using the all-pole signal model over very large bandwidths improves sharpness on isolated scattering centers, but cannot achieve resolution between closely spaced scattering centers that violate these fundamental limits.

Alternately, as discussed previously, one can process the frequency-time measurement data using the high-resolution spectral estimation techniques introduced in chapter 2 and eliminate the need for pulse compression altogether. The extracted observables $(R_n, \dot{R}_n, D_n)$, $n = 1, \ldots, N$ are then used to characterize the data block, rather than the compressed pulses or a range, range-rate image formed from the data block. Sequentially processing these extracted observables provides enhanced resolution over conventional processing and forms the basis of the majority of techniques developed in this text.

Finally, if the frequency dependence of the scattering center is strong and the bandwidth is sufficiently large, an all-pole model having complex poles can be used to estimate the diffraction coefficient frequency variation of each scattering center, for example, either decaying or growing exponentially, leading to better characterization of the scattering center. This situation is considered in the hypothetical example treated in section 12.3.1, and further developed in chapter 13 using ultrawideband processing.

12.3   One- and Two-Dimensional Data Extrapolation Examples

12.3.1   Example Targets

Two targets are introduced to illustrate the application of the techniques developed in this chapter. Target 1 illustrated in figure 12.2, is an ideal three-point scattering center target, having frequency-dependent scattering center diffraction coefficients. Each scattering center is characterized by a diffraction coefficient described by equation (12.2), exhibiting exponential frequency variation over the processing bandwidth. Two of the scattering centers are closely spaced relative to the resolution of the measurement bandwidth using Fourier-based processing. The objective using this target is twofold: to demonstrate the utility of single pulse EBWE to achieve better resolution on closely spaced scattering centers when using Fourier-based processing, and to demonstrate the utility of applying an all-pole signal model having complex poles to estimate each scattering center's diffraction component variation with frequency. Although much wider bandwidths than used in this example would be required to characterize the frequency behavior of most real target components , the example illustrates the application of the all-pole model in characterizing the exponential frequency behavior. The "measurement data" are represented using a GTD simulation model. Simulation data are generated over a larger bandwidth than the assumed measurement band, and used to check the accuracy of the extrapolation technique over the expanded region.

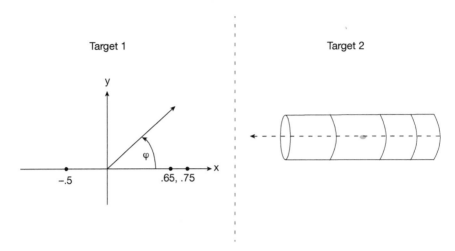

Figure 12.2 Example targets. Target 1 shows three-point scattering centers located on the *x*-axis having frequency-dependent diffraction coefficients defined by equation (12.2). The example provides a measure of the processing resolution relative to the closely spaced scattering centers as the angle of incidence changes. Target 2 shows the slotted cylindrical target described in section 6.4.1.

A geometrical theory of diffraction (GTD) signature model is used to develop the simulation data for each example. The second example target 2 uses simulated measurement data computed using the slotted cylindrical target introduced in chapter 6.

Target 1

Simulation data for target 1 is computed using a three-point scattering center all-poleGTD scattering model of the form

$$E(f,\varphi) = \sum_{n=1}^{3} D_n(f) e^{-j\frac{4\pi f}{c}\hat{k}\cdot r_n}, \tag{12.1}$$

where the frequency-dependent diffraction coefficients are exponential in form and defined by

$$D_n(f) = D_{0_n} e^{-\alpha_n(f-f_0)}, \quad n = 1, 2, 3. \tag{12.2}$$

The three scattering centers are located on the x-axis as illustrated in figure 12.2(a), so that the unit vector $\hat{k}$ lies in the (x, y) plane $\theta = \pi/2$. $\hat{k}$ is given by

$$\hat{k} = \cos\varphi\,\hat{x} + \sin\varphi\,\hat{y}, \qquad (12.3)$$

where φ is the standard spherical coordinate angle referenced to the x-axis. The values assumed for each α_n are tabulated in table 12.1. The measurement bandwidth is 1 GHz centered about center frequency $f_0 = 10$ GHz, which is a 10 percent bandwidth. However, simulation data are computed over a 4 GHz bandwidth to provide a truth data base for comparison to results obtained using the data extrapolation techniques. The rotational motion $\varphi(t)$ is assumed linear in time, and given by

$$\varphi(t) = \frac{2\pi}{T}(t - T/2), \quad 0 \le t \le T. \qquad (12.4)$$

A rotation period $T = 1$ s is selected, so that φ covers the $(-\pi, \pi)$ angular interval over the time interval $(0, 1$ s$)$. A sensor pulse repetition frequency equal to 1,000 Hz is used, resulting in time samples spaced $\Delta t = 0.001$ s. To illustrate the enhanced resolution realized using the 1D and 2D spectral estimation techniques, two of the scattering centers are closely spaced, separated by 0.1 m, much closer than the resolution limit of the 1 GHz bandwidth compressed pulse assuming 40 dB side-lobe weighting. For this weighting, the resolution limit $\Delta\rho$ is given by

$$\Delta\rho = \alpha\lambda/FBW, \qquad (12.5)$$

where $\alpha \sim 1$. The resolution limit at X-band having a 10 percent fractional bandwidth is given by $\Delta\rho \sim 0.3$m. These two scattering centers will not be resolved by the weighted 1 GHz BW compressed pulse.

The complete set of parameters used for the simulations are illustrated in table 12.1. Each set of complex amplitudes is unity weighted having phase selected from a random draw and then fixed over the $(0, 2\pi)$ rotational angle.

Note that for the two scattering centers located in close proximity, the maximum LoS range separation of the scattering centers occurs for incidence angle parallel to the x-axis, that is, $\varphi = 0$ and π, and decreases as φ deviates from these values. For $\varphi = \pi/2$, all of the range observables are identical as they lie perpendicular to the radar line of sight (LoS) at this incidence angle. Thus, the example provides a measure of achievable processing resolution relative to the closely spaced scattering centers as the angle of incidence changes.

Table 12.1   Simulation parameters

Parameter	Value
x_n, y_n, z_n	$(-0.5, 0, 0)$ $(.65, 0, 0)$, $(.75, 0, 0)$
$D_{0_1}, D_{0_2}, D_{0_3}$	$e^{j\psi_n}, \psi_n = rand\,(0, 2\pi)$
$\alpha_1, \alpha_2, \alpha_3$	$(-1, 0, +1) * 10^{-10}$
PRF	$1{,}000$
$f_o, BW, \Delta f$	10 GHz, 1 GHz, 0.0635 GHz

Target 2

Section 6.4.1 presents an overview of the slotted cylindrical target illustrated in figure 12.2. X-band data are simulated using the GTD scattering model described in section 6.4.1. Data over a 1 GHz subset of 4 GHz bandwidth is used to compare extrapolated data values to the truth over data extrapolation space.

12.3.2   Single Pulse Data Extrapolation: Target 1

The one-dimensional all-pole signal model used for single pulse data extrapolation applied to Target 1 is obtained using equation (2.13) and choosing $p_n = 1$ to obtain

$$E_0(k) = \sum_n E_n(k) = \sum_n D_{0_n} s_n^k, \tag{12.6}$$

where the poles s_n, $n = 1, 2, 3$ are extracted from the measurement data using the one-dimensional HRSE technique introduced in chapter 2 and described in appendix B. The D_{0_n} in equation (12.6) are determined by optimizing the fit of the signal model to the 1 GHz measured data, $E_d(k)$, over the set of n_f frequencies according to

$$\underset{(D_{0_n})}{Min}|\Delta(k)|^2, \quad k = 1, \ldots n_f, \tag{12.7}$$

where $\Delta(k)$ is defined by the difference

$$\Delta(k) = E_d(k) - E_0(k), \tag{12.8}$$

similar to the manner introduced in chapter 3.

As defined in section 2.3, the α_n characterize the component diffraction coefficient frequency variation over the measurement band according to

$$D_n(f) \sim D_{0_n} e^{-\alpha_n(f-f_0)}. \tag{12.9}$$

Assumed values for the α_n for each component associated with the three-point scattering center example are tabulated in table 12.1. The estimates of the α_n are obtained from the complex poles extracted from the measurement data using equation (2.19) applied to each pole estimate, according to

$$\hat{\alpha}_n = -\frac{1}{\Delta f} \mathrm{Re}(\ln s_n). \tag{12.10}$$

Once the poles s_n, $n = 1, 2, 3$ are extracted from the measurement data set, equation (12.6) is used to approximate the data over the entire extrapolation bandwidth, including the measurement band and the estimates of α_n are used to characterize the frequency response of each scattering center.

Figures 12.3, 12.4, and 12.5 each illustrate the compressed pulse shapes for differing look angles to target 1 for three cases: the 1 GHz data compressed pulse (dashed red); the $1 \rightarrow 4$ GHz EBWE synthetic data compressed pulse (red); and the 4 GHz truth data compressed pulse (blue). These are depicted for three angles of incidence: $\varphi = 2°$ (figure 12.3), 38° (figure 12.4), and 75° (figure 12.5), which correspond to a decreasing LoS range separation between the scattering centers. Also illustrated for each figure is the comparison of the frequency response of the synthetic data versus the truth data over the full 4 GHz band. Note that when all three scattering centers are clearly resolved in the 4 GHz data for $\varphi = 2°$, the frequency response of the synthetic data over the entire 4 GHz band is essentially identical to the 4 GHz truth data, and the diffraction coefficient estimates for the α_1, α_2, and α_3 frequency variations are very good. As the angle of incidences increases, for $\varphi = 38°$ the two closely spaced scattering centers (separated by 0.085 m along the radar line of sight) are barely resolvable at this incidence angle, but the frequency extrapolation is still quite good (excluding the poor estimate of α_2 for one of the closely spaced scattering centers). Finally, for $\varphi = 75°$ incidence, the closely spaced scattering centers (separated by 0.038 m along the radar line of sight) are not resolved even in the 4 GHz truth data, and the frequency response of the synthetic data gives a poorer match to the truth data outside the 1 GHz assumed measurement band.

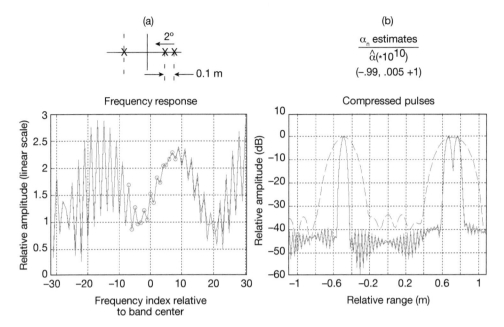

Figure 12.3   Target 1, three-point scattering center example. (a) Frequency response with extrapolation (red) versus truth (blue) comparison. (b) Compressed pulses with the 1 GHz data compressed pulse (dashed red), the $1 \to 4$ GHz EBWE synthetic data compressed pulse (red), and the 4 GHz truth data compressed pulse (blue) and alpha estimate. Angle of incidence is $2°$.

12.3.3   Single Pulse Data Extrapolation: Target 2

Now consider a comparison between synthetic data and truth for the slotted cylindri-cal target analogous to that presented in figure 12.3 for the three-point scattering cen-ter example. Figures 12.6, 12.7, and 12.8 illustrate the comparison of synthetic to measured data for three look angles to the target: $10°$, $40°$, and $60°$ relative to the target axis. The color legend is similar to that used in figure 12.3: for the frequency response of the target plotted to the left of the figure, a circle indicates the 1 GHz synthetic data, and blue solid lines indicate the 4 GHz data; for the pulse compressed data, the 4 GHz extrapolated data (red) is compared to the compressed pulse using the truth data over the 4 GHz band (blue) and the compressed pulse using the 1 GHz truth data (dashed red). For each look angle, all of the scattering centers are resolved over the 4 GHz bandwidth. Note the extrapolated data matches the truth data quite well,

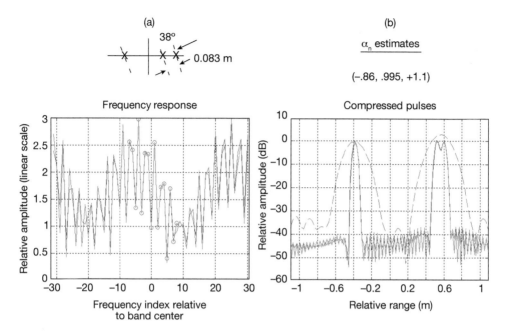

Figure 12.4   Target 1, three-point scattering center example. (a) Frequency response with extrapolation (red) versus truth (blue) comparison. (b) Compressed pulses with the 1 GHz data compressed pulse (dashed red), the 1 → 4 GHz EBWE synthetic data compressed pulse (red), and the 4 GHz truth data compressed pulse (blue) and alpha estimate. Angle of incidence is 38°.

even over four times the measurement bandwidth. As expected, given the good frequency response match, the corresponding compressed pulses using the 4 GHz synthetic data are well matched to those corresponding to the truth data.

12.3.4 Data Extrapolation Using 2D Block Processing

One-dimensional BWE techniques discussed in section 12.1 process only the frequency measurement data obtained from a single received pulse to achieve enhanced resolution. This section addresses the problem of data extrapolation and obtaining enhanced resolution in both range and cross range for a 2D frequency-time data block. The problem of achieving enhanced resolution relative to that obtained using conventional range-Doppler processing is also addressed in [105]. The technique developed there uses sequential applications of the 1D single pulse BWE technique described earlier to

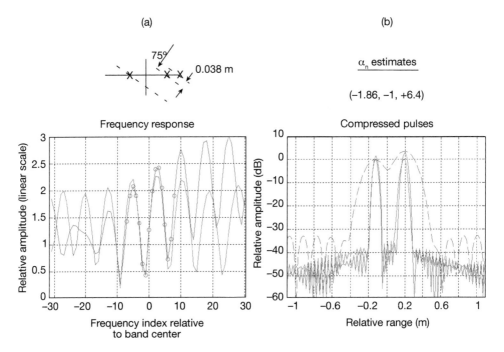

Figure 12.5　Target 1, three-point scattering center example. (a) Frequency response with extrapolation (red) versus truth (blue) comparison. (b) Compressed pulses with the 1 GHz data compressed pulse (dashed red), the 1 → 4 GHz EBWE synthetic data compressed pulse (red), and the 4 GHz truth data compressed pulse (blue) and alpha estimate. Angle of incidence is 75°.

both range and cross range to achieve enhanced resolution, first to enhance the cross-range resolution, followed by enhanced range resolution. To describe this technique, assume the data block consists of Q_0 pulses having time stamp $t = t_q$, $q = 1, \ldots, Q_0$. As described in [105], a three step procedure is implemented: first conventional range profiles are determined for each of the Q_0 pulses in the frequency-time data block, which are then used to form a conventional range-Doppler image; next, the BWE technique is applied to the Q_0 pulses across each range gate to achieve enhanced resolution in cross range (Doppler); finally, at each enhanced cross-range resolution cell, the range profile data in each cell is uncompressed and the conventional BWE technique applied to form an enhanced resolution range pulse for each enhanced cross-range cell. Conventional range-Doppler image processing applied to the new, larger data block then results in enhanced resolution in both range and cross range.

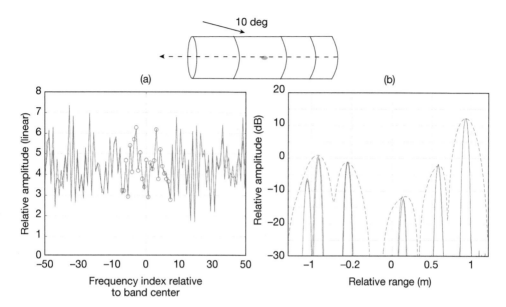

Figure 12.6   Target 2, slotted cylindrical target. Extrapolation versus truth comparison.
(a) Frequency response with extrapolation (red) versus truth (blue) comparison.
(b) Compressed pulses with the 1 GHz data compressed pulse (dashed red), the
$1 \rightarrow 4$ GHz EBWE synthetic data compressed pulse (red), and the 4 GHz truth data
compressed pulse (blue). Angle of incidence relative to target axis is 10°.

This section presents an alternative approach to this problem using a two-dimensional
signal model applied to the measurement frequency-time data block. Using a 2D signal
model applied to data measured over a block of multiple pulses provides significant
advantages:

1. It is more robust to noise than single pulse processing, due to the integration
   gain obtained by coherently processing a number of pulses.
2. It isolates closely spaced scattering centers in range, but which are separated in
   Doppler, providing enhanced resolution in range for these scattering center locations versus single pulse processing.
3. It permits very wideband extrapolation in frequency-time space.

The 2D signal model used is developed in chapter 3, characterized by equation (3.29)
with $p'_n = 1$, and is given by

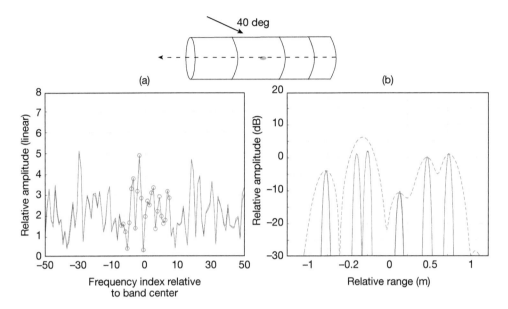

Figure 12.7   Target 2, slotted cylindrical target. Extrapolation versus truth comparison.
(a) Frequency response with extrapolation(red) versus truth (blue) comparison.
(b) Compressed pulses with the 1 GHz data compressed pulse (dashed red), the
$1 \rightarrow 4$ GHz EBWE synthetic data compressed pulse (red), and the 4 GHz truth data
compressed pulse (blue). Angle of incidence relative to target axis is 40°.

$$E_0(k, q) = \sum_n D_n s_n^k p_n^{q(1 + k\Delta f/f_0)}. \tag{12.11}$$

The coefficients D_n are determined using equation (12.7), with $\Delta(k, q)$ representing the best signal model to data fit over the 2D measurement observation space.

$$\Delta(k, q) = E_d(k, q) - E_0(k, q) \tag{12.12}$$

The pole pairs (s_n, p_n) are extracted from the measurement data block using the HRSE spectral estimation techniques as described in chapter 2. Of particular importance is that the 2D pole pair extraction provides a coherent gain in single pulse signal-to-noise ratio given by $10\log(Q_0)$, where Q_0 is the number of pulses in the 2D data block.

Although the extrapolation extent in frequency-time space can be chosen different for each dimension, it is convenient to choose the same expansion factor for both dimensions. Denote the bandwidth extrapolation factor by BWE_ex. Then a data block of size

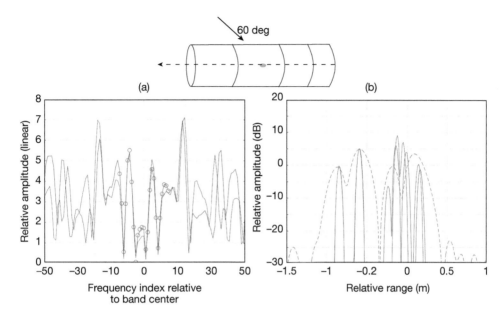

Figure 12.8 Target 2, slotted cylindrical target. Extrapolation versus truth comparison. (a) Frequency response with extrapolation (red) versus truth (blue) comparison. (b) Compressed pulses with the 1 GHz data compressed pulse (dashed red), the $1 \rightarrow 4$ GHz EBWE synthetic data compressed pulse (red), and the 4 GHz truth data compressed pulse (blue). Angle of incidence relative to target axis is 60°.

n_f frequencies by Q_0 pulses, that is, $[n_f \times Q_0]$ complex samples, results in data extrapolation over an expanded matrix block size of $[n_f \cdot BWE\_ex \times Q_0 \cdot BWE\_ex]$. For a single frequency-time data block, EBWE improves resolution over that achievable using single pulse BWE combining the enhanced resolution in frequency with the enhanced cross-range resolution in Doppler; that is, the EBWE data block is now extended synthetically by $BWE\_ex \cdot Q_0$ pulses. The enhanced resolution of a single EBWE range-rate image, scaled in cross range, is then given by $(\lambda/2 \cdot BWE\_ex \cdot FBW) \times (\lambda/2 \cdot \Delta\varphi')$, where $\Delta\varphi'$ denotes the angular extent covered by the $Q_0 \cdot BEW\_ex$ pulses used in the EBWE data block. By using synthetic data, the effective fractional bandwidth and angular extent are both increased relative to the original measurement parameters.

Figure 12.9 illustrates a typical example of the use of the 2D all-pole signal model applied to the three-point scattering example at radar line of sight of 53° to the horizontal axis using the simulation parameters in table 12.1. Figure 12.9(a) illustrates the amplitude of the 1 GHz frequency-time 16 frequencies by 16 pulses measurement data block samples as a subset of the extrapolated 2D data block using bandwidth and time expansion factors $BWE\_ex = 4$. This corresponds to an expanded block size of 64 frequencies by 64 pulses. Figure 12.9(b) compares the amplitude of the EBWE synthetic data over the 64×64 data block to the simulated (truth) data over this larger data block. The agreement is quite good.

Figure 12.10 illustrates the process of forming a range, range-rate image using EBWE beginning with the initial measurement data block of size 16×16 [figure 12.10(a)], progressing to the extrapolated data block of size 64×64 [figure 12.10(b), synthetic and truth data]. Figure 12.10(c) compares the range, range-rate images obtained using the 4 GHz, 64 pulse synthetic data and truth data to form the image. Because the signal model closely approximates the truth data over the extrapolation space, the agreement is quite good. The image results are essentially identical.

Figure 12.11 illustrates the EBWE processing methodology used for the three-point scattering center target applied to the slotted cylindrical target. The incidence angle is at $32°$ relative to the target axis. For this case the measurement frequency-time data block size consists of $n_f = 26$ frequencies by $Q_0 = 24$ pulses to form the 2D measurement data block. The EBWE results assume a frequency-time expansion factor of four, resulting in an extrapolated data block size of 104×96 data samples. As expected, because the 2D signal model used to extrapolate the data are inherently related to the GTD computational model used for the data simulation, the general trends in the variation in frequency-time space are consistent, and the impact on the resultant range, range-rate image comparison in figure 12.11 is minimal.

12.4   The Composite Target Space Mapping

As discussed in chapter 5, [105] introduces the concept of extended coherent processing (ECP) of the complex signature samples for targets exhibiting pure rotational motion by decomposing the three-dimensional representation of the correlation image function $I(x, y, z)$ expressed in equation (5.9) into piecewise segments consisting of blocks of range-Doppler images, properly scaled in cross range, rotated and coherently summed to provide a two-dimensional estimate $I(x, y)$ of the scattering center location. Coherent processing of the complex radar signature data in this manner is advantageous for fixed-point scattering centers having constant amplitude as a function of target rotation. However, it tends to suppress the contribution of the more practical case of scattering centers having amplitude and phase changes with rotation angle. This section incorporates a noncoherent application of the ECP technique developed in [105] into

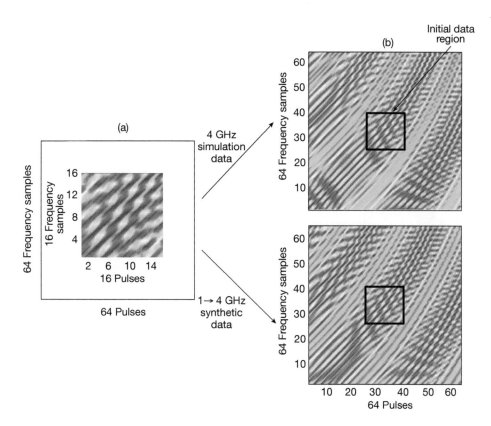

Figure 12.9   A typical example of the use of the 2D all-pole signal model applied to the three-point scattering example at radar line of sight of 53° to the horizontal axis using the simulation radar parameters in table 12.1. (a) The 1 GHz frequency-time measurement data block of 16 pulses as a subset of the extrapolated 2D data block using bandwidth and time expansion factors of four. (b) Comparison of the amplitude of the EBWE synthetic data over the 64 × 64 extrapolated data blocks to the simulated data over this larger data block.

the scattering center location estimation processing framework introduced in this text, which alleviates this limitation.

The observable to target space mapping techniques developed in section 5.2 can be applied to sequences of range, range-rate images to obtain composite, high resolution scattering center location estimates. It is useful to review the mapping techniques developed there and introduce variations of the fundamental mapping equations commonly

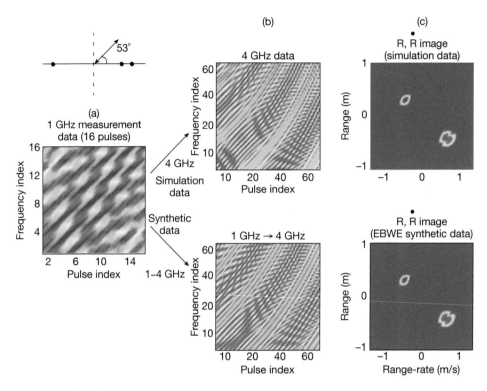

Figure 12.10 Illustrates the process of forming a range, range-rate image using EBWE. (a) The initial measurement data block. (b) The extrapolated synthetic data block. (c) Comparison of the range, range-rate image obtained from the synthetic data versus truth data using the 4 GHz, 64 pulse data.

used for processing Fourier-based image data blocks. Consider the fundamental mapping equations defined by equations (5.2) and (5.3), repeated here in slightly different form:

$$\hat{k}_b \cdot \underline{r}_n = R_{n,b},$$
(12.13)

$$\hat{\dot{k}}_b \cdot \underline{r}_n = \dot{R}_{n,b},$$
(12.14)

where the observables $(R_{n,b}, \dot{R}_{n,b})$ are extracted from the data at block time t_b. Assume the special case where the radar line of sight lies in the plane of the rotation motion of the target, for which case the target scattering center locations defined by $\underline{r}_n = (x_n, y_n, z_n)$ are projected to the (x, y) plane relative to the radar LoS. In this case, assuming the motion is known, the two equations (12.13) and (12.14) are directly invertible to

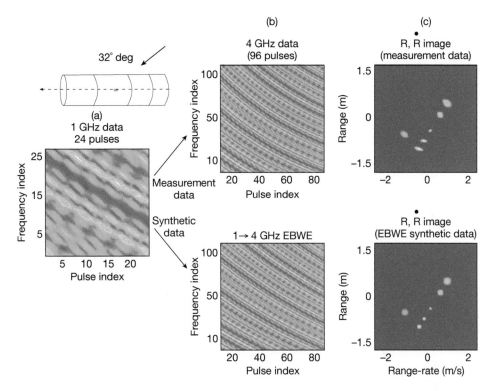

Figure 12.11 Illustrates the processing sequence parallel to figure 12.10 for the slotted cylindrical target for incidence angle of 32° relative to the target axis. For this case the measurement frequency-time data block size consists of $n_f = 26$ frequencies by $Q_0 = 24$ pulses to form the 2D data block. The EBWE results assume a frequency-time expansion factor of four.

provide estimates of each (x_n, y_n) scattering center location. Because a single block estimate of (x_n, y_n) has variance depending on the signal-to-noise ratio of the received signal, and because of shadowing of scattering centers on the target relative to the LoS of the radar to the target, it is common to noncoherently accumulate a collection of solutions for each (x_n, y_n) as the sensor viewing angle rotates about the target. Each single block scattering center location estimate can be noncoherently integrated by accumulating sliding block estimates covering a full 360° rotational motion onto a 2D target space grid. Applying this process, the composite collection of location estimates referenced to target space is referred to as the *composite target space mapping*.

Three techniques for obtaining the composite target space mapping, each based on applying the fundamental mapping equations (12.13) and (12.14), are developed in this section:

Method A

The first uses a sequence of Fourier-based range, range-rate images obtained from the baseline frequency-time measurement data block. For each image, defined at a specified block time t_b, the range, range-rate pair corresponding to *each pixel* of the image having amplitude greater than a specified threshold are used as the range, range-rate observables $(R_{n,b}, \dot{R}_{n,b})$ identified in equations (12.13) and (12.14) and mapped to target space. When this technique is applied, the index n in the observable corresponds to an image pixel location mapped to target space. Because the dimension of the image size is $(N_{fft} \times N_{fft})$, the range of the index n for a single image mapping to target space is large. Thus, the collection of target space locations obtained from a single image covers a broad region of target space. Now apply the target space persistence filter developed in section 5.6.3. Divide target space into bins and noncoherently integrate the target space mapping location bin counts (or amplitudes) accumulating in each bin as the target rotates 360 degrees. Those bins associated with the true scattering center locations contain mappings that persist and sum to a large value, resulting in a composite target space mapping. The resultant mapping contour, thresholded to a specified level below its maximum, is used to estimate the locations of the target scattering centers.

Method B

The second technique is identical to method A, except the target space mapping is applied to each pixel location having amplitude greater than a specified threshold obtained from the range, range-rate image *using the extrapolated EBWE frequency-time data blocks* as depicted in figure 12.11(c). Consequently, the collection of points in target space resulting from processing a single EBWE range, range-rate image is smaller due to the increased resolution contained in each image.

Method C

The third technique uses the composite target space mapping equations (12.13) and (12.14) applied sequentially to the set of HRSE range, range-rate observables extracted from each data measurement block using sequential estimation processing and is identical to the 2D composite mapping techniques developed in chapters 5 and 9. Compared to the composite mapping techniques in methods A and B shown previously, it can be viewed as the limiting case using Fourier-based spectral estimation on each expanded data block for very large block expansion sizes. In the limit of very large synthetic data blocks, each range, range-rate EBWE image reduces to a set of closely spaced image pixels surrounding the $(R_{n,b}, \dot{R}_{n,b})$ observables extracted using the HRSE technique applied to the measurement data block.

12.4.1   Target 1: Three-Point Scattering Center Example

Figure 12.12 illustrates the composite target space scattering center location estimates corresponding to each of the three techniques defined in methods A, B, and C applied to the three-point scattering center example: Figure 12.12(a) illustrates the composite target space mapping scattering center location estimates obtained applying the target space mapping methodology described previously in method A to the measurement data in sequential blocks covering the full rotation cycle of the motion. An initial data block size consisting of $Q_0 = 16$ pulses consistent with the single block example of figures 12.9 and 12.10 is used for each block time in the composite example. Figure 12.12(b) compares the scattering center location estimates obtained using the EBWE synthetic data as described in method B over the full rotational motion of the target compared to applying the same mapping to the truth data over the extrapolated data set region. An initial data block size consisting of $Q_0 = 16$ pulses, consistent with the single block examples of figures 12.9 and 12.10, is used for each block time in the composite image example. The synthetic bandwidth is chosen to be four times the measurement band. The location estimates are essentially identical for these synthetic and truth data cases, and the closely spaced scattering centers are resolved both for the 4 GHz synthetic data and the 4 GHz truth data. Finally, figure 12.12(c) illustrates the scattering center location estimates for method C obtained using the sequential HRSE range, range-rate observable sequence extracted from the 1 GHz measurement band. The increase in achievable resolution in the composite target space image using the range, range-rate observable sequence is striking. The location estimates are confined to just a few bins located at the scattering center locations, and are magnified in the figure to emphasize the sharpness of the location estimate.

12.4.2   Target 2: Slotted Cylindrical Target

Consider now the application of the techniques described previously to the slotted cylindrical target.

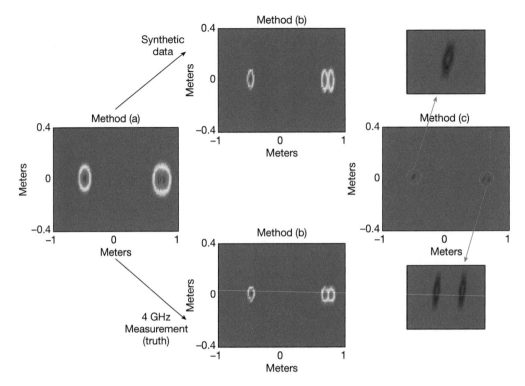

Figure 12.12   The composite target space mapping applied to target 1 using each of Methods A, B and C: (a) Method A: pixel threshold greater than 20 dB below each image maximum (b) Method B using the EBWE synthetic data over the extrapolated 4 GHz band compared to truth data over this same 4 GHz band. (c) Method C using the sequential HRSE range, range-rate observable sequence extracted from the 1 GHz measurement band.

Figure 12.13 illustrates the corresponding hierarchy of composite scattering center location estimates for the slotted cylindrical target using the same parameters used in the single image illustrated in figure 12.11. For figure 12.13(a), method A uses the target space mapping equations applied sequentially to the range, range-rate pairs corresponding to each pixel of the range, range-rate image obtained from the 1 GHz measurement data band using a 24 pulse data block. In figure 12.13(b), which shows method B, the comparison of the scattering center location estimates is obtained using the same methodology applied using the EBWE synthetic data over the extrapolated 4 GHz band using a block size expansion factor of 4 in both frequency and time.

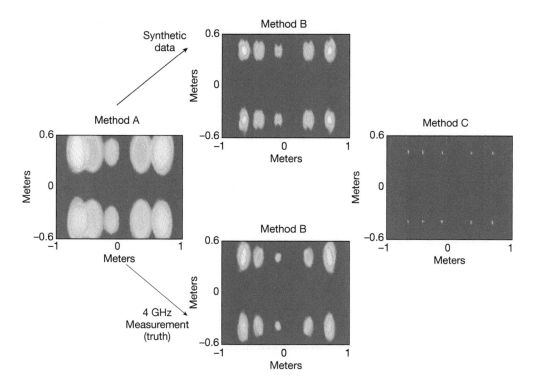

Figure 12.13   The composite target space mapping applied to target 2 using each of Methods A, B and C (a) Method A: pixel threshold greater than 20 dB below each image maximum. (b) Method B using the EBWE synthetic data over the extrapolated 4 GHz band compared to truth data over this same 4 GHz band. (c) Method C using the sequential HRSE range, range-rate observable sequence shown in figure 12.14 extracted from the 1 GHz measurement band data.

compared to truth data over this same 4 GHz. The location estimates are essentially identical using the measurement and synthetic data, and all the scattering centers are resolved both for the 4 GHz synthetic data and the truth data.

Now consider mapping the HRSE observables into target space using method C. Upon careful examination of the output of the 2D HRSE spectral estimation technique, at most six scattering centers will be visible to the sensor as the target rotates. Thus, an order $N=6$ is used in extracting the HRSE observables. Figure 12.14 illustrates the HRSE observable sequence $(R_{n,b}, \dot{R}_{n,b}), b=1,\ldots,B$ extracted from the 1 GHz measurement data block using block increments of two pulses as the target

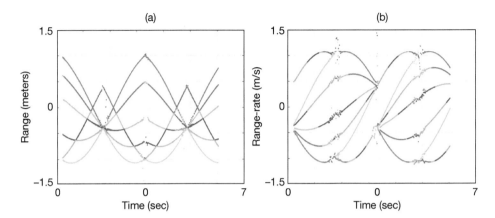

Figure 12.14   Extracted range, range-rate observable sequence for Target 2. Order = 6. Rotation period = 2πs (a) Range observables versus time. (b) Range-rate observables versus time.

rotates over full rotation angle assuming a rotation period of 2πs. The nonlinear signal model was used to estimate the observables, resulting in some smoothing in the extracted observable sequence as discussed in chapter 3.

Figure 12.13(c) illustrates the corresponding scattering center location estimates obtained applying the target space mapping equations to the observable sequence in figure 12.14. Note that in using the composite mapping to target space applied to the extracted HRSE observables, the scattering center location estimates are not particularly sensitive to over specifying the model order because the target space mapping acts as a filter to extraneous poles that occasionally show up (as are evident in figure 12.14) and only those bins corresponding to true persistent scattering centers add noncoherently in a given bin in target space. Essentially the same results would be obtained using a model order of five. The increase in sharpness using the HRSE observable sequence mapping relative to identifying the location of each scattering center on the target is clearly evident.

13

Colocated Sensors: Sparse Frequency Band Processing

Chapter 12 addressed the general problem of extrapolating the target signature in frequency-time space for a single sensor. In this chapter the problem of fusing data from multiple sensors operating independently over different frequency bands, as depicted in the two-sensor scenario in figure 11.1(b), is addressed. The incorporation of separate frequency bands offers the opportunity to cover ultrawide bandwidth if techniques can be developed to coherently fuse data from the multiple radars. The data extrapolation techniques introduced in chapter 12 provide a key step for compensating the data in each measurement band so that coherent processing is possible. The new data fusion techniques result in significantly improved scattering center characterization.

The simulation example presented in chapter 12, section 12.3.2, demonstrated that if the frequency dependence of the scattering center is strong, and the bandwidth is sufficiently large, a single sensor all-pole model can be used to estimate the diffraction coefficient frequency variation of each scattering center. As described in chapter 2, the all-pole signal model incorporates a characterization of the amplitude of the scattering center frequency behavior in the form $e^{-\alpha(f-f_0)}$, for example, either decaying or growing exponentially, where the parameter α is extracted from each scattering center complex pole using equation (2.19). However, the single sensor approach has some inherent limitations. Typically, the measurement bandwidth is less than ~10 percent, and the frequency behavior of most nonresonant type scattering center components exhibits little variation over such narrow bandwidths. However, appropriate processing of data obtained from multiple sensors operating over different frequency bands yields effective fractional bandwidths approaching 50 percent or greater (referred to as ultrawide bandwidths in this text). Measuring a target's ultrawideband radar signature can have significant payoff from a target signature modeling viewpoint. The amplitude of an individual scattering center can exhibit significant variation with frequency over the larger bandwidth. Spheres, edges, and surface joins are examples of scattering centers that exhibit such behavior. Not only would extremely fine range resolution be obtained, but the amplitude behavior of isolated scattering centers can help in typing the scattering center. Many canonical scattering centers are known to exhibit $f^{\tilde{\alpha}}$-type scattering behavior, where the parameter $\tilde{\alpha}$ is characteristic of specific scattering component types. For example, the radar cross section (RCS) of flat plates, singly curved surfaces (cone sections), and doubly curved surfaces (sphere) vary as f^2, f^1, f^0, respectively. The RCS of a curved edge varies as f^{-1}, whereas a cone vertex may be characterized with an f^{-2} RCS frequency dependence. The relationship of $\tilde{\alpha}$ to the parameter α can be determined by matching the slopes of the frequency variation of the two functional forms over the ultrawide bandwidth. Noting that α characterizes the amplitude of the scattering component, and $\tilde{\alpha}$ as defined previously the RCS of the component, it follows that α and $\tilde{\alpha}$ are related by the simple expression $\tilde{\alpha}/2 \approx \alpha f_0$. One of the advantages of ultrawideband processing is the ability to estimate the frequency-dependent terms using the measured data, and use the estimates to identify the scattering center type for incorporation into the radar signature model of the target.

The notional concept of synthetic ultrawideband processing has been discussed in chapter 1, section 1.4.2, and is illustrated in figure 1.15. To accomplish the data fusion between measurement sub-bands, two technical issues must be addressed. First, a method is needed to compensate for the lack of mutual coherence between the various radar sub-bands as discussed in chapter 11. For a two-sensor scenario, when individual all-pole signal models are fitted to the separate spectral bands, the results present independent estimates for the target's scattering center locations and can be visualized as two sets of poles in the complex plane. If the bands were completely coherent and the scattering centers were completely resolvable in each measurement band, the zero-noise case would result in separate pole sets that were identical, and a single pole set would provide the best target model representation over the multiple band data set. However, as described in chapter 11, equation (11.14), a phase offset $\Delta\Psi$ and delay error τ_0 must be compensated for in the data from band 1 (sensor 1) relative to band 2 (sensor 2) before data fusion.

Second, once the data in the measurement bands have been mutually cohered, an appropriate ultrawideband signal model must be used to fuse the sparse sub-band measurements. The fused signal model must be global in nature, accurately characterize ultrawideband target scattering and provide for meaningful interpolation between the measurement sub-bands.

This chapter discusses each of these steps and is organized as follows. In section 13.1, 13.2, and 13.3, a single pulse ultrawideband band fusion technique is developed and illustrated using a two-point scattering center example. Application of the technique applied to static range data is developed in [106].

13.1 Technique Development

The specific multiband fusion technique used in this text is adapted from [106]. The notation and development are modified from the reference to conform to the framework introduced in this text. The reader is encouraged to examine [106] for a more detailed development. An overview of the approach to ultrawideband coherent processing is illustrated in figure 13.1. The first step to achieving an ultrawideband target model is to maximize the degree of coherency between the separate signal bands. An estimate of the target's ultrawideband radar signature is then obtained by coherently combining the sparse sub-band measurements using a global all-pole signal model. While the figure illustrates ultrawideband processing for only two sub-bands, it is straightforward to apply the concept to an arbitrary number of sub-bands.

The process illustrated is divided into three major steps:

1. The time delay and phase differences between the separate radar sub-bands are estimated and the measurement data are adjusted so the bands can be processed coherently.

2. A single, global all-pole signal model is optimally fitted to the now mutually cohered sub-bands.

3. The resultant global (ultrawideband) signal model is used to synthetically represent the return signal over the vacant bandwidth of the received signals, and standard pulse compression methods are applied to provide a super-resolved range profile of the target.

Step 3 is straightforward while steps 1 and 2 are discussed subsequently in brief detail.

Step 1 is important when applying ultrawideband processing to field data collected by separate wideband radars. Time delays and phase differences between the radars result in mutually incoherent data in one band relative to the other. In essence, the dual band data must be adjusted so that it appears to the processor as if it was received from a single, dual band transmit source. To cohere the sub-bands, all-pole signal models are derived from the frequency domain data samples in each sub-band independently, and the sub-band models are adjusted as described in section 11.3.1 until they optimally match. The corresponding phase shift and time delay are then applied to the underlying data samples. A more detailed discussion of the mutual coherence processing is provided in section 13.2.

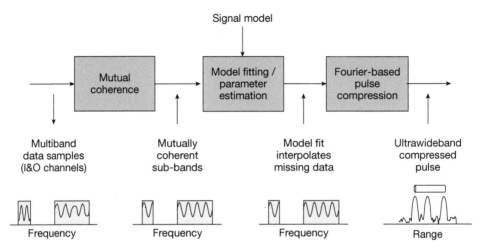

Figure 13.1   Ultrawideband processing methodology using sparse sub-band measurements to estimate the target's ultrawideband radar signature.

In step 2, a single global all-pole signal model is fitted to the complete set of adjusted sampled data across the multiple sub-bands. The model is then used for interpolation and/or extrapolation purposes. All-pole models are well-suited for ultrawideband processing as they accurately characterize the target by a superposition of discrete scattering centers, each with its own frequency-dependent term. While all-pole models are best matched to signal amplitudes that grow or decay exponentially fast with frequency, they can also be used to accurately characterize $f^{\tilde{\alpha}}$ scattering behavior over finite bandwidth intervals as discussed previously. This issue is further discussed in section 13.3 where a technique is developed that uses sparse sub-band measurements to estimate the global all-pole model parameters.

13.2   Mutual Coherence Processing

The methodology to cohere the sensors, termed mutual coherence processing, is now considered in detail. As described in chapter 11, mutual coherence discrepancies occur when the sub-band measurements are collected by separate wideband radars operating independently. The mutual coherence compensation technique allows one to apply ultrawideband processing across an arbitrary number of fielded radar platforms. This section illustrates the approach assuming two radars operating over sparsely separated bands viewing a target consisting of two scattering centers.

For illustration purposes, a simulation model is used to generate radar returns for a hypothetical target consisting of two discrete scattering centers. The scattering center closest to the radar has an $f^{\tilde{\alpha}}$-type amplitude behavior that decays with frequency, while the scattering center furthest from the radar has a scattering amplitude that grows with frequency. The simulated spectral signal samples are given by

$$E_k = 4\left(\frac{f_k}{f_1}\right)^{-1} e^{i\frac{\pi}{4}k} + \left(\frac{f_k}{f_1}\right)^{+1} e^{i\frac{\pi}{3}k}, \tag{13.1}$$

where the band center is chosen as $f_1 = 7.5$ GHz. Band 1 is assumed centered at 3.5 GHz and band 2 at 11.5 GHz, each having a bandwidth of 1 GHz. The ultrawideband spectrum covers the 3–12 GHz band and contains K frequency samples. Assume there are K_1 samples in band 1 and K_2 in band 2. The frequency sample index k ranges from $k = 0, \ldots, K_1 - 1$ for the lower sub-band and from $k = K - K_2, \ldots, K - 1$ for the upper sub-band. The frequency sampled phase terms correspond to a scattering center separation of 15 cm. White Gaussian noise was added to each signal sample with a signal-to-noise ratio of 20 dB.

The noisy signal frequency samples, E_k, are illustrated in figure 13.2. Figure 13.2(a) shows sparse multiband measurements of a target consisting of two closely spaced scattering centers; one has an amplitude that decays with frequency while the other's amplitude grows with frequency. The two sub-bands illustrated are not mutually coherent. Figure 13.2(b) shows the corresponding compressed pulses. The pulses do not line up in range because the sub-bands are mutually incoherent. The E_k signal samples in the lower sub-band (band 1) have been modulated by the function $e^{-i\frac{\pi}{9}k}$ to simulate the effects of sensor mutual incoherence, that is, the signal poles for the lower sub-band have been rotated 20° clockwise relative to the upper sub-band (band 2) signal poles. This is consistent with the two-sensor model introduced in chapter 11 where sensor incoherence is seen to be a consequence of the differing locations of the separate radars.

The mutual cohering process is initiated by modeling the spectral signals in each sub-band with a superposition of complex exponential functions using the all-pole signal modeling techniques introduced in chapter 2. Consider the application of the fusion techniques to a single radar pulse. Incorporating the notation introduced in chapter 11, a one-dimensional all-pole frequency domain signal model $M(f_k)$ of the form

$$M(f_k) = \sum_{n=1}^{P} a_n s_n^k, \tag{13.2}$$

is used to model the data in each sub-band, where P denotes the model order.

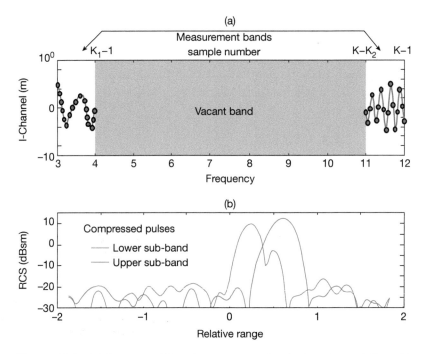

Figure 13.2 (a) Sparse multiband measurements of a target consisting of two closely spaced scattering centers. One has an amplitude that decays with frequency while the other's amplitude grows with frequency. The two sub-bands illustrated are not mutually coherent. (b) The corresponding compressed pulses. The pulses do not line up in range because the sub-bands are mutually incoherent.

Consider the sub-band modeling and mutual coherence process. This is done using a three-step process as follows:

1. Determine separately, the signal models $M_1(f_k)$ and $M_2(f_k)$ that best character- ize each sub-band. In general, the amplitudes and poles a_n, s_n, $n = 1, P$, can differ for each sub-band, and are extracted from each sub-band measurement as described in section 12.3.2, equations (12.6)–(12.8).
2. Use the sub-band signal models $M_1(f_k)$ and $M_2(f_k)$ to estimate the coherence parameters $\Delta\psi$ and τ_0 as shown subsequently in this section.
3. Using the estimates of $\Delta\psi$ and τ_0, adjust the data samples in sub-band 1 to make them more coherent with those in sub-band 2, as shown in equation (13.4).

In order to illustrate the effects of noise on the pole order estimates, the poles estimates were determined using an assumed model order $P=8$ (refer to [106–111] for a more detailed discussion on model order estimation). Figure 13.3 illustrates the estimates of the model order coefficients $|a_p|$, $p=1,\ldots,P$ (for $P=8$) for each sub-band data set illustrated in figure 13.2. It is clear from the figure that a pole order $P=2$ would be satisfactory for modeling this target, and the amplitudes for $p=1, 2$ correspond to the correct signal model order $P=2$ used in the simulation. For a less obvious result the Akaike information criterion or the minimum descriptive length model tests could be applied as described in [106].

The resulting pole estimates for the two sub-band data sets are illustrated in figure 13.4. The poles shown in figure 13.4(a) correspond to the lower sub-band (band 1) and those in figure 13.4(b) correspond to the upper sub-band (band 2). Notice that the signal poles in the lower sub-band do not line up with the signal poles in the upper sub-band. This is due to lack of mutual coherence.

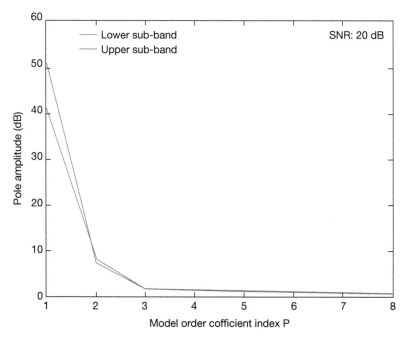

Figure 13.3   Pole amplitudes versus model order coefficient p for the two sub-band data sets illustrated in figure 13.2 for a model order $P=8$.

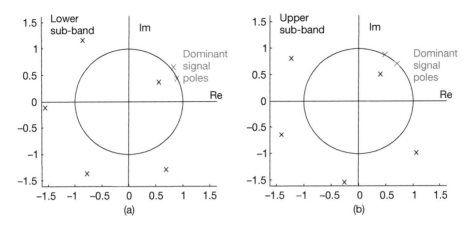

Figure 13.4   Pole estimates for the two sub-band data sets illustrated in figure 13.2. (a) Lower sub-band. (b) Upper sub-band. The dominant signal poles in the lower and upper sub-bands are shown in blue and red, respectively.

In step 2, to estimate $\Delta\psi$ and τ_0, the sub-band signal models $M_1(f_k)$ and $M_2(f_k)$ are adjusted to reduce the mismatch between them, and are determined by minimizing the cost function

$$C(\tilde{A}, \Delta\tilde{\theta}) = \underset{(A, \Delta\theta)}{Min} \sum_{k=0}^{K-1} \left| AM_1(f_k)e^{jk\Delta\theta} - M_2(f_k) \right|^2, \qquad (13.3)$$

with respect to the pole rotation angle $\Delta\theta = 2\pi\Delta f\tau_0$ and complex amplitude coefficient $A = |A|e^{j\Delta\psi}$. Because the sample index in equation (13.3) ranges over all the samples (including the vacant band) covering the entire ultrawide band, the cost function minimizes the mismatch between the two sub-band signal models extrapolated over the ultrawide bandwidth. The sub-band model alignment process determines the best adjustment to maximize the mutual coherence between the two sub-bands. Denote the solutions to 13.3 as $\tilde{A}$ and $\Delta\tilde{\theta}$. The model extrapolations, with and without coherence compensation are illustrated in figure 13.5.

Figure 13.5(a) illustrates the extrapolation of each sub-band signal model, without coherence compensation, into the vacant band. Because of the large sub-band separation, the two model extrapolations show considerable disagreement over the vacant band. The adjustment provided by $\tilde{A}$ and $\Delta\tilde{\theta}$ minimizes this difference.

In step 3, the optimal pole rotation angle $\Delta\tilde{\theta}$ and complex amplitude coefficient $\tilde{A}$ are applied to the lower sub-band signal model and corresponding data samples are replaced by the mutually cohered data samples $\tilde{E}_k$ given by

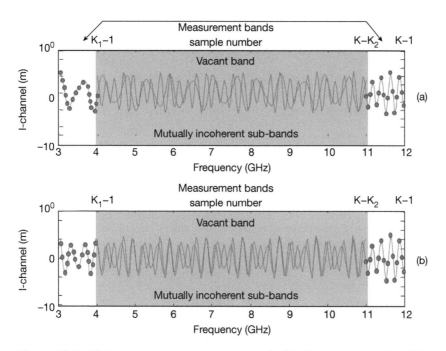

Figure 13.5 Mutual coherence processing applied to the sparse sub-band data set illustrated in figure 13.2. (a) Lower and upper sub-band models before coherence processing. (b) Models after mutual coherence processing.

$$\tilde{E}_k = E_k e^{j(k\Delta\tilde{\theta} + \arg(\tilde{A}))}, \text{ for } k = 0,\ldots,K_1 - 1. \tag{13.4}$$

Note that only the phase of $\tilde{A}$ is used for adjusting the data in the lower sub-band. The amplitude of $\tilde{A}$ is used in equation (13.3) only to place the two sub-band amplitudes on the same general level so as to aide in the accuracy of the phase and delay estimates. Figure 13.5(b) illustrates the extrapolation of the coherent sub-band signal models into the vacant band after mutual coherence adjustment. Observe from figure 13.5(b) that although the agreement of the two signal models over the vacant band is not perfect, the overall agreement is much closer than before mutual coherence compensation. Although the two signal models in figure 13.5(b) may not entirely agree, it is important to recognize that they have approximately the same signal poles. However, the corresponding all-pole model coefficients, a_n, for each band are different. The lower sub-band favors the decaying signal component whereas the upper sub-band favors the growing signal component.

Consider now the coherent combination of the two sub-bands using a single global all-pole signal model.

13.3   Ultrawideband Parameter Estimation and Prediction

Once the radar sub-bands have been mutually cohered, a global all-pole signal model is optimally fit to the adjusted measured data. Denote the global signal model as $\hat{M}(f_k)$ characterized by a refined set of poles $\hat{s}_n$ and pole amplitudes $\hat{a}_n$. The approach used determines those global all-pole model parameters that minimize the function J given by

$$J = \underset{(\hat{a}_n, \hat{s}_n)}{Min} \left\{ \sum_{k=0}^{K_1} \left| \tilde{E}_k - \hat{M}(f_k) \right|^2 + \sum_{k=K-K_2}^{K-1} \left| E_k - \hat{M}(f_k) \right|^2 \right\}. \tag{13.5}$$

Recall that the frequency sample index k ranges from $k = 0, \ldots, K_1 - 1$ for the lower sub-band and from $k = K - K_2, \ldots K - 1$ for the upper sub-band. The data samples in the lower sub-band now contains the adjusted, or mutually cohered data samples. The function J measures the total error between the global model

$$\hat{M}(f_k) = \sum_{n=1}^{\hat{N}} \hat{a}_n \hat{s}_n^k, \tag{13.6}$$

and the mutually cohered data samples in each sub-band. Note a model order $\hat{N} \geq P$ can be used in the global optimization to account for the case where closely spaced scattering centers not resolvable in either sub-band are present, but may be resolvable over the ultrawide bandwidth. There also may be cases that have scattering centers all of which are not common to every sub-band. For this simulation example $\hat{N} = 2$ because a common set of two scattering centers are resolved in each sub-band.

Minimizing J with respect to the complete set of global all-pole model parameters $\{\hat{a}_n, \hat{s}_n\}, n = 1, \ldots, \hat{N}$ is, unfortunately, a difficult nonlinear problem with no closed-form solution. One approach to solving this problem is to use the iterative approach illustrated in figure 13.6. Initial estimates of the all-pole model parameters are obtained using the lower and upper sub-band data, as described subsequently. Using the initial pole estimates, the set of amplitude coefficients $\{\hat{a}_n\}, n = 1, \ldots, \hat{N}$ can be determined using equation (13.5). These initial estimates are then iteratively optimized using a standard Newton-Raphson (NR) nonlinear least-squares technique found in many standard texts on numerical analysis [112]. If the initial parameter estimates are close to optimal, the NR technique will rapidly converge to those all-pole model parameters that minimize J.

There are two ways to obtain an initial estimate of the global all-pole model parameters $\{\hat{s}_n\}, n = 1, \ldots, \hat{N}$ using the HRSE spectral estimation technique developed in appendix B. One approach is to construct the multiband prediction matrix H given by

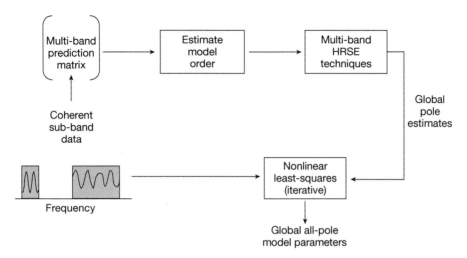

Figure 13.6   Ultrawideband parameter estimation. Initial global parameter estimates are obtained using high resolution spectral estimation (HRSE) techniques applied to the sub aperture or extended aperture Hankel matrix described by equation (13.7) or equation (13.11). The initial parameter estimates are iteratively optimized using a Newton-Raphson nonlinear least-squares technique.

$$H = \begin{bmatrix} H_1 \\ H_2 \end{bmatrix}. \tag{13.7}$$

The matrix H_1 has a special form; it is a Hankel matrix. Hankel matrices are associated with the transient response of linear time invariant (LTI) systems. Subspace decomposition methods exploit the eigenstructure of Hankel matrices to estimate the parameters of LTI signal models [107]. The Hankel matrices H_1 and H_2 correspond to the forward prediction matrices for the lower and upper sub-bands, respectively, as described by equation (B.3) in appendix B applied to a single pulse. Recall that there are K_1 samples in band 1 and K_2 samples in band 2. It is convenient to introduce the notation $E(k)$ to conform the HRSE development notation used in appendix B. The frequency data samples of the lower sub-band, $E_l(k)$, are then given by

$$E_l = [E_l(1) \quad E_l(2) \quad \dots \quad E_l(K_l)], \tag{13.8}$$

and for the upper sub-band, $E_u(k)$:

$$E_u = [E_u(K - K_2 + 1) \quad E_u(K - K_2 + 2) \quad \dots \quad E_u(K)]. \tag{13.9}$$

The Hankel matrix H_1 computed from the lower sub-band samples is given by

$$H_l = \begin{bmatrix} E_l(1) & E_l(2) & \cdots & E_l(L) \\ E_l(2) & E_l(3) & \cdots & E_l(L+1) \\ \cdots & \cdots & \cdots & \cdots \\ E_l(K_1-L+1) & E_l(K_1-L+2) & \cdots & E_l(K_1) \end{bmatrix}, \qquad (13.10)$$

where L denotes an averaging parameter and is heuristically chosen to be $L=[K_1/2]$; the brackets denote the largest integer that is less than or equal to the argument. The Hankel matrix H_2 computed from the entries of E_u defined by equation (13.9) can be obtained by simple extension of equation (13.10) using the upper band samples. It is important to mention that the vertical stacking of H_1 and H_2 requires that the two matrices have the same number of columns; therefore, the parameter $L=\min([K_1/2], [K_2/2])$ must be appropriately chosen.

The vertical stacking of the Hankel matrices H_1 and H_2 is referred to as sub-aperture processing. The approach combines the data samples from both sub-bands, providing the potential for robust parameter estimates from noisy data. This approach is most useful when the scattering centers are fixed on the target and somewhat resolved in each sub-band, and the frequency dependence of the scattering center over the ultrawide band is to be estimated. This is the case for the example considered in this section.

Using the poles obtained applying the HRSE technique to H defined by equation (13.7) as initial pole estimates to the nonlinear solution methodology outlined in figure 13.6 results in an adjustment of the upper and lower sub-band poles to jointly fit the dual band data. It essentially provides sharpness to the location estimate and a characterization of the ultrawideband frequency dependence for each scattering component.

It is also possible to obtain initial pole estimates by allowing for cross-correlation between the sub-bands, that is, by defining H as

$$H = [H_1 \quad H_2]. \qquad (13.11)$$

The horizontal stacking of H_1 and H_2 is referred to as extended aperture processing. This stacking topology requires the Hankel matrices to have the same number of rows chosen such that for $L_1=[K_1/2]$ and $L_2=[K_2/2]$, L_1 and L_1 are adjusted to satisfy the constraint $K_1-L_1=K_2-L_2$. The extended aperture processing provides the potential for true ultrawideband scattering center location resolution, although the resulting initial pole estimates are typically more sensitive to noise.

Figure 13.7 illustrates a plot of initial pole estimates for the sparse sub-band data set using sub-aperture processing. Note that by incorporating both H_1 and H_2 into the Hankel matrix, both signal poles are correctly identified and associated as f^{+1} and f^{-1} type pole behavior. The signal pole inside the unit circle corresponds to the f^{-1} scattering center; the signal pole outside the unit circle corresponds to the f^{+1} scattering center. These two signal poles are used to initialize the Newton-Raphson technique as discussed previously.

The NR technique uses the initial parameter estimates to find the global all-pole model parameters, $\hat{a}_n$ and $\hat{s}_n$, that locally minimize the cost function J defined in equation (13.5). The model order $\hat{N}$ remains fixed during this iterative process. The technique typically converges to a local minimum of J in only a few iterations. The approach is tested in the next few paragraphs by optimally fitting a global all-pole

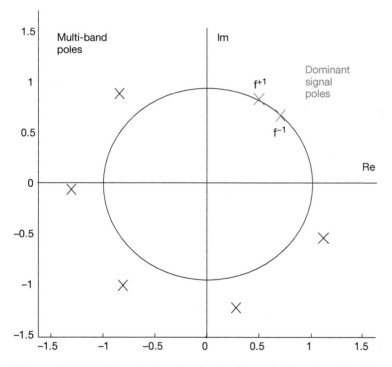

Figure 13.7 Multiband pole estimates for the mutually coherent sub-bands illustrated in figure 13.5(b) using the sub aperture Hankel matrix. The dominant signal pole inside the unit circle corresponds to the f^{-1} scattering center; the dominant signal pole outside the unit circle corresponds to the f^{+1} scattering center.

signal model to the two sub-bands data sets used in the two-point scattering center example introduced in section 13.2.

Figure 13.8(a) illustrates a comparison between the global all-pole signal model and truth. The all-pole model is in excellent agreement with truth over the entire ultra-wideband frequency range. The corresponding compressed pulses are shown in figure 13.8(b). The sparse sub-band compressed pulse uses the mutually coherent radar measurements within the two sub-bands and the global all-pole model in the vacant band. This approach clearly performs well; the two target points are well-resolved with the estimated ultrawideband response closely matching truth.

This example also demonstrates the potential for using all-pole signal models for accurately characterizing f^{α}-type scattering behavior over ultrawide bandwidths. In fact, the ultrawideband pole estimates can be transformed into equivalent estimates of the $\tilde{\alpha}$ coefficients for $f^{\tilde{\alpha}}$-type signal models. As discussed previously, by matching the slopes of the two functional frequency models, it is possible to find an $f^{\tilde{\alpha}}$ function

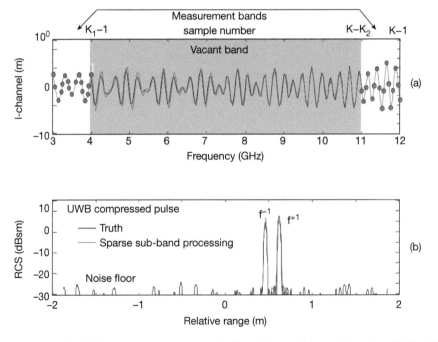

Figure 13.8 (a) Comparison between the fitted ultrawideband signal model (red line) and truth (black line). (b) The corresponding compressed pulses. The two scattering centers are well-resolved with the ultrawideband model closely matching truth.

that best matches the exponential behavior of the ultrawideband signal pole over the ultrawideband frequency range. One may also derive an approximate analytical relationship between the pole magnitudes and the corresponding $\tilde{\alpha}$ coefficients by matching the functions $f^{\tilde{\alpha}_n}$ and $|\hat{s}_n|^k$ at the lowest and highest ultrawideband frequencies. This relationship is given by

$$\tilde{\alpha}_n = \frac{(K-1)\log\left(|\hat{s}_n|^K\right)}{\log\left(1+(K-1)\dfrac{\Delta f}{f_1}\right)}, \tag{13.12}$$

where Δf and f_1 denote the spectral sample spacing and lowest ultrawideband frequency, respectively, and K denotes the total number of ultrawideband frequency samples. In the two scattering center example discussed previously, the two dominant signal poles $|\hat{s}_1|^K$ and $|\hat{s}_2|^K$ evaluated at the highest upper band frequency are determined to be

$$\begin{aligned} |\hat{s}_1|^K &= 0.992e^{j\pi/4} \\ |\hat{s}_2|^K &= 1.005e^{j\pi/3} \end{aligned} \tag{13.13}$$

By substituting these poles into equation (13.12), one obtains an estimate of the true $\tilde{\alpha}$ coefficients consistent with those used in the simulation. This information is useful for analyzing the details of targets with the viewpoint of constructing an accurate measurements-based model characterizing each scattering center.

14

Signature Modeling Using Sparse Angle Data

14.1 Introduction

In the previous chapter, techniques were developed for fusing data obtained by independent sensors operating in different frequency bands that were observing the target at the same viewing angle. In this chapter the problem of fusing data collected by sensors that observe the target from different viewing angles is considered. This fusion scenario is depicted in figure 11.1(c). It is assumed that the target is in the near field relative to the sensor separation distance L, so that $R_0 < 2L^2 / \lambda$, where L is perpendicular to the line of sight (LoS) between the sensors and R_0, the range to the target as discussed in section 11.1. As a consequence of the large separation, each sensor has a different view of the target geometry, and the fusion model for this scenario must account for the angular changes in the target signature.

As discussed in chapter 11, the signature model used for either fusion scenario is based on the geometrical theory of diffraction (GTD) scattering model defined by equation (2.6). Using this scattering model, the monostatic field scattered from the target is dependent on the frequency and angle of arrival of the incident wave registered in the target coordinate reference system, and is a function of the frequency and angle variation of the diffraction coefficients $D_n(f, \theta, \varphi)$ characterizing each scattering center on the target. When sparse band sensors are colocated, the viewing angles to the target are identical and the sensor fusion techniques need only account for the frequency behavior of the diffraction coefficients. This permits fusion techniques of the type developed in chapter 13 based on applying a global all-pole fusion model covering the entire sparse frequency bandwidth. Furthermore, because the look angle is the same for each sensor, the fusion techniques are independent of target motion and other target specific properties.

For the sparse-angle fusion problem considered in this chapter, the frequency bands of each sensor are assumed common. For fusion scenarios for which the variation of $D_n(f, \theta, \varphi)$ with angle can be approximated by an exponential growth or decay over the fusion observation space, a single global all-pole model having complex poles may be used to model the multisensor data. However, as the observation space increases, the variation of $D_n(f, \theta, \varphi)$ with angle is not typically modeled by an exponential behavior. In this case other models characterizing the variation of $D_n(f, \theta, \varphi)$ must be used. The challenge for the sparse-angle sensor fusion problem is to develop a suitable diffraction coefficient model that accounts for the angular variation of the measured data. In this chapter the angular behavior of $D_n(f, \theta, \varphi)$ is modeled using a set of parametric basis functions. Because diffraction coefficients are functions of viewing angle, a motion solution is also required when processing the time sequence of received radar data. In fact, as discussed in chapter 11, a joint multisensor motion solution is required in order to define the multisensor angle observation space.

The diffraction coefficient modeling techniques developed in this chapter for sparse-angle fusion are integrated into the methodology for developing a comprehensive computational signature model of the target in the following two chapters. The characteristics desired of the parameter-based diffraction coefficient models, and the rationale for the specific choices used in the sparse-angle fusion methodology, are addressed in this chapter.

In order to understand the differences in how frequency and angle measurement data are fused, it is useful to review the two types of signature models discussed in chapter 11, section 11.1. The two signature models used to characterize the scattering for each data fusion problem are summarized in equations (14.1a) and (14.1b):

$$E(f) \sim \sum_n D_n(f) e^{-j2\pi f \tau_n} \tag{14.1a}$$

$$E(\theta, \varphi) \sim \sum_n D_n(\theta, \varphi, c_1, c_2, \ldots c_m) e^{-j\frac{4\pi f}{c} \hat{k} \cdot \underline{r}_n}, \tag{14.1b}$$

where $\tau_n = \frac{2}{c} \hat{k} \cdot \underline{r}_n$ represents the propagation delay along the radar LoS for the nth scattering center referenced to the center of rotation (CoR) of the target. Only the explicit dependence of $E(f, \theta, \varphi)$ versus angle or frequency is emphasized in equation (14.1) for each respective signature model. Because $D_n(f)$ in equation (14.1a) is typically a slowly varying function of frequency (the exception being a resonant component), it is well approximated by an exponential behavior and the model defined in equation (14.1a) is identical to an all-pole model, having complex poles located off the unit circle as described in section 2.1. In contrast to equation (14.1a), sparse-angle fusion must account for a possibly strong variation of $D_n(\theta, \varphi)$ with angle as represented in equation (14.1b). Thus, a global all-pole model is typically inappropriate to use for this case, and a set of diffraction basis functions characterized by a parameter set $(c_1, c_2, \ldots, c_m)$ are incorporated into the model. The signature model is also an explicit function of the three-dimensional (3D) scattering locations defined by $\underline{r}_n$.

Application of a global model of the form defined in equation (14.1b) to the full multisensor data set presents a complex multiparameter nonlinear estimation problem. Along with the parameter set $(c_1, c_2, \ldots, c_m)$ associated with each scattering component, the sparse-angle model is dependent on the locations of the target scattering centers. As discussed in chapter 11.1, sparse-angle fusion generally requires that each sensor obtain a 3D location estimate for each of the target's scattering centers. Multisensor fusion, properly formulated, allows the noncoherent combination of these location estimates into a common three-dimensional target-centered coordinate system. Range and system delay mismatches between sensors result in a range misalignment of each location estimate, and sensor location estimation errors defining the baseline look angle of each sensor to the target result in a cross-range misalignment. It is only when the sensors are mutually cohered in both range and cross range to eliminate these misalignments that a true model of the target's signature based on equation (14.1) is obtained.

Along with differences in the structure of the fusion model, it is important to contrast the density of the data set relative to the available measurement space associated with each fusion problem. Figure 11.7 contrasts the sparsity dimension of the observation data for each of the two sensor scenarios. Interpolation in frequency is inherently a one-dimensional problem while interpolation in angle is a two-dimensional problem where the signature is a function of the polar angles (θ, φ), which in principle can vary over a full 4π steradians. A field or static range data collection is limited to a specific $(\theta(t)\phi(t))$ trajectory that typically covers only a small subspace of the total available observation space.

For the single sensor data extrapolation techniques developed in chapter 12, small extrapolations in frequency and angle centered about the measurement observation interval remain valid using a localized all-pole model. However, the all-pole model is inadequate when used to fuse angle data over large viewing angles due to three effects that are characteristic of the angular variation of $D_n(f, \theta, \varphi)$:

1. **Specular responses:** Speculars occurring in the angular domain are dual to resonance in the frequency domain. They occur when a continuous set of scattering centers lie on a line or surface reflecting directly back to the sensor. A single

all-pole model with constant amplitude coefficients represents such resonance phenomena poorly because of the large dynamic range of the response over small angular sectors. Multipole signature models might be used to model a specular response, where the distribution of the pole locations along the line or surface changes as the observation angle transitions through the specular. However, modeling the specular response by incorporating the diffraction coefficient angular variation associated with each contributing scattering center into the signature model is the preferred approach because it permits correlating the scattering model to a more unified framework using standard computational codes. The specular response might also be modeled using an estimate of the surface topology. This requires combining the GTD-based scattering model with a physical theory of diffraction scattering model, which uses the surface topology to compute the specular response. The attributes and differences of each approach are discussed in detail in chapter 15.

2. **Shadowing:** One basic premise of the GTD is that radiation or scattering to a particular point in space occurs only from scattering sources that are visible to the far-field observation point. Thus, when fusing signature data obtained at two different look angles, care must be taken to isolate the set of scattering centers that are visible to both sensors. As discussed previously, for arbitrarily positioned sensors, the integration of independent sensor data into a common signature model requires correlation to a composite set of scattering center locations referenced to a single, three-dimensional target-based coordinate system.

3. **Diffraction from discontinuities:** Diffraction from discontinuities on the target results in a component scattering pattern dependent on the particular geometry of the discontinuity. They are well characterized using parameter-based scattering models. Modeling the diffraction from a variety of discontinuities that might occur on the target is the subject of the next section.

14.2   Diffraction Coefficient Modeling: A Parameter-Based Basis Set

It is well known from GTD that the field scattered from a discontinuity on a scattering body is well modeled at high frequencies by a localized point source whose complex amplitude changes with frequency and angle. Using a two-dimensional example for simplicity, the field at a given polarization scattered from a discontinuity located at the coordinate (x_n, y_n) is characterized by

$$D_n(f, \varphi)e^{-jux_n}e^{-jvy_n}, \tag{14.2}$$

where u and v are as defined in chapter 11 for $\theta = \pi/2$. For a given look angle φ_0 to the target, the variation of $D_n(f, \varphi_0)$ with frequency is modeled by $e^{-\alpha(f - f_0)}$ as discussed in chapter 11, where the exponent α is dependent on scattering center type. This forms the rigorous basis for the general validity of the all-pole frequency model. However, for some resonant scattering centers (e.g., a resonant antenna on a conducting body) the variation of $D_n(f, \varphi_0)$ versus frequency is not characterized by $e^{-\alpha(f - f_0)}$, but by a tuned resonant circuit response (e.g., figures 16.10 and 16.11). In this case a different basis function set would be used for $D_n(f, \varphi_0)$. Fortunately, this resonance effect is *atypical* of the most common scattering feature types.

Unfortunately, for fixed frequency f_0, the *typical* variation of $D_n(f_0, \varphi)$ versus φ is analogous to a resonant frequency type scattering center behavior. This can be illustrated by examining the analytic expression characterizing $D_n(f, \varphi)$ for scattering from a two-dimensional wedge [113], as depicted in figure 14.1:

$$D(f, \varphi) = \frac{D_0}{\sqrt{f}} \left[D_1 + \frac{1}{\cos\left(\dfrac{\pi}{n_0}\right) - \cos\left(\dfrac{2\varphi}{n_0}\right)} \right], \tag{14.3}$$

where D_0 and D_1 are constants, $n_0 = 2 - \omega_a/\pi, \omega_a$ is the wedge angle, and transverse electric polarization incidence to the wedge is assumed. For $\varphi = \varphi_0$, $D(f, \varphi_0) \sim f^{-0.5}$, which is well modeled by an $f^{\tilde{\alpha}}$ frequency variation $\tilde{\alpha} = -0.5$. However, for fixed f_0 the variation of $D(f_0, \varphi)$ versus viewing angle φ is more complex. Observe that the wedge geometry illustrated in figure 14.1 generates a specular response (singularity) at two angles:

$$\varphi_1 = \frac{\pi}{2}$$
$$\varphi_2 = \frac{3\pi}{2} - \omega_a = n_0\pi - \frac{\pi}{2}. \tag{14.4}$$

Using equation (14.4), it is possible to rewrite equation (14.3) in the form

$$D(f, \varphi) = \frac{D_0}{\sqrt{f}} \left[D_1 + \frac{1/2}{\sin\left(\dfrac{1}{n_0}(\varphi - \varphi_1)\right) \sin\left(\dfrac{1}{n_0}(\varphi - \varphi_2)\right)} \right], \tag{14.5}$$

which illustrates the singularities occurring at $\varphi = \varphi_1, \varphi_2$. Thus the variation of $D_n(f_0, \varphi)$ versus angle contains singularities analogous to a frequency resonance response.

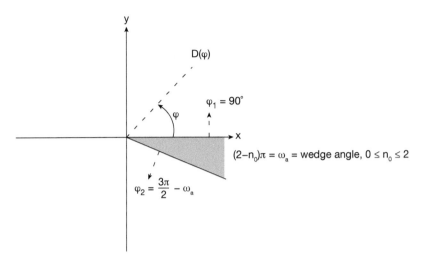

Figure 14.1 Geometry for wedge diffraction. Observe that the wedge geometry generates a specular response (singularity) at angles φ_1 and φ_2.

Observe that in the regions adjacent to both sides of a given specular, $D_n(f_0, \varphi)$ is well modeled using a general expansion of the form

$$D_n(f_0, \varphi) \approx c_1 + \frac{c_2}{\sin(\varphi - \varphi_S)}, \qquad (14.6)$$

where φ_S is the specular angle, and c_1 and c_2 are parameters that can be used to optimize the data fit when modeling a measured diffraction coefficient. Figure 14.2 illustrates the variation of $D_n(f_0, \varphi)$ for a 90° wedge, and the approximation in equation (14.6) that is used in later sections as a data fusion basis function.

The functional form for $D_n(f_0, \varphi)$ given by equation (14.6) is typical for localized scattering centers that contribute to a specular response. Identifying this type of scattering is important because it provides general information about the nature of the surface topology, for example, flat or wedge-like types of surfaces. Other types of scattering centers are characterized by a functional form for $D_n(f, \varphi)$ that is a generalization of equation (14.6). One example is a ridge or channel set on a curved conducting surface; in that case, the functional form for $D_n(f, \varphi)$ versus angle would be characterized by modeling the angular variation of the field scattered using the ridge or channel dimensions set on the surface topology.

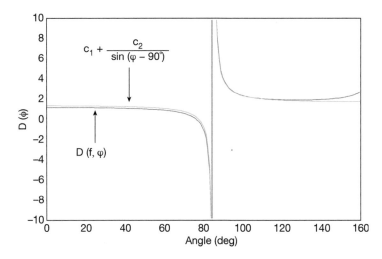

Figure 14.2 Diffraction coefficient behavior from 90° wedge. Theoretical (blue) versus approximation using equation (14.6) (green).

Thus a general parameter-based choice for the diffraction basis expansion set must be robust enough to cover a variety of diffraction phenomena. Figure 14.3 illustrates some useful parametric functional forms for several types of diffraction.

Although these examples are limited in scope, they provide a class of examples characterized by a more general form for $D_n(f_0, \varphi)$, given by

$$D_n(f_0, \varphi) \approx c_1 + c_2 \frac{\sin^{K_1}(\varphi - \varphi_{s_1})}{(\varphi - \varphi_{s_1})} + \frac{c_3}{\sin^{K_2}(\varphi - \varphi_{s_2})}, \tag{14.7}$$

where K_1 and K_2 are constants that depend on the specific scattering center types. Such parametric models for $D_n(f_0, \varphi)$ are useful for forming a measurements-based signature model and discussed more extensively in chapter 16 (e.g., section 16.3).

14.3 Diffraction Coefficient Modeling Example: The Two-Dimensional Strip

Perhaps the simplest deviation from the constant amplitude two-point scattering center example is the two-dimensional perfectly conducting strip illustrated in figure 14.4. A constant amplitude two-point scattering model approximating this geometry was

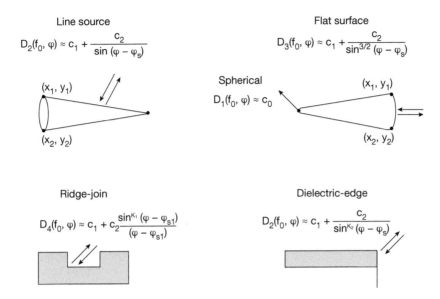

Figure 14.3   Example parameter-based diffraction basis functions for diffraction coefficient modeling.

used in chapter 11 to contrast sensor mutual coherence compensation requirements for sparse-angle data fusion to those for sparse band fusion. Now consider the specific diffraction basis functions that characterize the diffraction from each edge of the strip.

Relative to the three scattering mechanisms discussed previously, the strip exhibits specular and diffraction effects but no shadowing. Consider, for example, the two-dimensional strip of width $2a$ oriented on the x-axis of the x-y coordinate system illustrated in figure 14.4. Denote the backscatter from the strip as $E(f, \varphi)$, where φ is the standard azimuthal angle variable. Neglecting mutual coupling between the ends of the strip, the GTD scattering model for the strip can be characterized by scattering centers located at each end of strip, using the appropriate GTD diffraction coefficient defined by equation (14.3), with $n_0 = 2$. The backscattered field $E(f, \varphi)$ can be written in the form

$$E(f, \varphi) = D_1(f, \varphi)e^{+jua} + D_2(f, \varphi)e^{-jua}, \tag{14.8}$$

where $u = \dfrac{4\pi f}{c}\cos\varphi$. $D_1(f, \varphi)$ and $D_2(f, \varphi)$ are given by

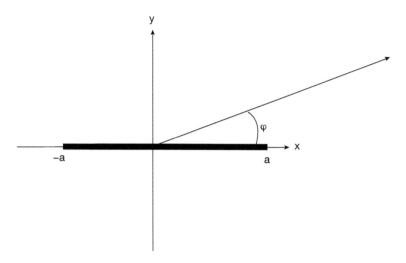

Figure 14.4   Backscatter from a two-dimensional strip of length 2a.

$$D_1(f, \varphi) = \frac{D_0}{\sqrt{f}} \left[1 + \frac{1}{\cos\varphi} \right],$$

$$D_2(f, \varphi) = \frac{D_0}{\sqrt{f}} \left[1 - \frac{1}{\cos\varphi} \right].$$

(14.9)

The sign difference for $D_1(f, \varphi)$ relative to $D_2(f, \varphi)$ in equation (14.9) arises due to the difference in incidence angle referenced to each end of the strip in applying equation (14.3).

Consider the frequency and angle variation of the field scattered from the strip using equation (14.8) for two cases: First examine the frequency behavior $E(f, \varphi_0)$ for a given look angle $\varphi = \varphi_0$, and second, examine the angular behavior of $E(f_0, \varphi)$ at a fixed frequency $f = f_0$. The real part of the variation of $E(f, \varphi_0)$, assuming $\varphi_0 = 45°$ and strip width $2a = 2$ m is shown in figure 14.5. Observe that the envelope of $E(f, \varphi_0)$ is characteristic of a typical $f^{\tilde{\alpha}}$ type frequency behavior. For the second case, fix the frequency at $f_0 = 1.5$ GHz and examine the angular variation of the magnitude of $E(f_0, \varphi)$ illustrated in figure 14.6. The monostatic scattering versus angle (normalized to unity at the specular angle) exhibits a large dynamic range as the angular variation passes through the specular region, analogous to the frequency resonance response of a high Q component. As a consequence of this large dynamic range, a single all-pole

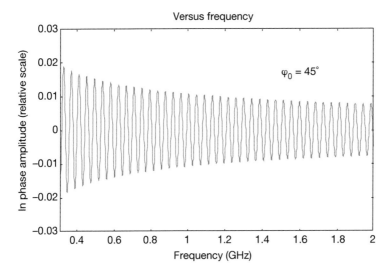

Figure 14.5   Scattered field from two-dimensional strip width $2a = 2$ m as function of frequency at 45° incidence angle (relative amplitude scale). The real part of the variation of $E(f, \varphi_0)$ is shown.

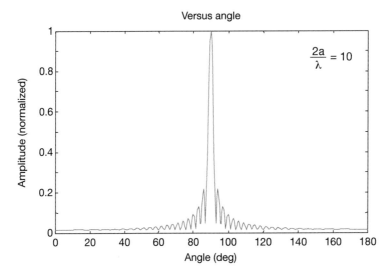

Figure 14.6   Scattered field from two-dimensional strip as function of aspect angle (relative amplitude scale). Fixed frequency at $f_0 = 1.5$ GHz.

model cannot be used to characterize the field variation over broad angular regions. However, note that a single pole model having an exponential variation of the form $e^{-\beta_n \varphi}$ can be used to model the field variation over small localized angular sectors.

14.4   Sector Fusion Processing Example: A Comparison of Two Fusion Methods

Assume the conducting strip illustrated in figure 14.4 represents the target, and simulate narrowband angle data samples using equation (14.8). Relative to the mutual coherence discussion of section 11.2.1, assume two sensors lie in the plane of rotation of the target so that the time to angle mapping is defined by equations (11.6) and (11.7) and $\kappa = 90°$. The observation space is then one-dimensional, characterized by the rotational angle φ as depicted in figure 14.4. Furthermore, assume the baseline look angle to the target for each sensor, φ_{0_1} and φ_{0_2}, are correctly estimated. Assume for this example that narrowband observation data at $f_0 = 1.5$ GHz is collected over the two sparsely separated and noncontiguous viewing sectors illustrated in figure 14.7(a), centered at $\varphi_{0_1} = 40°$ and $\varphi_{0_2} = 80°$, respectively, each having a sector width of $12°$.

For *each sector* of simulated angle data, the angle extent of each is such that a separate all-pole model having complex poles is adequate to model the data. If the two viewing angle sectors were closely spaced, a global all-pole model could be used to model the data over the composite observation space and sparse-angle data fusion could be carried out using fusion techniques dual to those developed in chapter 13. When the sectors are widely separated, as in this example, the fusion model must be generalized using parameter-based basis functions to model the diffraction coefficients. To motivate the utility of incorporating the parameter-based diffraction coefficient basis functions in the data fusion model, this section considers using two models to fuse the data over the composite observation space: first assuming an all-pole model having complex poles, and subsequently using the diffraction-based fusion model.

14.4.1   Data Fusion Using the All-Pole Signature Model

First consider the form of the complex poles characterizing each $12°$ sector of angle data used for the sparse-angle data fusion. It was demonstrated in chapter 11 that the location of each scattering center can be isolated in cross range using a narrowband all-pole model applied to limited sectors of data, from which the constant amplitude poles take the form of equation (11.20). Recall that when a time sequence of narrowband pulses are processed over a given sector of angle data, the range-rate $\dot{R}$ extracted from the time sequence of data is mapped to cross range using the estimated motion of the target according to $R_{XR} = \dot{R}/\dot{\varphi}$, as defined in equation (6.25). It is useful to generalize the expression for the narrowband poles for constant amplitude diffraction coefficients that lie on the unit circle (see figure 11.5) to include poles displaced from the unit circle corresponding to scattering centers that have diffraction coefficients characterized by an exponential variation versus angle. The analysis is similar to that developed in section 11.3.2, except now the effects of edge diffraction are included.

Consistent with equation (2.9), approximate the angular variation of each diffraction coefficient $D(f_0, \varphi)$ defined in equation (14.8) over the localized region about each sensor's look angle, $\varphi = \varphi_{0_m}$, $m = 1, 2$, by an exponential behavior defined by

$$D(f_0, \varphi) = D_0 e^{-\beta(\varphi - \varphi_{0_n})}, \tag{14.10}$$

The value of β depends on the proximity of each viewing angle to the specular location. Using equation (14.10), the field variation over each angular sector about $\varphi = \varphi_{0_m}$ resulting from the edge scattering center located at $(x_n, 0)$ can be expressed in the form

$$E_n(f_0, \varphi) \approx D_0 e^{-\beta_n(\varphi - \varphi_{0_m})} e^{-j\frac{4\pi f_0}{c} x_n \cos\varphi}. \tag{14.11}$$

To determine the complex pole p_n associated with each edge of the strip, consider data samples equally spaced in angle $\Delta\varphi$ so that $\varphi = \varphi_{0_m} + q\Delta\varphi$, and assume $q\,\Delta\varphi \ll 1$. After some manipulation, equation (14.11) can be expressed in the form of an all-pole model defined by

$$E_n(f_0, \varphi_{0_m} + q\Delta\varphi) \approx D_0 e^{-j\frac{4\pi f_0}{c} x_n \cos\varphi_{0_m}} (p_n)^q. \tag{14.12}$$

where the complex pole p_n is given by

$$p_n = e^{-\beta_n \Delta\varphi} e^{j\frac{4\pi f_0}{c} x_n \Delta\varphi \sin\varphi_{0_m}} \tag{14.13}$$

The pole p_n appropriate to each sector can be expressed in terms of the cross-range dimension R_{XR} introduced in chapter 11, equation (11.21). Define R_{XR_n} according to

$$R_{XR_n} = x_n \sin\varphi_{0_m}, \tag{14.14}$$

The complex pole p_n can be written in the more descriptive form

$$p_n = e^{-\beta_n \Delta\varphi} e^{j\frac{4\pi f_0}{c} \Delta\varphi R_{XR_n}}. \tag{14.15}$$

Compare equation (14.15) to the poles characterized by equation (11.21) for the constant amplitude two-point scattering center example illustrated in figure 11.4. The amplitude of the poles for the constant amplitude, two-point scattering example correspond to $\beta_n = 0$, whereas β_n in equation (14.15) characterizes the local approximation about each sector of $D(f_0, \varphi_{0_m})$ for each edge of two-dimensional strip. The edge diffraction coefficient modeled by $e^{-\beta_n(\varphi - \varphi_0)}$ results in a value of β_n considerably larger over angular regions approaching the specular location, and the complex poles will be displaced a considerable distance off the unit circle.

Consider first the model-to-data fit in each angular sector using the all-pole model having complex poles. The data-to-model comparison for each sector gives the results illustrated in figures 14.7(b) and 14.7(c). Because the data are noise free, the comparison of data to model for each individual sector is indistinguishable from the plots: the values for β_n appropriate to each sectors' all-pole model are clearly quite adequate to model the data in each sector. Note, in particular, the rapid exponential growth associated with the data in sector 2 as the sector region approaches the specular location.

Now consider the model-to-data fit using a *global* two-pole model having complex poles with the goal of interpolating the sector data over the entire observation region containing the two sectors (34° to 86°). The comparison of model to data for this case is illustrated in figure 14.7(d). The comparison of model (red) to data (blue) obtained by applying a two-pole exponential model over this very large sector is considerably degraded and favors the sector nearest the specular location where the scattered field is strongest.

14.4.2 Data Fusion Using a Specular-Based Diffraction Basis Set

The sparse-angle fusion technique described previously using a global all pole model required no knowledge of the target scattering center locations. Fusion was accomplished using the dual of the sparse-band fusion methodology developed in chapter 13, and it was evident that the global all-pole model having exponential variation over the very wide angle angular sector was inadequate to model the data. Now consider using the global signature model defined in equation (14.1b) using the set of basis functions appropriate for the strip geometry applied to the same sparse-angle fusion problem. Assume for simplicity that the locations of the target end point scattering centers, $(x_1, x_2) = (-a, a)$, are precisely located using the narrowband scattering location estimation techniques developed in chapter 6. Using equation (14.6), the constants $(c_{1,n}, c_{2,n})$ and a specular location estimate $\hat{\varphi}_s$ are required to model each scattering center diffraction coefficient $D_n(f_0, \varphi)$ over the entire region connecting the two sectors. Begin with the general expansion for $E(f, \varphi)$ determined using the GTD scattering model:

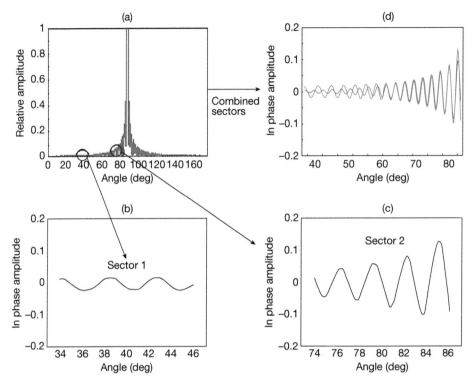

Figure 14.7 (a) Scattering amplitude and two angular sectors used for two-dimensional strip data fusion example. (b and c) All-pole model having complex poles with model fit versus data applied separately to sector 1 and sector 2 centered at $\varphi_{0_1} = 40°$ and $\varphi_{0_2} = 80°$. The model to-data comparisons are indistinguishable from the graphs. (d) Global all-pole model having complex poles applied to the data over the combined (34° to 86°) angular sectors. Blue indicates data and red shows the model. Note the model to data fit degrades as the global all-pole model (red) favors the data fit in the larger amplitude sector.

$$E(f_0, \varphi) = \sum_{n=1}^{2} D_n(f_0, \varphi) e^{-j u x_n}. \qquad (14.16)$$

Using the basis set defined by equation (14.6), the expression for $E(f_0, \varphi)$ becomes

$$E(f_0, \varphi) = \sum_{n=1}^{2} \frac{1}{\sqrt{f_0}} \left[c_{1,n} + \frac{c_{2,n}}{\sin(\varphi - \hat{\varphi}_s)} \right] e^{-j u x_n}. \qquad (14.17)$$

Equation (14.17) is general enough to characterize the field variation over broad angular regions for either constant amplitude or specular-dominant scattering. Assume for now that the location of the specular is known. Applying equation (14.17), figure 14.8(b) illustrates the model match to the data over the entire $34°$ to $86°$ sector at $f_0 = 1.5$ GHz using the expansion defined by equation (14.17). This is to be compared to figure 14.8(a) using a global all-pole signature model, having complex poles. The data match illustrated in figure 14.8(b) using the diffraction-basis set in equation 14.17 essentially overlays the data over this broad angular region.

There is clearly a drawback to using the expansion basis set defined by equation (14.17): the location of the specular and the scattering center locations must be accurately estimated. Estimating the scattering center locations is straightforward for this example using the techniques developed in chapters 5 and 6. However, the specular location appears as a nonlinear parameter and is not generally known if the target topology is unknown. There are several solutions to this dilemma: the first is to estimate φ_s by iteration, fitting the model to the observation data set to obtain the best match; the second is to estimate φ_s from other information, such as using field data collected on a target rotating through a large angular spread, where the angular location of the specular amplitudes can be extracted from the data using the motion solution; and the third would be to determine the specular directions from the geometry of the target if its surface topology is known.

As an example of the first approach, it is possible to develop a first-order estimate of the specular location using the exponential variation of the complex pole characterizing a particular scattering center. Consider the local approximation fitting the pole model to the diffraction basis set over a limited angular sector.

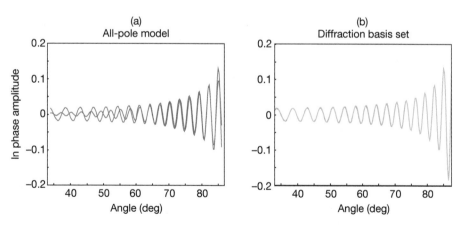

Figure 14.8 (a) Global all-pole model (red) having complex poles applied to data (blue) over $34°$ to $86°$ angular sector. (b) Diffraction basis set model defined by equation (14.17) applied to data over $34°$ to $86°$ angular region. Model and data are indistinguishable.

$$D_0 e^{-\beta_n \varphi} \approx c_{1,n} + \frac{c_{2,n}}{\sin(\varphi - \hat{\varphi}_s)}. \tag{14.18}$$

Assuming the second term dominates the exponential growth of the all-pole model, an estimate of φ_s is given by

$$\hat{\varphi}_s \approx \varphi_0 + \tan^{-1}\left(\frac{1}{\beta_n}\right), \tag{14.19}$$

where φ_0 denotes the center of the angular sector under consideration and β_n is determined from the real part of the complex all-pole model. The closer φ_0 is to φ_s, the more correct the estimate. Once the estimate for φ_s is obtained, equation (14.17) leads to a linear set of equations for the remaining coefficients. An iterative technique can be developed for estimating the parameters in equation (14.17). The reader is referred to [114] for a more detailed discussion.

IV

Measurements-Based RCS Signature Modeling

15

An Integrated Predictive/Measurements-Based RCS Signature Model

15.1   Introduction

The results of chapter 14 demonstrate that the use of a parametric diffraction coefficient model, except for limited special cases, is necessary for sparse-angle sensor fusion. In this chapter the parametric model is used to develop a comprehensive computational signature model for the target. This signature model is, in fact, more robust for data extrapolation than the techniques introduced in chapter 12. Diffraction coefficient modeling represents a key element in creating a signature model for the target using the geometrical theory of diffraction (GTD) scattering model characterized by equation (2.6). The choice of diffraction coefficient basis functions becomes particularly important when integrating the diffraction coefficients extracted from measurement data into a comprehensive target model characterized by the diffraction coefficient basis sets.

The signature modeling techniques that employ the diffraction coefficient basis functions can be used to fuse data received from any number of sensors, provided that the data from each sensor are registered to a common target-centered coordinate system having origin located at the target center of rotation (CoR). In this chapter techniques are developed for integrating high pulse repetition frequency data obtained by a single sensor over specified time intervals into the signature model. Each data collection time interval covers scattering from the target over the angular sector characterized by the motion solution, and the composite of these angular sectors define the measurement observation space. The computational signature model developed using these measurements is referred to as a measurements-based model (MBM) of the target radar signature. The methodology for extracting the 3D scattering center locations and associated diffraction coefficients from the measurement data, and their integration into the measurements-based signature model are addressed. The resultant signature model presents a good match to the measurement data over the observation space, and is predictive for extrapolations outside the observation region.

As an introduction to the measurements-based modeling methodology, it is useful to review the conceptual approach introduced in chapter 5 for developing a signature model of the target based on measured data. Referring to the discussion related to figure 5.1, the model development depends on obtaining solutions to the joint target-motion estimation problem, and extracting the signature model parameters from the data. The sequential estimation technique introduced in chapter 2 is applied to B sliding blocks of frequency-time data to obtain the observable sequence $(R_{n,b}, \dot{R}_{n,b}, \ddot{R}_{n,b}, D_{n,b})_{b=1,\ldots B}$ for each of the N scattering centers contributing to the target signature. This sequence is then used in a joint target-motion space estimation framework to obtain the motion solution $\theta(t_b)$, $\varphi(t_b)$, $b=1,\ldots, B$. Using the motion solution, the set of observable estimates is mapped into a fixed target-space coordinate system yielding 3D scattering center location estimates. The complex amplitudes for each of the N target components, $D_n(\theta_b, \varphi_b)$, $b=1,\ldots, B$, viewed as a function of angle, characterize the measured diffraction coefficients, averaged over the frequency bandwidth (assuming $FBW \leq 10$ percent). These initial motion and scattering center location estimates are used to develop the signature model using the iterative framework introduced in section 15.4.3.

For the formulation illustrated in figure 5.1, the scattering from the target is modeled using a GTD-based signature model. The measurements-based model developed in this chapter integrates two computational models: a fast-running physical theory of diffraction (PTD) computational model to characterize the general topology of the target and "easy" to model components, and a GTD measurements-based computational model to characterize specific components on the target. The fixed-point scattering center locations on the target are time invariant (except for their disappearance caused by shadowing). A special class of time varying scattering center locations caused by edges and smooth surfaces require special treatment. The GTD-based approach described using the simulation example developed in section 2.5 provides one way of modeling these types of scattering centers, and was also used to model the slotted cylindrical target introduced in chapter 6. Incorporating an estimate of the target topology directly into the signature model using a PTD prediction code provides a second approach for characterizing the non-fixed-point scattering centers. Both approaches are consistent with the formulation developed in this chapter.

The chapter begins by presenting a brief overview of different analytical and numerical techniques used to predict the electromagnetic scattering from targets. Some deficiencies in these techniques and how to alleviate them by incorporating target measurements are discussed. An integrated approach to forming the signature model using a combination of electromagnetic codes, and either static range or field measurements of the target, is developed. Examples illustrating the application of the technique using simulated measurement data are presented.

Following this development, those factors that most affect the fidelity of the signature model, and the demands placed on model fidelity for specific use cases are examined. The chapter concludes with a discussion of three techniques based on the new 3D scattering center location estimation framework developed in the text that could be implemented to exploit static range measurement data for estimating three-dimensional target information.

15.2   An Overview of RCS Prediction Techniques: The MBM Signature Model

An excellent overview of the different radar cross-section (RCS) prediction techniques along with their attributes and limitations is presented in a series of IEEE lectures by O'Donnell [115], which are available on the web. Figure 15.1 is adapted from those lectures, with additional material added to provide an understanding of how measurements are incorporated into the computational model. Figure 15.1 illustrates a comparison of exact computational techniques (i.e., those based on a rigorous formulation of Maxwell's equations applied under a variety of situations) versus prediction techniques based on approximations to Maxwell's equations applied to specific regions of the frequency spectrum. The acronyms delineated in figure 15.1 are listed here as notional only, to illustrate the wide variety of prediction techniques available to the analyst, and a complete description can be found in [115]. The techniques listed under exact techniques are divided into three categories: classical analytical solutions that can only be obtained for a few canonical target geometries; strictly numerical techniques, which in general are computationally slow and have limited application at higher frequencies; and hybrid techniques. The class of approximate techniques based on high-frequency approximations to Maxwell's equations are computationally fast relative to the rigorous numerical methods, but are still limited to modeling only certain types of components and materials.

In this chapter attention is focused on a modified geometrical theory of diffraction (GTD) model that combines PTD with GTD. It is computationally fast and has the flexibility to accommodate measurements-based component features. The GTD scattering model provides a direct correlation to the high-resolution spectral estimation (HRSE) processing introduced in chapter 2 for extracting information about the target from the measurement data. In developing the signature model of the target using

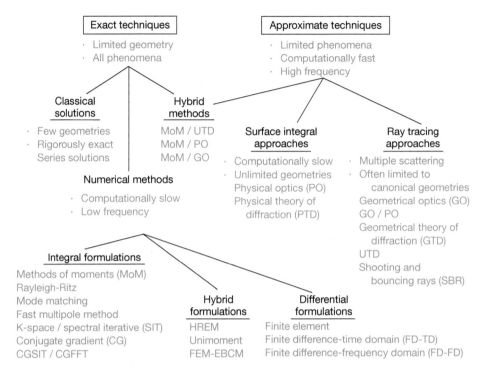

Figure 15.1   A compendium of RCS prediction techniques (adapted from [115]). The compendium is extensive, and included here only to delineate the extensive nature of the many techniques developed over the past years appropriate to electromagnetic scattering predictive codes. The reader is referred to [115] for a more complete description of the techniques listed in the figure.

GTD-based techniques, two options are available. The first represents all the scattering centers on the target by their measured GTD diffraction coefficients correlated to their corresponding scattering center locations. This approach was used to model the scattered field for the two-dimensional perfectly conducting strip example treated in section 14.3 as well as the slotted cylindrical target. Incorporating measured diffraction coefficients eliminates the specular singularities that often characterize the theoretical GTD diffraction coefficients, and allows for a comprehensive solution using only GTD. The second option models only selected components on the target using their measured GTD diffraction coefficients, and augments the GTD model with a PTD contribution that characterizes the topology and the remaining easy to model components. This is most useful for targets having difficult-to-model components. Both options use the observable sequence $(R_{n,b}, \dot{R}_{n,b}, \ddot{R}_{n,b}, D_{n,b})_{b=1,\dots B}$ extracted from the measurement data to estimate the target topology, scattering center locations, and component diffraction coefficients in order to provide insight into the overall character of the target and scattering components.

In theory, GTD and PTD are complementary: GTD characterizes the component scattered field well, but the theoretical diffraction coefficients generally degrade in the specular and shadow regions. PTD characterizes the specular region well, but adds a GTD diffraction coefficient to correct the physical optics solutions far away from the specular region.

When incorporating PTD into the signature model, the GTD measured diffraction contribution is coherently added to the baseline PTD prediction. A rigorous theoretical justification of this measurements-based approach is not presented in this text, but follows from a logical extension of the foundation of GTD in which the scattered field is approximated as the coherent sum of fields emanating from a number of individual scattering centers. The PTD contribution is equivalent to a sum of coherent scattering centers distributed over the surface of the target that are visible to the sensor, referenced to the origin of the target model coordinate system. The GTD contribution is formed from additional components on the target, having locations referenced to the same target model coordinate system and coherently added to the PTD scattered field. Examples are presented that compare predictions of the signature using the MBM model to simulated measurement data. The approach is also consistent with the comparative analysis of GTD versus PTD developed by Knott and Senior [116], where it is shown that the GTD model in the specular regions (caustics) can be corrected using the physical optics solution.

Using the previous rationale, the general form of the measurements-based signature model, for a specified polarization of incident and scattered field, is defined by a generalization of equation (2.6)

$$E(f, \theta, \varphi) \sim E_{PTD} + \sum D_n(f, \theta, \varphi)e^{-j\frac{4\pi f}{c}\hat{k}\cdot\underline{r}_n} \tag{15.1}$$

where $\underline{r}_n = (x_n, y_n, z_n)$ denotes the location of the nth scattering center and $D_n(f, \theta, \phi)$ the associated diffraction coefficient. The second term represents the GTD contribution determined from measurement data on the target. The determination of the PTD and GTD contributions in equation (15.1) is dependent on one of two possible conditions:

1. The target is known, that is, its topology and component locations can be deter-mined by examination, but the component scattering response may be difficult to model, and thus is extracted from the measurement data set.

2. The target general topology is known, but the component locations and behav-ior are unknown; for example, it may have a cylindrical structure, but little is known about the precise locations and character of the components on the target.

For case 1, where the scattering center locations $\underline{r}_n$ are known, the PTD contribu-tion is determined using the topology of the target, and the diffraction coefficients are estimated from the measurement data. For case 2 both the scattering center locations and diffraction coefficients must be estimated from the measurement data. If the measurement data set spans a robust observation space, the general topology of the target can be fine-tuned by isolating the specular responses, which correspond to a dis-tribution of range, cross-range poles lying along the surface of the target. Errors in estimating $\underline{r}_n$, the scattering center locations, degrade the fidelity of the resulting model, particularly for narrowband signature modeling. The consequences of this error are addressed in more detail in sections 15.5 and 16.2.1.1.

15.3   The Rationale for a Measurements-Based Model

Table 15.1 identifies some limitations of the computational codes discussed in fig-ure 15.1, and shows how using a measurements-based model alleviates these limita-tions. Four specific areas are highlighted that typically dominate the use of the computational codes: computational run time, modeling of complex components, tar-get fabrication imperfections, and trajectory dynamics.

The measurements-based model overcomes these limitations by using a fast-running PTD code to model the basic target topology and integrates a measurements-based component model extracted from measurements on the target. The PTD model char-acterizes the specular returns from extended surfaces of the target (e.g., a cylindrical or spherical topology) while the GTD model is used to model isolated components on the target that contain complex features and fabrication imperfections. Because the GTD diffraction coefficients are extracted from measured data on the target, they include all attributes of the component as well as target fabrication imperfections. The resulting computational model is fast running and can be used in near-real-time appli-cations, such as in interactive dynamic scenarios where the motion of the target can-not be defined a priori.

Table 15.1   Code modeling limitations versus a measurements-based modeling approach

Existing Code Limitations	Measurements-Based Modeling Attributes
• Significant run time	• PTD/GTD-based computations
– Days/weeks/etc.	– Real time/near-real time
• Can't accurately model some essential target scattering components	• Target component characterization extracted from measurements
– Resonant structures, dielectrics, etc.	
• Real targets have fabrication imperfections not a priori included in model	• Fabrication imperfections included in component characterization
• Run time limits motion space and trajectory dynamics	• Parameter-based component characterizations
– Body dynamics limited to a priori defined motion	
– Not suitable for interactive Monte Carlo applications	– Readily incorporates nonnominal dynamics
	– Facilitates interactive Monte Carlo analysis

One approach to mitigate the run time limitation in a motion dynamics simulation has been to generate a database of target signatures using predictive codes, where the motion of the target is specified. This approach is suitable for a fixed scenario, but is cumbersome when the simulation scenario changes in real time in response to a sequence of interactive events. An alternate approach might be to compute the database over the full 4π steradian sphere of possible look angles to the target. A simple analysis demonstrates the infeasibility of this approach as the operational frequency increases. Denote T_0 as the code run time at a given look angle, frequency, and polarization. For a target having characteristic dimensions D_1 and D_2, Nyquist sampling in angle requires angle samples on the order of $\lambda/(4D_1)$ and $\lambda/(4D_2)$ for each (θ, φ) look angle to the target. For a number of frequencies spaced Δf over a bandwidth BW, the computational time, T_c required to generate the 4π steradian database is approximated by

$$T_c \approx T_0 64\pi \left(\frac{D_1 D_2}{\lambda^2} \right) \left(\frac{BW}{\Delta f} \right) \quad \begin{matrix} \rightarrow \sim 200 T_0 D_1 D_2 BW/\Delta f \text{ at UHF} \\ \rightarrow \sim 2.2 \times 10^5 T_0 D_1 D_2 BW/\Delta f \text{ at X-Band.} \end{matrix} \quad (15.2)$$

The computational time required to form the database becomes intractable as the operational frequency increases. Using the measurements-based model, the signature can be generated in real time consistent with the simulation parameters for any given motion dynamics.

15.4   An Integrated Approach: The Measurements-Based Computational Model

15.4.1   Conceptual Overview

Figure 15.2 provides a conceptual overview of the formation and application of a wideband measurements-based signature model applied to simulate a compressed pulse, using the slotted cylindrical target as an example. A purely GTD-based signature model is illustrated and assumes wideband data are available.

An estimate of the target scattering attributes is initially obtained from solutions to the joint target-motion estimation problem as described in section 15.1. Typically, this results in a composite target space mapping identifying the locations of the dominant

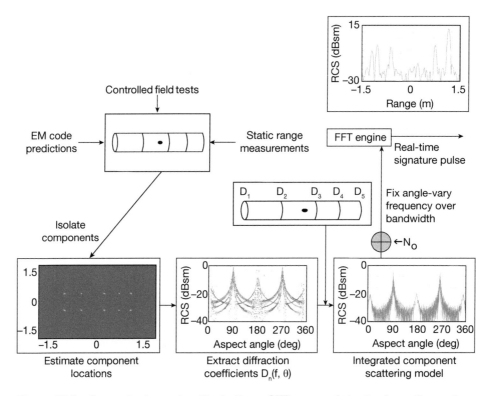

Figure 15.2   Conceptual overview illustrating a GTD approach for the formation and application of a wideband measurements-based signature model applied to simulate a compressed pulse

scattering centers on the target referenced to target space. Electromagnetic prediction codes might also be employed to validate a particular component model identified from the measurement data. Because modern RCS measurement ranges generally cover a very broad range in frequencies, static range measurements provide very high resolution of closely spaced scattering centers on the target, which may not be cleanly resolved over the operational sub-band of interest. The high resolution, with its associated sharpness, facilitates high precision estimates of the component locations on the target to be included in the MBM. This is particularly important to maintain fidelity of the resulting target model when used in narrowband applications. The diffraction coefficients characterizing each component are extracted from the measurement data and stored in a library to represent $D_n(f, \theta, \phi)$. Typically, this can be accomplished using the parametric model approximation $D_n(f, \theta, \phi, c_1, c_2, \ldots, c_m)$ to $D_n(f, \theta, \phi)$ as discussed in section 14.1, where the parameters $(c_1, c_2, \ldots c_m)$ are determined by fitting a functional model to the measurements over the observation interval. Using a purely GTD scattering model eliminates the need for the PTD term to model of the topology of the target as discussed previously. The scattering center locations, the target topology (if available), and corresponding diffraction coefficients are integrated into an MBM signature model that takes the form of equation (15.1).

15.4.2   Static Range versus Field Data Measurements

The methodology for developing the MBM is dependent on the amount and type of target information available, which in turn depends on the target measurement environment. As described in figure 15.2, the model development begins with an initial estimate of the scattering center locations. The initial estimates of the 3D scattering center locations can be determined in either of two measurement environments: (1) static range measurements and (2) field measurements.

1. **Static range data:** This is the case for generating an MBM from static range measurements on a known target having well-defined scattering components. For static range data, the time-to-angle mapping of the received radar data is carefully controlled and known. Furthermore, most measurement ranges are capable of covering very wide frequency bands, and the data collected has very high signal-to-noise ratio. Consequently, the resultant MBM is of very high fidelity. Using the known motion solution (the pylon rotation), the extracted observable sequence can be associated with a corresponding observation angle history. If the target is simple, initial estimates of the 3D target scattering component locations $\underline{r}_n = (x_n, y_n, z_n)$ can be obtained from mechanical drawings of

the target. Note that, as discussed previously, most standard static range measurement techniques collect target measurements using fixed roll cuts which provide a limited angle observation space (e.g., figure 5.2) and are not amenable to estimating 3D scattering center locations. As an alternative, static range instrumentation techniques described in section 15.6 can potentially be used to obtain data suitable for estimating the 3D locations.

2. **Field data:** For field data, the radar signature data are collected as a function of time, and the target-motion solution, topology, and scattering center locations must be estimated from the measurement data. Some characteristics of the target may be known, and the motion may be estimated independently. Typically, the sensor used in the data collection is bandwidth limited, and the signal-to-noise ratio is considerably lower than with static range data. The degradation of the MBM fidelity as a result of these limitations is discussed in section 15.5. An initial estimate of the motion solution can be obtained using the techniques introduced in chapters 8–10. Using this motion solution, estimates of the 3D target scattering center locations, component diffraction coefficients and refinements to the initial motion are obtained using the estimation techniques introduced in the next section.

15.4.3   Signature Model Development

For either static range or field data, errors in the estimates of the 3D scattering center locations used in the MBM model will affect the fidelity of the model. This section describes an optimization metric that uses the initial motion and 3D location estimates obtained from the joint target-motion solution, and iterates on these initial estimates to provide the best target signature model match to the measurement data. Consider first the iteration on the scattering center location estimates using the initial motion solution. Corrections to the initial motion solution can be obtained repeating the procedure described, but iterating on the parameters characterizing the motion solution.

The primary goal for the development of the MBM signature model is to optimize the match between the signature predicted using the MBM model and the measurement data. To accomplish this optimization, construct a target model observable sequence, denoted as $\hat{R}_n(t_b)$, $\dot{\hat{R}}_n(t_b)$, and $\hat{D}_n(t_b)$ defined at times $t = t_b$, $b = 1, \ldots, B$ as developed in this section. An indication of target model fidelity is available by comparing the HRSE model observables extracted directly from the measurement data to the target

model observables obtained using the target model scattering center location estimates. Using the target model scattering center location estimates derived from the
joint target-model solution, the range and range-rate target model observable sequences
are determined using the following relations:

$$\hat{R}_n(t_b) = \hat{k}(t_b) \cdot \hat{\underline{r}}_n$$
$$\hat{\dot{R}}_n(t_b) = \hat{\dot{k}}(t_b) \cdot \hat{\underline{r}}_n,$$

$$(15.3)$$

where the estimates of the scattering center locations are denoted as $\hat{\underline{r}}_n$, $n = 1, \ldots, N$.
Because the motion is assumed known, using $\hat{k}(t)$ and $\hat{\dot{k}}(t)$ in equations (15.3) determines the range, range-rate target model observable sequences. Recall that equations
(15.3) have also been used extensively in previous chapters of the text to develop truth
values for comparison to the HRSE observables extracted from simulation data using
the known values of the simulation motion and true locations $\underline{r}_n$, $n = 1, \ldots, N$ [e.g.,
equations (3.36) and (3.37) and figures 3.5 and 3.8].

The target model amplitude sequence is determined by replacing the HRSE poles
extracted from each measurement data block with a corresponding set of target model
poles estimated using the target model range, range-rate sequences. Denote the target
model pole sequences as $\hat{s}_n(t_b)$, $\hat{p}_n(t_b)$, $b = 1, \ldots, B$. Then, by analogy to equations
3.26 and 3.27, $\hat{s}_n(t_b)$, $\hat{p}_n(t_b)$ are determined by the expressions

$$\hat{s}_n(t_b) = e^{-j\frac{4\pi\Delta f}{c}\hat{R}_n(t_b)};$$
$$\hat{p}_n(t_b) = e^{-j\frac{4\pi\Delta t}{\lambda}\hat{\dot{R}}_n(t_b)};$$

$$(15.4)$$

Consider a block of measurement data $E_d(k, q)_{t=t_b}$ centered at block time $t = t_b$. As
described in chapter 2, the HRSE extracted amplitudes, $D_n(t_b)$, are determined applying the linear signal model defined by equation (2.13) to $E_d(k, q)_{t=t_b}$, imposing a least-
squares fit of the 2D all-pole model to $E_d(k, q)_{t=t_b}$. In a similar manner, the target
model amplitudes $\hat{D}_n(t_b)$ are determined using the same functional form of equation
(2.13), but replacing the HRSE signal model poles in equation (2.13) with the target
model poles defined in equation (15.4). Denote the target model prediction of the data
over the data block as $\hat{E}(k, q)_{t=t_b}$, where

$$\hat{E}(k, q)_{t=t_b} = \sum_n \hat{D}_n(t_b)\hat{s}_n^k \hat{p}_n^q.$$

$$(15.5)$$

The target model amplitudes $\hat{D}_n(t_b), n = 1,\ldots N$ in equation (15.5) are determined using the best target model prediction to measurement data match determined by the optimization

$$\underset{(D_n)}{Min}\,|\Delta(k, q)|^2,\, k = 1,\ldots n_f,\quad q = 1,\ldots,Q_0,$$

where $\Delta(k, q) \equiv E_d(k, q) - \hat{E}(k, q)$, and Q_0 and n_f denote the number of pulses and frequencies, respectively, in the data block. Comparing the target model amplitudes $\hat{D}_n(t_b)$ to the HRSE signal model extracted amplitudes, $D_n(t_b)$, both extracted from the same data block but using different sets of pole pairs, provides a robust check on the accuracy of the location estimates as well as the estimated motion.

It is important to contrast the information contained in the HRSE signal model extracted observables with the target model extracted observables. For the HRSE extracted observables, the data block $E_d(k, q)_{t=t_b}$ is directly input to the two-dimensional HRSE spectral estimation processor, and the $(R_n, \dot{R}_n, D_n), n = 1,\ldots N$ observables are the direct output, independent of the signature model. The extracted amplitudes, although indicative of each component scattering behavior, are not necessarily correlated to a particular scattering center, and provide no information as to the scattering center locations. The location estimates are obtained using the sequence of range, range-rate observables as input to the specific joint target-motion estimation solution technique employed.

The target model observables are based on signature model predictions using the target scattering center locations and associated diffraction coefficients incorporated into the MBM model, and are thus indicative of the target model fidelity. The target model amplitudes are directly correlated to each component on the target. Comparing the target model observable sequence $(\hat{R}_{n,b}, \hat{\dot{R}}_{n,b}, \hat{D}_{n,b})_{b=1,\ldots B}$ to the HRSE signal model observable sequence, $(R_{n,b}, \dot{R}_{n,b}, D_{n,b})_{b=1,\ldots B}$, properly associated over a time span of B_0 data blocks, provides a high fidelity check on the accuracy of the 3D scattering center location estimates used in the measurements-based target model, as well as the motion solution. The accuracy of the component locations used in the MBM model can be quantified by introducing the optimization metric C_1, which minimizes the distance between the HRSE signal model extracted complex amplitude sequence and target model predicted complex amplitude sequence, properly associated over a time span of B_0 data blocks, as defined in equation (15.6):

$$C_1 = \underset{(x, y, z)}{Min}\left\{\sum_n \sum_b \left(|D_{n,b} - \hat{D}_{n,b}|^2\right)\right\},\quad b = 1,\ldots,B_0. \tag{15.6}$$

As discussed previously, the metric in equation (15.6) can also be applied to the motion estimate, fixing the scattering center locations, and using the parameter-based characterization of the motion as optimization variables.

A sub-optimal application of the metric in equation (15.6) can be applied "visually" by overlaying the amplitudes of the sequences $|\hat{D}_n(t_b)|, n = 1, \dots N$ and $|D_n(t_b)|$, $n = 1, \dots N$, without correlation, on a common graphical scale versus block time. The visual association effectively acts as the correlation agent. Including the phase of the amplidudes in equation (15.6) provides a more precise metric. This is because the optimization metric is an implicit function of all three range, range-rate and amplitude observables, as well as the motion estimate, as indicated by equations (15.4) and (15.5). It has been found that using the sub-optimal metric based on the amplitudes provides a sensitive measure of the accuracy of the motion estimate, as well as the scattering center locations applied to a wideband MBM model. The optimization defined in equation (15.6) is highly nonlinear, and is typically used to obtain small corrections to the initial scattering center locations and/or the motion estimate. It has proved extremely valuable in improving the target model signature fidelity.

Once the sequence of target model amplitudes, $\hat{D}_n(t_b)$, are determined for each scattering component, the phase correction defined by equation (3.25) is applied.

$$\hat{D}_n(t_b)_{GTD} = \hat{D}_n(t_b)e^{j4\pi\hat{R}_n(t_b)/\lambda},$$

Using the motion to angle space mapping, the angular dependence of the model-based GTD diffraction coefficients are obtained in the form $(\hat{D}_n(\theta_b, \varphi_b))_{GTD}$. The diffraction coefficient basis set appropriate for the MBM model is then obtained using a parametric fit of the chosen basis set, $D_n(\theta, \varphi, c_1, c_2, \dots, c_m)$, to $(\hat{D}_n(\theta_b, \varphi_b))_{GTD}$. The extracted model GTD diffraction coefficients and model component location estimates are jointly used in equation (15.1) to form the GTD contribution to the MBM signature model.

The inclusion of a PTD contribution to the MBM model follows a similar procedure. The main difference is that the total field from both PTD and GTD is used in equation (15.5) to optimize the MBM model to data fit. In this case, equation (15.5) is generalized to the form

$$E(k, q)_{t = t_b} = E_{PTD}(k, q)_{t = t_b} + \sum_n \hat{D}_n(t_b)\hat{s}_n^k \hat{p}_n^q. \tag{15.7}$$

However, when the PTD and GTD contributions are combined, it is important to register the separate pieces to a common target coordinate reference system. This can be accomplished in numerous ways. One would be to overlay the target topology defining the PTD contribution onto the composite map of the estimated component scattering center locations in the target-based reference system, and adjust the position of the topology to fit these locations. As a final check on the quality of the target model, for static range data having high signal-to-noise ratio, a reconstruction of the temporal evolution of the narrowband signature can be generated using the MBM model and compared to the measurement data, similar to that illustrated in figure 15.8. The target registration and topology can then be adjusted to determine whether the data fit could be improved. Because the PTD model includes the specular response(s), this also provides a robust check on motion solution because the specular location can typically be located precisely in the time history of the signature. The resultant MBM can then be used for extrapolation and prediction outside the measurement observation space.

It is important to point out that the signature model characterized by the right side of equation (15.7) is defined relative to a target-based coordinate system defined by the location of the CoR at a particular origin on the target associated with the measurement system. For static range measurements, this location corresponds to the positioning of the target on the measurement pylon; for field data, the CoR is located at the position of the center of mass of the target, characterized by the mass distribution of the target. As discussed in chapter 2 and appendix F, the location of the measurement coordinate system CoR is typically different from the location of the CoR used in the computational signature model, which has a location chosen for the convenience of the analyst. Integration of the measured diffraction coefficients into the computational signature model is discussed in detail in appendix F.

As a step beyond using the MBM signature model developed and characterized using the optimization metric defined in equation (15.6), the analyst could construct a very high-fidelity signature model using one of the more accurate electromagnetic modeling techniques described in figure 15.1. One can then validate the model relative to how well it predicts the measurement data by using the higher fidelity model as a standard, and generate frequency and angle simulation data over the angle observation space characterized by the motion solution. Extracting an observable sequence from the high-fidelity simulation data and applying equation (15.6) to compare the measurement observable sequences provides an additional measure of model fidelity.

15.4.3.1   Example: Cylindrical topology having a ridge component

At this point it is instructive to consider an example to illustrate the application of the MBM methodology to develop a hybrid PTD/GTD model. Consider a long cylinder of length $L = 2\,\text{m}$ and radius $a = 0.03\,\text{m}$ to which an azimuthally symmetric ridge has been added at the center as indicated in figure 15.3.

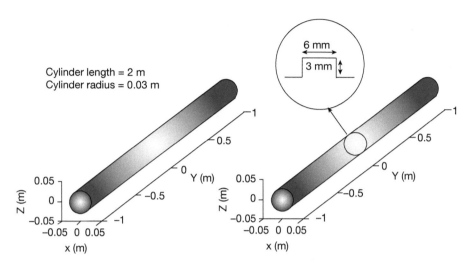

Figure 15.3   Isolated cylinder (PTD) and cylinder with protruding ridge (GTD). Cylinder has length $L = 2\,\mathrm{m}$ and radius $a = 0.03\,\mathrm{m}$ to which an azimuthally symmetric ridge has been added at the center.

Choose an operational frequency at X-band (10 GHz) having a 1 GHz bandwidth, corresponding to a frequency regime well within the limits of the high frequency approximation models for this target. Define the following experiment: to represent the measurement data, use a highly accurate method of moments (MOM) prediction code to compute the monostatic scattered field for the target, including the cylinder and ridge. For the MBM model choose the PTD contribution to be the scattered field of the cylinder without the ridge; and use the GTD contribution derived from the measurement data to model the ridge component. The target geometry and the associated PTD and GTD decomposition are indicated in figure 15.4(a).

Figure 15.4(b) compares the PTD scattered field contribution (at band center) versus angle for the cylinder compared to the total (cylinder plus ridge) scattered field. Most notable is the difference in the scattering at $\varphi = 0$ and $\varphi = \pi$ between the PTD and MOM calculations. In these regions the MOM prediction includes the interaction of the ridge and the cylinder end, whereas this interaction is not present in the PTD cylinder model.

Now determine the GTD contribution to the measurements-based model. The target is characterized by three scattering center locations located at $x = (-1, 0, +1)$ on the x-axis. The ridge component is located at $x = 0$. Figure 15.5 compares the target model amplitude and phase sequence and HRSE extracted measurement amplitude and phase

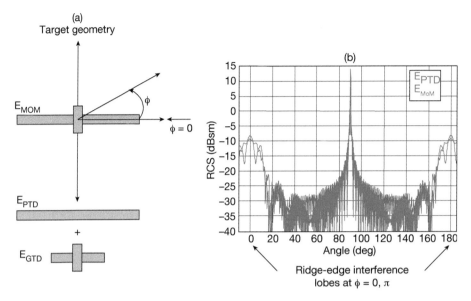

Figure 15.4   (a) Decomposition of the PTD and GTD model components. (b) Scattered field RCS for the E_{ptd} contribution (red) versus the total scattered field E_{mom} computed using an exact MOM code (blue).

sequence for the ridge component, that is, $\hat{D}_n(t_b)$, $b = 1, \ldots, B$ versus $D_n(t_b)$, $b = 1, \ldots,$ B, extracted from the MOM data over the $(0, 180)$ degree observation region. Because the component locations are assumed exactly known, the target model and HRSE extracted complex amplitude observable sequences for the ridge component are identical for this example. To examine the sensitivity of the metric defined in equation (15.6) to the scattering center location accuracy, assume a location estimate error of $\Delta x = \lambda$ is introduced in the assumed ridge component location and is used to determine the target model poles. Figure 15.6 compares the ridge RCS amplitude and phase sequence extracted from the MOM model prediction data to that extracted using the target model poles for the ridge component in equation (15.4), but containing the positioning error. Note that the extracted amplitudes are still identical, but the extracted phase using the target model poles has changed considerably. This phase error will have a direct impact on the fidelity of a narrowband signature model as discussed in the following section and chapter 16. Also note that using the sub-optimal application of equation (15.6) comparing HRSE and target model amplitudes only indicates that the location error $\Delta x = \lambda$ would have little effect on a wideband model.

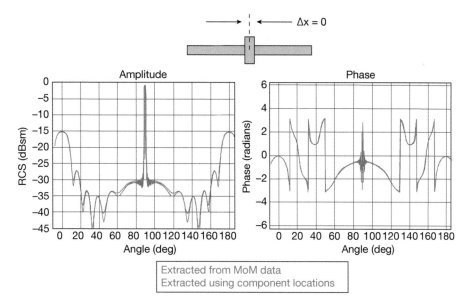

Figure 15.5 The target model amplitude sequence for the ridge component (known location) and HRSE extracted measurement amplitude sequence. Red indicates the HRSE measurement observable sequence extracted from MOM data. Blue indicates target model observable sequence.

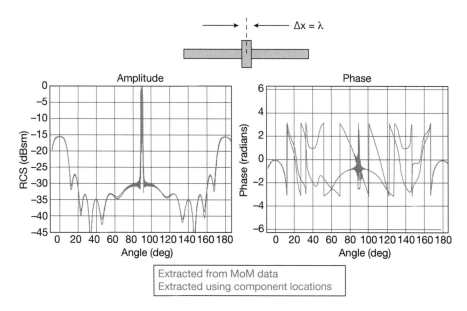

Figure 15.6 Extracted HRSE measurement and target model observable sequences for the ridge component. Red indicates the HRSE sequence extracted from MOM data. Blue indicates target model sequence. Location error $\Delta x = \lambda$.

It becomes apparent that comparing the model and truth sequences using the metric defined by equation (15.6) provides a good check on the accuracy of the component location estimate.

Figure 15.7 illustrates the target model diffraction coefficient amplitudes for all three scattering components, that is, the center ridge and two edges, obtained for $\Delta x = 0$. Note the complementary nature of the two edge component diffraction coefficients, which indicates the strong response of the edge most visible to the radar line of sight.

Figure 15.8 illustrates the comparison of the MOM data and MBM model at band center frequency assuming $\Delta x = 0$. Note the comparison is very good, which demonstrates the utility of combining the PTD code with the GTD model to form the composite MBM model.

15.5   Model Fidelity Requirements

One driving question when developing a measurements-based signature model from data collected on a target is what degree of fidelity can be achieved from the signature

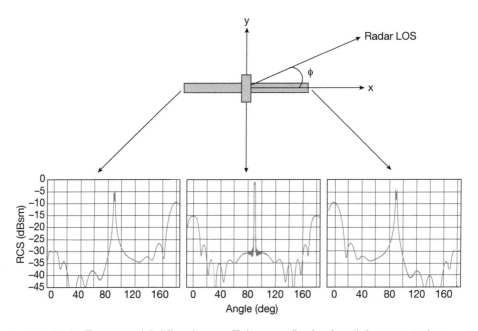

Figure 15.7   Target model diffraction coefficient amplitudes for all three scattering components: the center ridge and two edges extracted for $\Delta x = 0$.

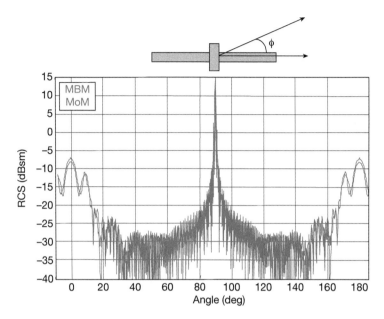

Figure 15.8   Comparison of the MOM data and MBM model at band center frequency assuming Δ*x* = 0. MOM data (red) versus MBM model (blue)

model? Some general statements regarding those factors that most affect the fidelity of the model can be made . These factors are dependent on the application of the model to either a wideband or narrowband application.

Consider first the fidelity requirements for a narrowband signature model. For the narrowband model, the signature is formed from the coherent sum of the scattering from each individual scattering center. Thus, the signature behavior as a function of angle and frequency is primarily dependent on the location accuracy of each scattering component, and to a lesser extent, the amplitude and phase of each scattering center component diffraction coefficient. To characterize the tolerable location uncertainty for a high-fidelity narrowband model, consider the phase change resulting from a location error, ΔR, for any particular scattering center. The phase error for the scattering center, defined along the range line of sight to the target, is given by

$$\Delta(phase) = (4\pi/\lambda)\Delta R.$$

This phase error, applied randomly to each scattering center, results in a change in narrowband signature as determined by equation (15.1). For $\Delta(phase)$ to present a negligible contribution of the scattering center contribution to the narrowband signature, constrain the phase error to be $\Delta(phase) < \pi/2$. This results in a bound on the tolerable scattering center location error

$$\Delta R < \lambda/8.$$

This presents a severe location error requirement, particularly at the higher frequencies. For example for an X-band sensor, $f_0 = 10$ GHz, ($\lambda = 0.03$ m), the range location requirement becomes $\Delta R < 0.004$ m. Such a tight bound often exceeds the dimensions of some components present on the target. Thus, for the narrowband target model, it is often sufficient to settle for modeling the envelope of the narrowband scattering response versus angle, or to average the narrowband prediction over a number of adjacent frequencies.

Now consider the fidelity requirements of a wideband signature model. Typically, the wideband signature model is incorporated into a dynamic field exercise that models the waveforms used by specific sensors in the simulation. Often, the objective is to model the compressed pulse at the output of the matched filter. For this application, the accuracy of the scattering location estimates used in equation (15.1) need only match the resolution achievable by the waveform used in the simulation, and are more relaxed relative to the narrowband modeling requirement. Of most interest is the accuracy of the wideband RCS model at the peaks of the compressed pulse as a function of angular changes in the look angle to the target, and this measure is primarily dependent on the amplitude of the scattering center diffraction coefficient. This accuracy is governed by imposing the fit of the wideband MBM model to the data after pulse compression. Depending on the signal signal-to-noise ratio, the application of the resultant wideband MBM to a narrowband application would be appropriate for modeling the envelope of the narrowband scattering response versus angle.

The dependence of model fidelity on the measurement signal-to-noise ratio cannot be overemphasized. This can be illustrated by examining the signal-to-noise ratio per frequency sample in the received uncompressed wideband waveform, which represents the measured target frequency response. Consider the single pulse processing example illustrated in figure 1.8 in section 1.3.2. Each target in the compressed waveform at the output of the matched filter has nominally a 25 dB signal-to-noise ratio, and the matched filter compression gain relative to the 6,000 frequency samples is 38 dB. The signal-to-noise ratio per sample of the uncompressed waveform, that is, each narrowband frequency component of the target signature response represented by $E'(\omega_k, \phi)$ in figure 1.8 has a signal-to-noise ratio of -13 dB, much too low to develop a high-fidelity narrowband MBM model over a limited viewing angle space. One can increase the signal-to-noise ratio per sample by processing the uncompressed 512 samples after applying the time window filter, characterized by the signature response $E(\omega_k, \phi)$, but this still places the signal-to-noise ratio per frequency sample at −2 dB, which is an inprovement but still much too low to develop a high fidelity narrowband signature model using conventional techniques. However, referring to the linear signal model representation of the target frequency response for this same example illustrated in figure 1.9, the frequency response characterized by the linear signal model derived from the noisy data set compared to the actual target response is replicated quite well. Because the signal model is defined in the uncompressed frequency domain, the integration gain provided using the HRSE technique applied to the noisy data block provides a narrowband signature estimate incorporating the 6,000 pulse integration gain. Thus the HRSE framework offers a narrowband signature modeling technique capable of compensating for the low signal-to-noise ratio per frequency sample in the uncompressed data. As discussed subsequently, the limit on signal-to-noise ratio can be further overcome by processing the data over very wide-angle regions, for which higher resolution estimates of the target scattering center locations can be obtained, resulting in a more accurate MBM model for use in estimating the narrowband signature.

The accuracy of the location estimate is governed by the sensor data collection scheme: either static range or field data. For either case, the location accuracy is dominated by three factors, namely measurement sensor bandwidth, data signal-to-noise ratio, and the angular extent of the measurement observation space. The accuracy of the estimate of the scattering center locations is governed by the fundamental limitations on range and cross-range resolution. The location error is dictated by the range resolution $\Delta\rho_R$, and cross-range resolution $\Delta\rho_{XR}$ developed in chapter 5. Applied to a single 2D frequency-time data block, the location accuracy is a function of wavelength λ, fractional bandwidth FBW, and angular extent $\Delta\phi$ according to

$$\Delta\rho_R \sim \lambda/FBW;\ \Delta\rho_{XR} \sim \lambda/2\Delta\phi. \tag{15.8}$$

The accuracy in scattering center location favors higher bandwidths at the higher frequencies, and collecting data over wide angular sectors. For example, using an X-band sensor (10 GHz) having a bandwidth of 1 GHz and a data block spanning a 5° angular sector, equation (15.7) results in a location uncertainty of $\Delta\rho_R \sim 0.27$ m in range and $\Delta\rho_{XR} \sim 0.18$ m in cross range. Noncoherent integration of a number of location estimates covering a larger angular sector, along with using HRSE processing, can reduce these values significantly.

The fidelity achieved using the two data collection schemes can be summarized as follows:

Static range data collections provide data collection over a 360° rotation period at ultrawide bandwidths at very high signal-to-noise ratios. The combination of coherent and noncoherent processing gain over the 360° pylon rotation angle can lead to very precise scattering center location estimates. However, as discussed in the next section, 3D scattering center location estimates using a rotating pylon cannot generally be extracted using standard static range data collection schemes, and one must complement the 2D measurement estimates with multiple roll cuts and mechanical drawings of the target.

Field data collections are typically limited by the bandwidth of the data collection sensor and are signal-to-noise ratio limited due to the range of the target. However, they do offer a more robust angular observation space, which often permits 3D scattering center component characterization. However, as discussed previously, the resultant scattering center location estimate accuracy is typically bandwidth limited, and often it is difficult to form a high-fidelity narrowband signature model that requires the precise scattering center location accuracy tolerances discussed previously.

15.6   3D Target Characterization Using Static Range Data

Because of the low cost, high signal-to-noise ratio, and extremely wide frequency bands achievable on a static range, the data from static range collections are most desirable for developing an MBM. For these reasons, it is useful to develop techniques for extracting 3D estimates of the target component locations and scattering properties using the static range data collection environment. Of prime interest in this text is integration of the static range measurements into a computational model for efficiently generating the target's signature. This requires proper association of the static range measurement system angles to those of the computational model coordinate system. This association is discussed in some detail in appendix F, and is incorporated into the 3D static range measurement techniques discussed in this section.

Fortunately, the scattering center location estimation techniques developed in chapter 5, 7, and 9 can be adapted to extract 3D location estimates from data obtained with standard sensor instrumentation typically found on the static range using the standard rotational pylon configuration. The techniques developed in the previous chapters for extracting three-dimensional target information from field data can be directly applied to static range data by slightly modifying the data collection scheme. The static range data collection schemes for three such cases are illustrated in figure 15.9.

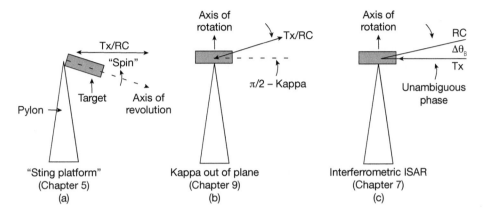

Figure 15.9 Static range data collection schemes providing 3D target scattering center location estimates. (a) Single transmit/receive sensor, spin axis sting platform. (b) Single transmit/receive sensor, radar line of sight elevated out of target rotation plane. (c) Separate transmit/receive sensors using interferometric processing.

Consider each case:

1. **Figure 15.9(a):** In this case the target is mounted on a sting pylon that permits the target to be spun around its axis of symmetry. This static range setup corresponds to the spin only target-motion simulations treated in chapter 5. For the sting-like platform, the viewing aspect angle is fixed at an angle to the target reference z-axis, and the target revolves 360° about the azimuthal axis defining the roll angle. The sting platform allows the target z-axis to be fixed in aspect angle relative to the incident wavefront. Observe that the measurement angles are described in a spherical coordinate system allowing for direct association to that typically used for the computational model. This configuration is suitable for 3D target characterization but is difficult to implement for large or heavy targets, and provides only forward views of the target. Static range measurements using this technique, not discussed in this text, have demonstrated good results for 3D scattering center characterization.

2. **Figure 15.9(b):** For the typical static range roll cut measurement, the transmit/receiver instrumentation is set up in the plane orthogonal to the rotation axis of the pylon. This measurement setup corresponds to simulating a tumbling target in a field measurement scenario having a value of $\kappa = 90°$. The measurement is implemented by fixing the target atop the pylon and rotating the pylon 360°.

As positioned on the pylon, the target roll angle, $\varphi = \varphi_0$, defined relative to the target z-axis and referenced to a target model coordinate system having the CoR located on the z-axis, is fixed, and the pylon rotational measurement angle φ' (as defined in the measurements coordinate system), characterizes the pylon rotation angle, where $0 \leq \varphi' < 2\pi$. As discussed in chapter 9, this setup is only capable of 2D scattering center characterization. The location estimate of each scattering center is projected to the plane of rotation. In an attempt to extract 3D target information, multiple roll cuts $[\varphi_0, \varphi_1, \ldots]$ are used. However, when implementing multiple roll cuts, the target must be physically rolled in discrete roll angle increments, and a full set of measurements repeated over a full rotation of the pylon for each roll angle. In physically displacing the target to its new roll angle, the location of the target CoR on the pylon is (typically) slightly different for each roll cut. This adds a rotationally dependent phase uncertainty to the data when attempting to coherently combine one roll cut data set to another. Furthermore, complete roll characterization of the target requires Nyquist angle samples in roll, which requires very close roll cut spacings, resulting in a prohibitive number of cuts to cover a broad angular observation space. Consequently, coherent processing of static range data covering multiple roll cuts is typically limited to a single roll angle cut.

The roll angle limitation can be overcome by displacing the transmit/receiver out of the plane of target rotation corresponding to a value $\kappa < 90°$. This corresponds to the pure-tumble observer scenario treated in chapter 9 where the look angle to the target is raised out of the tumble plane, simulating a look angle $\kappa < 90°$ to the target angular momentum. When the look angle is raised out of the tumble plane, and a single transmit/receive sensor location is used, the target space mapping equations introduced in section 9.3 can be used to extract the 3D locations of the scattering centers. This requires including the acceleration observable $\ddot{R}_n$ in the sequence $(R_{n,b}, \dot{R}_{n,b}, \ddot{R}_n)_{b=1,\ldots,B}$ in applying the target space mapping equations. For a given roll angle orientation of the target, $\varphi = \varphi_0$, the diffraction coefficient associated with each scattering center is extracted in the form $D_n(\theta'_b = \kappa, \varphi'_b, \theta_b, \varphi_b)_{b=1,\ldots,B}$, where $0 \leq \varphi'_b < 2\pi$. Association of the static range measured diffraction coefficient to the coordinate system used in the computational model requires associating the static range roll cut measurement angles θ'_b, φ'_b to the computational model angles θ_b, φ_b. This association is described in appendix F. As discussed there, for $\kappa < 90°$, both θ_b and φ_b change as a function of pylon rotation angle as governed by equations (2.23) and (2.24), which extends the angle observation space for use in the computational model to differing views of the target. An example of the observation space covered by θ_b and φ_b as the pylon rotates for $\kappa = 70°$ is illustrated in figure F.4.

3. **Figure 15.9(c):** Applying the instrumentation setup for the interferometric case, the transmit sensor remains in the plane orthogonal to the pylon's rotation axis, and a bistatic receiver is elevated out of that plane. The bistatic angle is constrained small enough to preserve the bistatic phase, as discussed in chapter 7 on interferometric processing. Data are collected for rotation angles $0 \leq \varphi' < 2\pi$ and target reference roll angle $\varphi = \varphi_0$. The extraction of the scattering center diffraction coefficients and scattering center locations proceeds in the manner described in chapter 7.

For some static range configurations, it may be difficult to introduce a second sensor located out of the target plane of rotation. This would occur in particular for a compact static range measurement environment where the far-field plane-wave incident on the target is implemented by placing the measurement pylon in the near field of a large parabolic reflector having a transmitter located at its focal point. However, for this case the monostatic look-ahead technique introduced in section 7.3 can be implemented to extract the 3D scattering center locations. For conventional IF-ISAR processing as described in section 7.2, one requires two closely spaced views of the target with the constraint that these views must provide orthogonal (independent) information. As described in section 7.3, for some specialized motion types using a single sensor, one can look ahead in time and search for a look angle to satisfy the 3D orthogonality condition described there. For implementation on the static range, the look-ahead condition can be replicated by a simple rotation of the target in roll corresponding to the bistatic angle, and repeating the 360° rotational measurement. The data are processed as described in the following examples. Because the target must manually be repositioned in roll angle, subject to repositioning errors, this results in a data collection that must be phase compensated to assure both 360° rotational measurements are aligned to the same CoR located on the target. Static range measurements made using this technique, not discussed in this text, have demonstrated good results for 3D scattering center characterization.

In order to illustrate the similarity of the two IF-ISAR techniques discussed in case 3, consider simulating monostatic and bistatic static range measurements using the simplistic fixed-point scattering center target described in figure 9.3. This target was used in chapter 9 to illustrate the technique for estimating 3D scattering center locations using the observable sequence $(R_{n,b}, \dot{R}_{n,b}, \ddot{R}_n)_{b=1,\ldots B}$ for a single monostatic sensor placement elevated from the target tumble plane. For this example, modify the target so that the scattering centers displaced above the (x, y) plane are at a height of 0.25 m versus the 0.125 m assumed in chapter 9. Assume the target is placed on the static range pylon such that the monostatic sensor lies in the (x, y) plane of the target, and assume the same radar operating parameters listed in table 9.1. Consider two bistatic scenarios for estimating the 3D scattering center locations using this simulated data measurement scenario:

1. A separate bistatic receiver, positioned at an angle 0.5° elevated from the target plane of rotation, is used to measure the bistatic field reflected from the target as the pylon rotates.

2. Monostatic data versus rotation angle is collected from the target for each of two target positionings on the pylon: the reference roll angle position, and a repositioning of the target with a roll angle displacement of 0.5°.

For each scenario, the measurement data are recorded versus viewing angle $0 \le \varphi' < 2\pi$. Each scattering location estimate uses dual frequency-time data block pairs, one for the monostatic received signal, and for scenario 1 the dual data block corresponding to the bistatic received signal; and for scenario 2 the dual bistatic data block is from the mono-static measurement at the same viewing angle obtained after repositioning the target. Figures 15.10 and 15.11 illustrate simulation results estimating the 3D scattering center locations for each of the two measurement scenarios, respectively. The results illustrated for each scenario are obtained using a typical dual data block pair selected from a set of viewing angles to the target as the pylon rotates, using the IF-ISAR processing developed in chapter 7. The location estimates would be identical for different data block pairs viewed at a different pylon position assuming no scattering centers are obstructed. Figure 15.10(a) illustrates the range, range-rate image (scaled using a 1 s rotational motion) for the monostatic sensor located in the plane of rotation, identifying the scattering center range, range-rate observables extracted from the data block used in the IF-ISAR processing; figure 15.10(b) illustrates the corresponding image phase difference extracted from the dual range, range-rate images corresponding to the monostatic and bistatic sensors. The locations of the scattering centers extracted from the monostatic range, range-rate image in figure 15.10(a) are indicated in red on the phase image contour. Figure 15.10(c) illustrates the 3D location estimates (red dots) of the scattering center locations compared to the actual target scattering center locations (circles) using the processing techniques described in section 7.3. The agreement is excellent, as evidenced by the coincidence of the red dots inside the circles.

Figure 15.11 represents the corresponding 3D scattering center location estimates for the simulation described earlier by scenario 2. The measurement scenario assumes

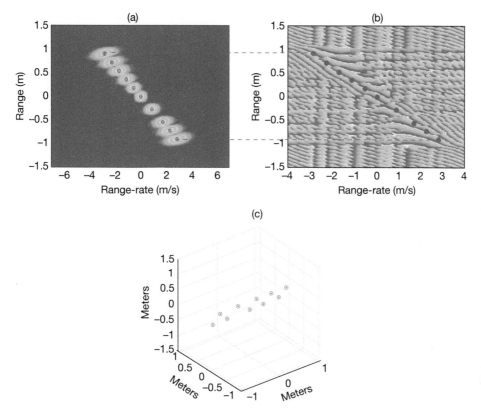

Figure 15.10   Static range data 3D scattering center estimation scheme using a bistatic receiver for the simulation target described in figure 9.3. Bistatic angle of 0.5°. (a) Typical range, range-rate image contour for transmitter in plane of rotation. (b) Image phase difference (image scale ±π) between bistatic and monostatic receivers using the dual data blocks corresponding to (a). (c) The 3D scattering center location estimates for a typical dual data block pair where a red dot indicates estimated, and a circle indicates true.

the same monostatic sensor. The dual range, range-rate images are obtained using the same viewing angle to the pylon before and after rotating the target 0.5° in roll, assuming no displacement error in repositioning the target.

Note that figure 15.11(a) is identical to figure 15.10(a), because the same monostatic sensor is used in both simulations. However, as discussed in section 7.3, the phase values in figure 15.11(b) differ from those in figure 15.10(b). Although the general shape of the image phase contours appears identical, the phase values in figure 15.11(b) are double those for the simulation using a bistatic receiver, as discussed in section 7.3.

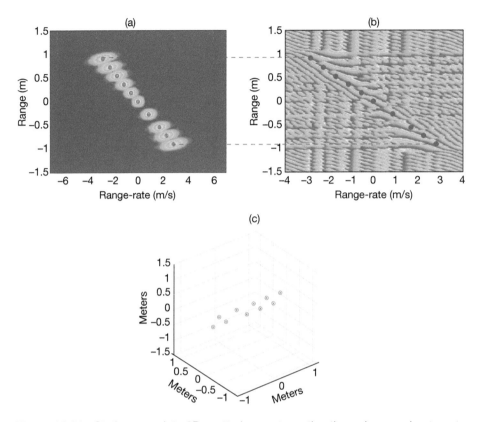

Figure 15.11   Static range data 3D scattering center estimation scheme using target rotation and monostatic receiver. Simulation target described in figure 9.3. Rotation angle of 0.5°. (a) Typical range, range-rate image contour for transmitter in plane of rotation. (b) Image phase difference (image scale $\pm\pi$) between "bistatic" (after target roll angle displacement) and monostatic receivers using dual data blocks paired from each pylon rotation. (c) The 3D scattering center locations estimates for a typical dual data block pair where a red dot indicates estimated, and a circle indicates true.

The agreement of estimated versus true scattering center locations illustrated in figure 15.11(c) for this noise-free example is excellent.

It is instructive to note that for the simulation target used in the previous example, each scattering center present on the target is uniquely identified in the reference coplanar monostatic range, range-rate image. Thus, the phase change for each scattering center is uniquely characterized by the bistatic measurement. However, this will not

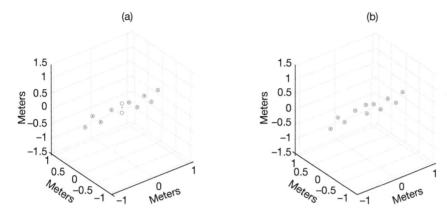

Figure 15.12   Static range data 3D scattering center estimation scheme using target rotation and monostatic receiver. (a) Estimated scattering center locations for the target described in figure 9.3 with additional scattering center added to the target at location (0,0,0.3); (b)) Estimated scattering center locations for the target described in figure 9.3 with additional scattering center added to the target at location (0,0,0.3) after target rotation angle of 45°.

always be the case. For example, consider the addition of a scattering center in simulation case 2, located at position (0,0,0.3) m, placing two scattering centers at the $(x, y) = (0, 0)$ location in the coplanar image plane, one located 0.3 m in height above the other. The phase difference between the monostatic and bistatic image contours for these particular scattering centers for the assumed viewing angle to the target now becomes ambiguous. The results of the processing for this situation are illustrated in figure 15.12(a). Because only a single scattering center is detected in the coplanar monostatic reference image at the (0,0) location, only a 3D single scattering center location estimate is obtained using the IF-ISAR processing, and this location estimate is positioned midway between the two actual scattering centers.

This ambiguous phase occurrence can be rectified by rotating the target in roll angle to a different viewing angle, and repeating the measurement described in scenario 2. For example, figure 15.12(b) illustrates the 3D scattering center location estimates obtained by repositioning the target on the pylon at a roll angle of 45° and repeating each roll angle measurement, before and after 0.5° target rotation. For this target-sensor orientation, each of the 11 scattering centers are resolved in the in-plane monostatic reference image, and the out-of-plane component of the scattering center locations is estimated correctly.

16

Component Modeling Using Measurement Data

16.1 Introduction

A key feature of the measurements-based target signature modeling process is the ability to use measurement data to characterize the scattering characteristics of components on the target. This chapter presents examples illustrating this feature using two sets of examples. One emphasizes the validity of the procedure using simulation data for which the components have hypothesized diffraction coefficients representing the components. The second example set illustrates how difficult-to-model components for which theory is not available can be characterized using parameter-based diffraction coefficient models.

For the first example, a simple target consisting of two point-like scattering centers having a hypothesized component behavior defined by a parameter-based form of the Keller diffraction coefficient defined in equation (14.5) is developed. The measured data is represented using simulation data. The MBM target model diffraction coefficients are extracted from the simulation data using the target model parameters defined in equation (15.5) as a best model fit to the data, assuming the scattering center locations are known, and compared to truth. The fidelity of the resultant MBM signature model is then examined for this simple example by introducing an error in the target scattering center location estimates.

16.2 Diffraction Coefficients Extracted Using HRSE Processing Compared to Simulation Truth Values

16.2.1 A Two-Point Scattering Center Example

Consider the example illustrated in figure 16.1, where the target consists of two scattering centers, located at $x = \pm a$ on the x-axis, having diffraction coefficients, $D_1(\varphi)$ and $D_2(\varphi)$, respectively.

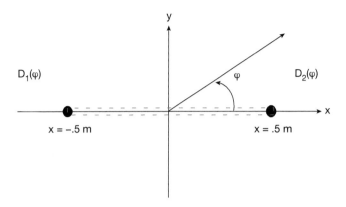

Figure 16.1   Two-point scattering center example having hypothesized diffraction coefficients, $D_1(\varphi)$ and $D_2(\varphi)$.

Choose a parametric-based diffraction coefficient model for each scattering center, characterized by the general form

$$D_1(\varphi) = D_{01} \left\{ 1 + \frac{c_1}{\sin(\varphi - \pi/2 + \varepsilon)\sin(\varphi - 3\pi/2 + \varepsilon)} \right\}$$

$$D_2(\varphi) = D_{02}e^{j\xi_0} \left\{ 1 + \frac{c_2}{\sin(\varphi - \pi/2 + \varepsilon)\sin(\varphi + \pi/2 + \varepsilon)} \right\},$$

(16.1)

where ξ_0 is a random phase. The complex parameter ε is introduced to control the amplitude of the specular response. For $|\varepsilon|$ small, the specular response occurs at angles near $\varphi = \pi/2$ and $\varphi = 3\pi/2$. In this example assume $D_1(\varphi)$ and $D_2(\varphi)$ are functions only of angle and independent of frequency.

The backscattered field from the target is modeled by equation (15.1) with $E_{PTD} = 0$. $D_1(\varphi)$ and $D_2(\varphi)$ in the GTD portion model are specified by equations (16.1). The simulation assumes an operational frequency at X-band (10 GHz) having a bandwidth of 1 GHz. The scattering centers are located assuming $a = 0.5$. The parameters in equations (16.1) are given by $\xi_0 = 0.44\pi$, $D_{01} = 0.01e^{-j\pi/2}$, $D_{02} = 0.005$, $c_1 = 0.5e^{-j\pi/2}$, $c_2 = 2$, and $\varepsilon = 0.1j$. For the simulation, assume a rotational target motion and observation samples such that each received pulse corresponds to viewing angle samples of the target spaced at $0.25°$ increments. Figure 16.2 illustrates the scattered field versus rotation angle at the center of the band.

Figure 16.3 (red) illustrates the magnitude of the angular variation of each hypothesized diffraction coefficient for the parameters chosen. The specular response occurring at $\varphi = 90°$ and $270°$ have amplitudes controlled by the parameter ε. Choosing ε to be a small complex parameter preserves the location of the specular, while adding a continuous phase shift to the diffraction through the specular transition. The parameter values for c_1 have been chosen to be different than those for c_2 to allow an easier comparison of the extracted data to the underlying truth.

Now consider the simulated scattered field as representing actual noiseless measurements of the target, and form a measurements-based model of the signature. First, the scattering center locations are assumed precisely known. The effects of location error are examined in section 16.2.1.1. Given the locations, the target model diffraction coefficients are extracted from the measurement (simulation) data as described in section 15.4.3. A sliding data block 12×32 in size, consisting of 12 frequencies covering the 1 GHz bandwidth and 32 pulses is used to extract the target model range, range-rate, and complex amplitude observables from sequential data blocks that cover the complete 360° rotation. Each block increment is 10 pulses and corresponds to a 2.5° increment in angle.

Figure 16.3 illustrates the magnitude of the target model diffraction coefficients (blue) extracted from the simulation data using known scattering center locations,

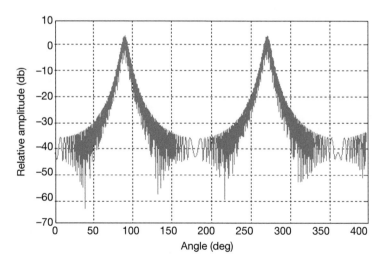

Figure 16.2   Backscattered field (relative amplitude) as function of rotation angle at band center for the two-point scattering center example illustrated in figure 16.1

compared to truth (red). As described in section 15.4.3, the target model amplitudes $\hat{D}_n(t_b)$ extracted directly from the data block at time $t = t_b$ require phase correction to determine the GTD diffraction coefficients used in the measurements-based GTD model. This relationship is given by

$$\hat{D}_n(t_b)_{GTD} = \hat{D}_n(t_b)e^{+j4\pi\hat{R}_n(t_b)/\lambda} \tag{16.2}$$

where $\hat{R}_n(t_b)$ denotes the target model range observable sequence. Figure 16.4 illustrates the corresponding phase for the GTD target model diffraction coefficients (truth vs. estimated) after applying the phase correction in equation (16.2).

Applying the motion to angle mapping, and inserting the estimated GTD diffraction coefficients and known scattering center locations into equation (15.1) determines the MBM model for this simple example.

Consider first the MBM signature prediction model compared to the measurement (simulation) data over a single data block of 32 pulses centered at $\varphi = 134°$ aspect.

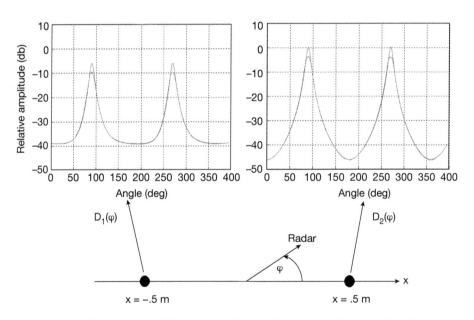

Figure 16.3 Target model diffraction coefficients (blue) extracted from the simulated measurement data using known scattering center locations, compared to truth defined by equation (16.1) (red).

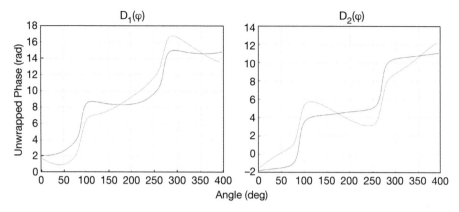

Figure 16.4   Extracted target model diffraction coefficient phase (blue) compared to truth (red) after phase correction applied using equation (16.2).

Figure 16.5 compares the MBM model to simulation data over this data block for both narrowband and wideband cases. Figure 16.5(a) illustrates the difference between the MBM model evaluated at each frequency over the 1 GHz bandwidth and the corresponding simulation data. The model to data difference is plotted versus pulse number for the 32 pulses over the data block. The difference between the MBM model and data versus pulse number is shown normalized to the data block maximum value, and the results superimposed for each frequency. Figure 16.5(b) compares the narrowband MBM model to data versus pulse number at the center band frequency. Note that because the exact locations of the scattering centers were used for extracting the target model diffraction coefficients, the agreement is quite good. Figure 16.5(c) illustrates a wideband comparison of model compressed pulses versus data compressed pulses for the pulse at the center of the data block. The wideband model versus data compressed pulses are essentially identical.

16.2.1.1   Narrowband fidelity considerations

As discussed in chapter 15, the fidelity of the narrowband signature model depends strongly on being able to accurately estimate the location of each component scattering center. In the MBM model to data comparison examples discussed in figure 16.5 the exact scattering center locations were used in developing the MBM model. Now examine the effect of scattering center location error on the fidelity of the model. Use

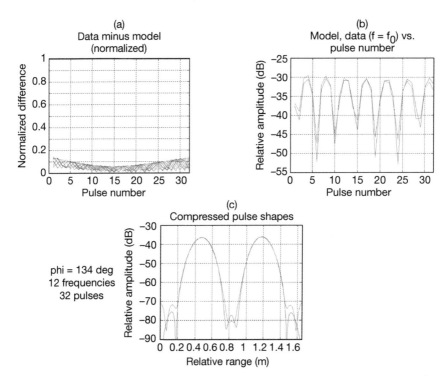

Figure 16.5   Data versus MBM model comparison for a data block centered at 134° look angle to the target. (a) Data minus MBM model difference as function of pulse number superimposed for each frequency over the 1 GHz bandwidth. (b) MBM model compared to data versus pulse number at band center frequency. (c) Wideband MBM model and data compressed pulses for the pulse at the center of the data block.

the same 32 pulse data block at aspect 134° treated in figure 16.5, but now examine the narrowband measurements-based signature model assuming an error, Δx, in the scattering center location used to form the target model poles defined in equation (15.4). Typically, if the target scattering centers are estimated in the presence of noise, or the scattering centers are very closely spaced, the location of each scattering center can be measured to a wideband resolution uncertainty in range on the order of $res \sim \lambda/2FBW$, where FBW denotes the fractional bandwidth. Using this resolution as a baseline location error, figure 16.6 illustrates the normalized error between the narrowband MBM model and data versus pulse number superimposed for each frequency over the data block for three cases: $\Delta x = res/8$, $\Delta x = res/4$ and $\Delta x = res/2$, compared to $\Delta x = 0$. The

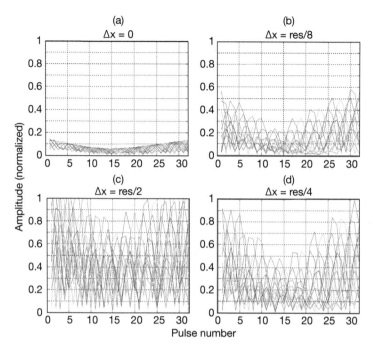

Figure 16.6   Normalized difference between narrowband measurements-based model and data for differing values of Δx. Δx is defined relative to *res* = λ/(2 * *FBW*). Data block centered at 134° look angle to the target.

baseline result in figure 16.6(a) corresponds to figure 16.5(a). The deterioration of the narrowband model illustrated in figures 16.6(b)–(d) is clearly evident as the location error increases.

Figure 16.7 illustrates the comparison of the MBM model (assuming the known scattering center locations) to the simulated measurement data at band center as a function of the rotation angle plotted over a complete rotation cycle.

16.3   Diffraction Coefficients for Difficult-to-Model Components

16.3.1   Dielectric Components

Consider now a more general case for components for which the theoretical GTD diffraction coefficients are difficult to model. For the example considered here, it is of

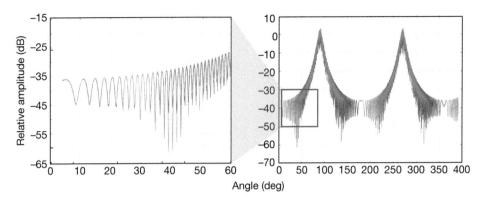

Figure 16.7   Narrowband MBM model (red) versus simulated measurement data(blue) versus rotation angle at band center, for $\Delta x = 0$.

interest to examine the behavior of a dielectric coated wedge having the same geometry as the perfectly conducting wedge illustrated in figure 14.1. Because the theoretical diffraction coefficient for the perfectly conducting wedge is well known, the parametric model is based on a perfectly conducting wedge and, using the results from [117], can be expressed in the form

$$D(f, \varphi) = \frac{D_0}{\sqrt{f}} \left[2c_1 \cot \frac{\pi}{2n_0} + c_2 \cot \frac{\pi - 2\varphi}{2n_0} + c_3 \cot \frac{\pi + 2\varphi}{2n_0} \right]. \qquad (16.3)$$

For $c_1 = c_2 = c_3 = 1$, it can be shown that equation (16.3) closely approximates the Keller diffraction coefficient defined by equation (14.5). Adjusting the parameters (c_1, c_2, c_3) in equation (16.3) provides a generalized parameter-based function that can be used to characterize the coated wedge. One choice would be to simply choose (c_1, c_2, c_3) to best fit the data over the measurement observation space. Another would be to develop a physics-based model related directly to the parameters of the coated dielectric. This provides a more general form of equation (16.3) that allows the parameters (c_1, c_2, c_3) to vary with frequency and angle. Consider, for example, modeling the diffraction coefficients characterizing the base of a coated cylindrical target exhibiting double diffraction from the base edges, each modeled by a 90° coated wedge. Introduce the plane-wave reflection coefficients at each junction of the rear base edges, $R_1(f, \varphi)$ and $R_2(f, \varphi)$, and modify the parameters (c_1, c_2, c_3) according to [119].

$$c_1 = [1 + R_1^2(f, \varphi)]/2, c_2 = c_3 = R_1(f, \varphi) \quad \text{for single edge illumination}$$
$$c_1 = R_1(f, \varphi)R_2(f, \varphi), c_2 = R_1(f, \varphi), c_3 = R_2(f, \varphi) \quad \text{for both edges illuminated.}$$

$$(16.4)$$

For fixed frequency, the constants for scattering by a perfectly conducting wedge are replaced with the corresponding plane-wave reflection coefficients for the face of the coated wedge, based on the thickness and permittivity of the coating. For fixed angle, the plane-wave reflection coefficients as a function of frequency are used.

By appropriately modifying the (c_1, c_2, c_3) parameters defined in equation (16.4), one can determine parameter-based models that adequately characterize the measured diffraction coefficients for dielectric coated targets under a variety of test conditions, and provide a good approximation to the measured results. Examples illustrating the application of equation (16.4) to static range measurements for the uncoated base edge are illustrated in [118]. Applications of equation (16.4) to charactering the angular variation of the extracted diffraction coefficients for a dielectric coated conducting base edge, not presented in this text, have demonstrated good results. Figure 16.8 illustrates the application of the parameter-based formula defined by equations (16.3) and (16.4) to published data characterizing a 60° impedance wedge as a function of angle [119]. Results are shown for various values of the impedance.

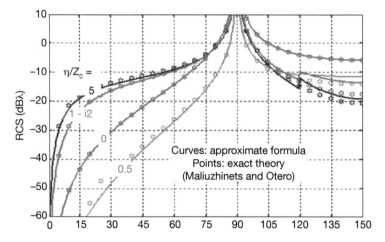

Figure 16.8   RCS versus incidence angle for an impedance wedge having 60° wedge angle.

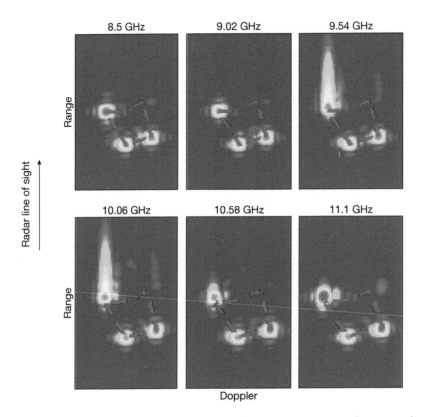

Figure 16.9 Sequence of range-Doppler images versus center frequency for each image data block for the circumferential slot. HH polarization 19° angle of incidence.

16.3.2 The Frequency Resonant Slot on a Cylindrical Structure

Another example of a frequency dependent component, briefly discussed in chapter 6, figure 6.1, is the frequency resonant circumferential slot fabricated on a cylindrical structure illustrated in figure 16.10. It exhibits, along with the responses to the other scattering centers on the structure, large delay responses in range emanating from the slot for excitation using wideband waveforms. Figure 16.9 illustrates a sequence of range-Doppler images for the circumferential slot at 19° angle of incidence relative to the cylinder axis, for horizontal transmit, horizontal receive (HH) polarization. The center frequency of each image data block ranges from 8.5 to 11.1 GHz. The strong response from the slot over the region from 9.54 to 10.06 GHz is evident, and decays away from this narrow 500 MHz bandwidth.

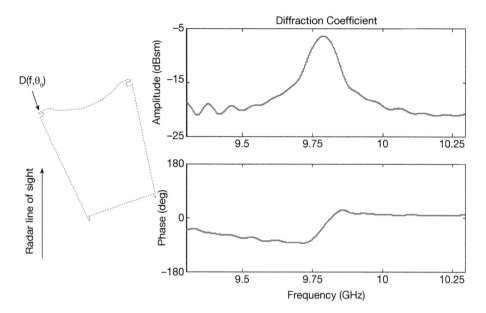

Figure 16.10 Measurements-based diffraction coefficient response, $D(f, \theta_0)$, for the circumferential slot extracted from the measurement data over the band 8.5 to 11.1 GHz.

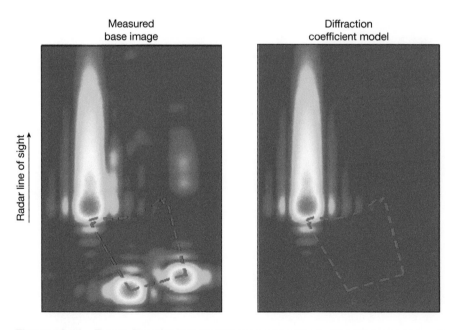

Figure 16.11 Range-Doppler image data reconstruction for the resonant circumferential slot compared to measurement.

Figure 16.10 illustrates the measurements-based characterization of $D(f, \theta_0)$ for the slot extracted using 2D ESPRIT (see appendix C) from the measurement data over the band 8.5 to 11.1 GHz.

Using the measurements-based estimate of $D(f, \theta_0)$, the result can be applied to the circumferential slot target geometry, referenced to the scattering center location of the slot, to reconstruct the measurement data. The data reconstruction compared to measurement is illustrated in figure 16.11. It demonstrates a clean isolation of the diffraction coefficient of the slot from those of the other scattering components.

Additional examples illustrating the diffraction coefficients extracted from static range data are presented in [118].

Acknowledgments

The authors are indebted to many of our colleagues at Lincoln Laboratory (LL) for their collaboration and technical contributions to the measurements-based modeling initiative at the laboratory. Special thanks go to Drs. Bing Chang and Keh-Ping Dunn for suggesting this material be integrated into a text for the LL book series. For a project that evolved slowly over time, numerous LL staff have contributed formative ideas leading to the development of the concepts and processing tools incorporated in this book. Special thanks are due to the major contributors to material in the text: Drs. Jean Piou, Michael Burrows, Kevin Cuomo, Audrey Dumanian and Martin Tobias for helping integrate ideas and techniques developed relating to several diverse research areas into a comprehensive analysis framework. Early years of the project benefited from technical support from Dr. Dieter Wilner, Dr. Mohamed Abouzahra, and Russ Evans. Collecting and resurrecting static range data from a variety of measurement ranges, as well as data formatting and processing, is due to Drs. Mohamed Abouzahra, Audrey Dumanian, Professor Henry Helmkin and Shirin Kubat. Thanks are due to Kathy Nassif, Simon Qiu, Dr. Thierry Copie, and Dr. John Wilkinson for challenging technical discussions, as well as processing support related to data obtained from numerous field tests. Thanks also for the continued interest and comments of Dr. Bing Chang, Dr. James Ward, and Richard Mozzicato while moving the text through publication cycle. The authors appreciate the management and technical support from Drs. Eric Evans, Melissa Choi, Justin Brooke, Marc Bernstein, Hsiao-Hua Burke, Kathy Rink, Keh-Ping Dunn, Dan O'Connor, and Sung-Hyun Son. The authors are indebted to Allison Loftin for her continued perseverance and dedication to the preparation of this text over the past years. The authors would also like to thank the detailed comments and suggestions to improve the text from each of the two reviewers of the manuscript selected by the MIT Press.

This author, in particular, would like to acknowledge the vast amount of mentoring received from numerous colleagues during his career both at the Lincoln Laboratory, industry and academia: Dr. Ronald Fante, one of the brightest analysts at Avco Corp. in the late 1960s time frame; Dr. Donald Thorn at the University of Akron; Dr. Leon

Ricardi, Dr. Alan Simmons, and Bill Cummings in the Satellite Antennas Group at LL in the 1970s time frame; Bill Lemnios at LL for the opportunity to learn radar technology from extensive site experience at the Kwajalein Missile Range in the 1980s and early 1990s; and too many technical discussions to mention with my colleagues at the laboratory during my formative years after returning from Kwajalein in learning the techniques of radar data analysis and radar signal processing: Gerald Augeri, Joe Dituri, Drs. Dieter Wilner, Benny Sheeks, Keh-Ping Dunn, Gary Jones, Robert O'Donnell, as well as the major contributors to the text mentioned previously.

At a personal level, I would like to acknowledge the continuing patience and support from my friend, mentor, and spouse of 37 years, Stella Mayhan. Her constant love and caring are a blessing to all who know her.

Appendix A: Characterization of Torque-Free Euler Rotational Motion

A.1 Geometrical Considerations and Motion Variables

For a target constrained by Newtonian forces referenced to inertial space, the motion of the target can be decomposed into two components: translation of the target's center of mass, and rotation about the target's center of mass (a more extensive discussion of this motion decomposition and its importance in processing field measurement data is presented in appendix F.) The spherical coordinates $\theta(t)$, $\varphi(t)$ introduced in section 2.2 describe the viewing angle from the target to the radar resulting from the rotational component of the target motion, referenced to a target-fixed coordinate system having origin located at the target's center of rotation (CoR). The objective of this appendix is to develop a parameter-based characterization of the rotational motion component constrained by Euler's equations as viewed in the radar's frame of reference. The characterization is valid as long as changes in the target's angular momentum vector as viewed from the radar-fixed frame of reference are small compared to the effects of the rotational motion, and the motion is torque free. For different observation intervals along the trajectory, the torque-free rotational motion referenced to the target angular momentum vector remains unchanged; however, the radar line of sight (LoS) to the target angular momentum vector defined by the angle κ will have changed. In appendix F, the case where κ varies along the trajectory of the target is considered in detail (see figure F.5).

It will be shown that over the limited time span for which the LoS between the radar and the target's angular momentum vector is unchanged, the rotational motion $\theta(t)$, $\varphi(t)$ is fully determined by the functional mapping introduced in chapter 2:

$$F\{\kappa,\ \theta_p,\ \psi,\ \phi_p,\ t\} \rightarrow [\theta(t),\ \varphi(t)], \tag{2.22}$$

where κ is the localized aspect angle from the radar to the target motion angular momentum vector, θ_p is the precession cone angle relative to target motion angular momentum vector, ϕ_p is the precession rotation angle, and ψ is the spin-angle.

The mapping defined by equation (2.22) is valid over the time span for which κ remains constant. For this assumption, the aspect and roll angles $\theta(t)$, $\varphi(t)$ defined by the mapping shown previously are given by equations (2.23) and (2.24) introduced in chapter 2:

$$\cos\theta = \cos\kappa\cos\theta_p + \sin\kappa\sin\theta_p\sin\phi_p \qquad (2.23)$$

$$\varphi = -\psi + \tan^{-1}\left(\frac{\cos\kappa\sin\theta_p - \sin\kappa\sin\phi_p\cos\theta_p}{\sin\kappa\cos\phi_p}\right). \qquad (2.24)$$

Development of equations (2.23) and (2.24) is the subject of this appendix; they represent a general characterization of the rotational motion for torque-free Euler motion and are valid for constant κ. For targets that are nearly rotationally symmetric about the target's z-axis such that the target's x and y moments-of-inertia I_x and I_y are approximately equal, $I_x \approx I_y$, the precession angle θ_p is nearly constant, and the motion can be further simplified by introducing the precession frequency and phase, f_p, α_p, and the spin frequency and phase f_s, α_s. In this case, the Euler angles ϕ_p and ψ become linear in time $\phi_p = 2\pi f_p t + \alpha_p$ and $\psi = 2\pi f_s t + \alpha_s$, and the rotational motion is characterized by the parameters θ_p, f_p, α_p, f_s, α_s, and κ, which is the angle between the radar LoS and the target's angular momentum vector.

For a given applied impulse torque, the target undergoes a well-defined motion characterized by a unique set of Euler parameters. For $I_x \approx I_y$ the parameters θ_p, f_p, f_s are characteristic of the motion and remain constant over the entire trajectory of the center of mass. The parameters α_p, α_s, κ change depending on the change in look angle, κ, to the angular momentum vector $\underline{J}$. Of primary interest in this text is the problem of estimating that set of Euler parameters characterizing the motion of the target. These can then be used to determine $\theta(t)$, $\varphi(t)$ defined relative to the target-fixed coordinate system, that is, the motion solution. For a given look angle κ, the rotational motion is completely defined by the six parameters θ_p, f_p, α_p, f_s, α_s, κ, and consequently, for this special case the problem can be posed as a $6 + 3N + 1$ parameter estimation problem where N denotes the number of scattering centers on the target and "+1" corresponds to a range bias after removing the translational motion.

A.2 Functional Characterization of $\theta(t)$, $\varphi(t)$ Constrained by Euler's Dynamical Equations

Consider a coordinate system with axes (x_0, y_0, z_0) having its orientation fixed in space (the inertial system) and its relationship to a target-centered system, the axes (x, y, z) of which are fixed to the body. The rotational motion of the target relative to the angular momentum vector $\underline{J}$ can be characterized by three angles, θ_p, ϕ_p, ψ, representing the precession and spin properties of the object as defined in section A.1. It can be shown [120, 121] that the relationship between the target-fixed system, (x, y, z), and the target-inertial system (x_0, y_0, z_0) are related by a simple transformation

$$\begin{pmatrix} x_0 \\ y_0 \\ z_0 \end{pmatrix} = \begin{pmatrix} a_{11} & a_{12} & a_{13} \\ a_{21} & a_{22} & a_{23} \\ a_{31} & a_{32} & a_{33} \end{pmatrix} \begin{pmatrix} x \\ y \\ z \end{pmatrix} \equiv A \begin{pmatrix} x \\ y \\ z \end{pmatrix} \qquad (A.1)$$

where A is the unitary matrix defined by

$$\begin{aligned}
a_{11} &= \cos\psi \cos\phi_p - \sin\psi \sin\phi_p \cos\theta_p \\
a_{12} &= -\cos\phi_p \sin\psi - \sin\phi_p \cos\psi \cos\theta_p \\
a_{13} &= \sin\theta_p \sin\phi_p \\
a_{21} &= \cos\psi \sin\phi_p + \cos\theta_p \cos\phi_p \sin\psi \\
a_{22} &= -\sin\phi_p \sin\psi + \cos\phi_p \cos\theta_p \cos\psi \\
a_{23} &= -\cos\phi_p \sin\theta_p \\
a_{31} &= \sin\theta_p \sin\psi \\
a_{32} &= \sin\theta_p \cos\psi \\
a_{33} &= \cos\theta_p
\end{aligned} \qquad (A.2)$$

Consider now the orientation of the target as viewed in (x_0, y_0, z_0) inertial space. Because the angular momentum vector, $\underline{J}$, is fixed in this inertial space, it is conventional to align (x_0, y_0, z_0) so that $\underline{J}$ is aligned with z_0. Consider a radar observing the target having its coordinate system registered with the target's inertial coordinate system. Assume that over the observation interval for which κ is constant, the (x_0, y_0, z_0) system is aligned such that the radar lies in the $y_0 = 0$ plane. Then, as viewed in (x_0, y_0, z_0) inertial space the radar line of sight to the angular momentum vector $\underline{J}$ (z_0-axis) is defined by the unit vector $\underline{\hat{r}}_0$.

$$\underline{\hat{r}}_0 = \sin\kappa \hat{x}_0 + \cos\kappa \hat{z}_0, \qquad (A.3)$$

where κ is the radar viewing angle relative to the z_0, ($\underline{J}$) axis. Referenced to the target-fixed coordinate system, the θ, φ spherical-coordinate look angles to the target expressed in the target-fixed (x, y, z) system are given by the respective projections onto the (x, y, z) axes:

$$\cos\theta = \hat{\underline{r}}_0 \cdot \hat{z} \tag{A.4}$$

$$\tan\varphi = \hat{\underline{r}}_0 \cdot \hat{y} / \hat{\underline{r}}_0 \cdot \hat{x}. \tag{A.5}$$

First consider equation (A.4). Using equation (A.1), and noting that because A is a unitary matrix, $A^{-1} = A^T$, where A^T denotes the transpose of A, $\hat{z}$ can be expressed in terms of $\hat{x}_0$, $\hat{y}_0$, and $\hat{z}_0$ in the form

$$\hat{z} = a_{13}\hat{x}_0 + a_{23}\hat{y}_0 + a_{33}\hat{z}_0, \tag{A.6}$$

so that equation (A.4) yields

$$\cos\theta = a_{13}\sin\kappa + a_{33}\cos\kappa. \tag{A.7}$$

Substituting for a_{13} and a_{33} from equation (A.2) it follows that

$$\cos\theta = \cos\kappa\cos\theta_p + \sin\kappa\sin\theta_p\sin\phi_p. \tag{A.8}$$

Equation (A.8) is solved for $\theta(t)$ using the inverse cosine function. Note that in constructing motion solutions to equation (A.8) using κ and θ_p, there is an inherent ambiguity because equation (A.8) is invariant to interchanging κ and θ_p. This two-fold ambiguity, when coupled with the two-fold ambiguity encountered using the inverse cosine function in solutions to equation (A.8), leads to a four-fold ambiguity.

In a similar manner, using equation (A.5), it can be shown that

$$\tan\varphi = \frac{\cos\kappa\sin\theta_p\cos\psi - \sin\kappa\cos\phi_p\sin\psi - \sin\kappa\sin\phi_p\cos\psi\cos\theta_p}{\cos\kappa\sin\theta_p\sin\psi + \sin\kappa\cos\phi_p\cos\psi - \sin\kappa\sin\phi_p\sin\psi\cos\theta_p} \tag{A.9}$$

Equation (A.9) can be used directly to solve for φ. However, a simple observation yields a much nicer analytic form. Note that equation (A.9) can be written in the form

$$\tan\varphi = \frac{c\cos\psi - d_1\sin\psi - d_2\cos\psi}{c\sin\psi + d_1\cos\psi - d_2\sin\psi}, \tag{A.10}$$

where

$$c = \cos\kappa\sin\theta_p$$
$$d_1 = \sin\kappa\cos\phi_p \tag{A.11}$$
$$d_2 = \sin\kappa\sin\phi_p\cos\theta_p.$$

Equation (A.10) can be expressed as the ratio of two shifted functions.

$$\tan \varphi = - \frac{N(\psi)}{N(\psi + \pi/2)}, \tag{A.12}$$

where $N(\psi)$ is defined by the numerator in equation (A.10). Comparing equations (A.10) and (A.12), $N(\psi)$ must then take the form

$$N(\psi) = A \sin(\psi + B)$$

$$N(\psi + \pi/2) = A \cos(\psi + B), \tag{A.13}$$

from which the solution for φ becomes

$$\varphi = -(\psi + B). \tag{A.14}$$

The phase angle B is readily determined to be

$$B = - \tan^{-1}\left(\frac{c - d_2}{d_1}\right). \tag{A.15}$$

Using equations (A.14) and (A.15)

$$\varphi = - \psi + \tan^{-1}\left(\frac{\cos\kappa \sin\theta_p - \sin\kappa \sin\phi_p \cos\theta_p}{\sin\kappa \cos\phi_p}\right). \tag{A.16}$$

Note that, in general equation (A.16) shows that the spin and precession components are decoupled as they affect the roll angle φ. Furthermore, observe that even if the target has no Euler spin component, the roll angle will change with time. If one considers a second sensor positioned relative to the first, equation (A.3) must be generalized. Assume this sensor is located at spherical coordinate angles (κ_2, β_2) relative to the angular momentum vector J. It can readily be shown that equations (A.8) and (A.16) remain valid with ϕ_p replaced by $\phi_p - \beta_2$, where β_2 is the rotational angle of the radar line of sight relative to J.

It is important to note that the assumptions regarding the alignment of the radar and target coordinate systems leading to equation (A.3) are not generally valid when dealing with field measurements processed over long portions of the target translational motion. In this case it is essential to account for changes in κ and proper registration of the data in the target-based coordinate system.

Appendix B: 2D Spectral Estimation: A State-Space Approach

B.1   Development

The one-dimensional (1D) and two-dimensional (2D) high resolution spectral estimation (HRSE) techniques described in this appendix are adapted from [122] and
[124] and incorporate the notation introduced in this text. Due to the extensive development, only an overview of the technique is presented. The interested reader should
refer to the references for a more detailed development. MATLAB code implementa-
tions of the techniques, along with illustrative examples are presented at the end of
this appendix. The formulation for the 2D technique is presented in sufficient detail
for correlation to the 1D and 2D versions of the MATLAB codes presented, where the
1D case is treated as a subset of the more general 2D case.

The 2D all-pole signal model is characterized by equation (2.13) based on the geometrical theory of diffraction scattering center model discussed in chapter 2. For the
development treated here, it is assumed the signal is corrupted with white noise data
samples, $w(k, q)$, so that equation (2.13) can be expressed in the form

$$E(k, q) = \sum_{n=1}^{N} D_{0n} s_n^k p_n^q + w(k, q) \quad k = 1, \ldots, K; \quad q = 1, \ldots, Q, \tag{B.1}$$

where D_{0n}, apart from a phase shift, represents the diffraction coefficient associated
with the nth scattering center having complex pole pair (s_n, p_n). The objective here is
to develop a technique mapping the $[K \times Q]$ frequency-time data samples contained
in $E(k, q)$ to the N pole pairs (s_n, p_n, D_{0n}), $n = 1, \ldots N$, that is

$$[s_n, p_n, D_{0n}] = F_{HRSE} \{E(k, q)\}, \quad n = 1, \ldots, N \tag{B.2}$$

where $F_{HRSE}\{\ldots\}$ represents the functional mapping. The MATLAB code listed at the
end of the appendix is identical to the one used throughout the text applied to the
simulation examples.

In matrix notation, the data samples defined by E(k, q) in equation (B.2) are given by

$$
E(k,q) =
\begin{bmatrix}
E(1,1) & E(1,2) & E(1,3) & \cdots & E(1,Q) \\
E(2,1) & E(2,2) & E(2,3) & \cdots & E(2,Q) \\
\vdots & \vdots & \vdots & \cdots & \vdots \\
E(K-1,1) & E(K-1,2) & E(K-1,3) & \cdots & (K-1,Q) \\
E(K,1) & E(K,2) & E(K,3) & \cdots & E(K,Q)
\end{bmatrix}. \quad \text{(B.3)}
$$

The qth column of the data matrix denotes the data frequency samples for the received pulse sampled at time $t = t_q$. The kth row of the data matrix denotes the data time samples for the received pulse referenced to the frequency component at $f = f_\kappa$. The two-dimensional HRSE technique coherently processes all the data samples contained in $E(k, q)$. The 1D version of the HRSE technique, either in frequency or time, are obtained by processing either a single row $(k = 1)$ or column $(q = 1)$ of the data matrix.

The 2D HRSE technique described in this appendix is based on introducing two separate multi-input–multi-output (MIMO) systems [122, 123]: the first is derived from a MIMO system applied to a row-enhanced data matrix H^{row} formed by combining sub-Hankel matrices H^{row}_κ with $k = 1, \ldots, K$ as entries. The second set of matrices is obtained from a MIMO system derived from a column-enhanced data matrix H^{col} by combining sub-Hankel matrices H^{col}_q; $q = 1, \ldots, Q$ as matrix elements. These two systems are coupled using an eigenvalue modal decomposition technique to obtain the desired 2D HRSE technique. Note that for 1D processing only a single-input–single-output system model is used, either in frequency $(q = 1)$ or time $(k = 1)$, and eigenvalue modal decomposition is not required.

On examination of equation (B.3), one can form Hankel matrices from the one-dimensional data sequences associated with every row and column of the data matrix. For example, the Hankel matrices associated with the kth row and qth column are given by

$$
H^{\text{row}}_k =
\begin{bmatrix}
E(k,1) & E(k,2) & \cdots & E(k,L) \\
E(k,2) & E(k,3) & \cdots & E(k,L+1) \\
\vdots & \vdots & \cdots & \vdots \\
E(k,Q-L+1) & E(m,Q-L+2) & \cdots & E(k,Q)
\end{bmatrix}
k = 1, \ldots, K; \quad \text{(B.4)}
$$

and

$$H_q^{col} = \begin{bmatrix} E(1,q) & E(2,q) & \cdots & E(J,q) \\ E(2,q) & E(3,q) & \cdots & E(J+1,q) \\ \vdots & \vdots & \cdots & \vdots \\ E(K-J+1,q) & E(K-J+2,q) & \cdots & E(K,q) \end{bmatrix} q = 1,\ldots,Q, \quad \text{(B.5)}$$

respectively. Parameters L and J that appear in equations (B.4) and (B.5), respectively, denote the correlation windows in the time and frequency directions. They are heuristically chosen here to be $L = [Q/2]$ and $J = [K/2]$, where the brackets denote the smallest integer less than or equal to the inserted quantity.

The first MIMO system uses the set of Hankel matrices defined by H_k^{row}, $k = 1,\ldots,K$. A row-enhanced data matrix is defined by stacking the K Hankel matrices described by equation (B.4) into a column matrix such that

$$H^{row} = \begin{bmatrix} H_1^{row} \\ H_2^{row} \\ \vdots \\ H_K^{row} \end{bmatrix}. \quad \text{(B.6)}$$

Following the development in reference [122], an expanded form of the enhanced Hankel matrix, denoted by H_e^{row}, is formed from the set of H_k^{row} according to

$$H_e^{row} = \begin{bmatrix} H_1^{row} & H_2^{row} & \cdots & H_J^{row} \\ H_2^{row} & H_3^{row} & \cdots & H_{J+1}^{row} \\ \vdots & \vdots & \cdots & \vdots \\ H_{K-J+1}^{row} & H_{K-J+2}^{row} & \cdots & H_K^{row} \end{bmatrix} \quad \text{(B.7)}$$

The decomposition of H_e^{row} into a product of two matrices is given by

$$H_e^{row} = \Omega\Gamma, \quad \text{(B.8)}$$

In linear system and control theory, Ω and Γ are known as the observability and controllability matrices, respectively. By computing the singularvalue decomposition of the enhanced Hankel matrix H_e^{row} and its low rank truncation, the following N rank reduction of H_e^{row} is obtained:

$$\tilde{H}_e^{\text{row}} = U_{sn} \Sigma_{sn} V_{sn}^*. \tag{B.9}$$

In equation (B.9), U_{sn} denotes the signal components of the left-unitary matrix; Σ_{sn} is a diagonal matrix with the signal singular values of H_e^{row} arranged in decreasing order as entries on its main diagonal. Furthermore, V_{sn} is the signal component of the right-unitary matrix, and * denotes conjugate and transpose. Therefore, the observability and controllability matrices corresponding to the low rank truncation are given by

$$\Omega = U_{sn} \Sigma_{sn}^{1/2} \tag{B.10}$$

and

$$\Gamma = \Sigma_{sn}^{1/2} V_{sn}^*. \tag{B.11}$$

Referring to [122] and [124], a set of complex matrices (A_r, B_r, C_r) describing the first MIMO system may be derived using either Ω or Γ. The derivation of these matrices in this appendix is based on Ω. Referring to the development in [124], one can show that

$$A_r = (\Omega_{-rl}^* \, \Omega_{-rl})^{-1} \Omega_{-rl}^* \, \Omega_{-r1} \tag{B.12}$$

$$B_r = (\Omega_K^* \, \Omega_K)^{-1} \Omega_K^* H^{\text{row}} \tag{B.13}$$

$$C_r = \Omega(1 : Q - L + 1, :), \tag{B.14}$$

where

$$\Omega_{-r1} = \Omega(Q - L + 2 : (K - J + 1) \, (Q - L + 1), :), \tag{B.15}$$

$$\Omega_{-rl} = \Omega(1 : (K - J)(Q - L + 1), :), \tag{B.16}$$

and Ω_K is a full observability matrix that is defined by

$$\Omega_K = \begin{bmatrix} C_r \\ C_r A_r \\ \vdots \\ C_r A_r^{K-1} \end{bmatrix}. \tag{B.17}$$

In equations (B.14)–(B.16), the MATLAB notation $\Omega(m_1 : m_2,:)$ denotes the matrix obtained by keeping rows m_1 to m_2 of Ω. The eigenvalues of the matrix A_r are used to characterize the poles s_n, $n = 1, \ldots, N$.

Now consider the development of the second MIMO system characterizing the poles p_n, $n = 1, \ldots N$. A column-enhanced data matrix is formed by stacking the Q Hankel matrices described by equation (B.5) into a row vector. Mathematically,

$$H^{\text{col}} = \begin{bmatrix} H_1^{\text{col}} & H_2^{\text{col}} & \cdots & H_Q^{\text{col}} \end{bmatrix}. \tag{B.18}$$

The second MIMO system is derived from H^{col} by following steps similar to those described by equations (B.7)–(B.11). The set of complex matrices (A_c, B_c, C_c) follow the derivations of (A_r, B_r, C_r) as described by equations (B.12)–(B.17).

The two MIMO systems are ready to be coupled leading to the desired 2D HRSE technique. Consider the extraction of the range pole s_n from A_r and the range-rate poles p_n from A_c. The range poles are determined from the eigenvalues of A_r and the range-rate poles from the eigenvalues of A_c. Note that the poles extracted from each system are not necessarily paired as corresponding to the same scattering center. A pole pairing technique must be developed to reorder the poles of one system to the other to accomplish this pairing. The pole pairing technique used here is based on the eigenvalue decomposition of the open-loop matrix A_r of the first MIMO system described by equation (B.12). The modal matrix of A_r will be used to reorder the eigenvalues of A_c. The eigenvalue decomposition of A_r is written as

$$\Lambda_r = M_r^{-1} A_r M_r, \tag{B.19}$$

where Λ_r is a diagonal matrix with the eigenvalues of A_r defined by

$$\gamma\{A_r\} = \{s_1, \quad s_2, \quad \ldots, \quad s_N\}, \tag{B.20}$$

as entries on its main diagonal. In equation (B.19), M_r denotes the modal matrix of A_r and is written

$$M_r = \begin{bmatrix} v_1 & v_2 & \cdots & v_N \end{bmatrix}, \tag{B.21}$$

where v_n represents the nth eigenvector associated with the pole s_n.

To accomplish the pole pairing, a matrix A_{rc} is formed and is defined by

$$A_{rc} = M_r^{-1} A_c M_r. \tag{B.22}$$

The important elements of A_{rc} are its diagonal entries, expressed in the form

$$diag\{A_{rc}\} = \{A_{rc}(1, 1), A_{rc}(2, 2), \ldots, A_{rc}(N, N)\}. \tag{B.23}$$

The eigenvalues of the second MIMO system are computed according to

$$\gamma\{A_c\} = \{\gamma_1, \quad \gamma_2, \quad \ldots, \quad \gamma_N\}. \tag{B.24}$$

To associate the poles γ_n with their corresponding s_n, the elements of $\gamma\{A_c\}$ are reordered with respect to the position and strength of the entries of $angle(diag(A_{rc}))$, denoted by $\{\gamma\{A_c\}\}^{angle(diag(A_{rc}))}$. The MATLAB operation $angle(y)$ extracts the phase of the complex variable y. For example, if $angle(diag(A_{rc}(k, k)))$, $1 \leq k \leq K$ is the strongest element of $angle(diag(A_{rc}))$, the nth entry of $\gamma\{A_c\}$ must be the element that exhibits the highest strength. The poles associated with equation (B.20) obtained from equation (B.24) are reordered according to

$$\{\gamma_1, \gamma_2, \ldots, \gamma_N\}^{angle(diag \ (A_{rc}))} = \{p_1, p_2, \ldots, p_N\}. \tag{B.25}$$

Equations (B.20) and (B.25) then provide the correct pole pairs (s_n, p_n) associated with the nth scattering center.

The coupled MIMO (CMIMO) technique for extracting the pole pairs (s_n, p_n, D_n) is summarized in the following steps:

Form the row (column) enhanced data matrix H_e^{row} (H_e^{col}) by using equation (B.7) and compute the complex matrix $A_r(A_c)$ given in equation (B.12).

Compute the eigenvalue decomposition of A_r using equation (B.19) to obtain the complex pole s_n and the entries of $diag(A_{rc})$ according to equations (B.20) and (B.23), respectively.

Compute the complex poles p_n paired with s_n according to equation (B.25).

The amplitude coefficients D_n are then determined using the signal model poles (s_n, p_n) in equation (C.1) and by fitting the signal model to the data using a least-squares error metric.

B.2   MATLAB Code and Data Examples

B.2.1   One-Dimensional MATLAB Code

```
function [sn_poles,Awt]=State_Space1D(data,order);
%function [pn_poles,Awt]=State_Space1D(data,order);
```

```
%++++++++++++++++++++ Input +++++++++++++++++++++++++++++++++++++
++++++++++++++
% data is a row vector of frequency or time sampled data
% order= estimated number of scattering centers in the data

%+++++++++++++++++++++ Output +++++++++++++++++++++++++++++++++++
++++++++++++
% sn_poles / pn_poles= range or range-rate poles sn or pn;
Awt = signal model amplitudes;
%-------------------------
N=length(data);
p=fix(N/2);

  H=zeros(N-p+1,p);

  for n=1:N-p+1
    H(n,:)=data(n:n+p-1);
  end

[k,n]=size(H);
[U,S,V]=svd(H,0);
V=V';

Cont=sqrt(S(1:order,1:order))*V(1:order,:); % controlability
matrix
        A=Cont(:,2:n)/Cont(:,1:n-1);

sig_poles=eig(A);
sn_poles=-imag(log(sig_poles));

Kf=[-length(data)/2+1:length(data)/2];

F=[];F1=[];F11=[];
for n=1:order
  F1(:,n)=exp(-j*sn_poles(n)*Kf');
  F11(:,n)=reshape(F1(:,n),length(data),1);
  F=[F,F11(:,n)];
end

Edata=reshape(data.',length(data),1);

Awt=inv(F'*F+1.e-5*eye(order))*F'*Edata;
```

1D HRSE Example

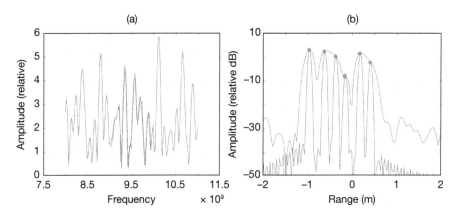

Figure B.1   1D HRSE MATLAB code applied to a typical single pulse data set using slotted cylindrical target. (a) Blue shows the amplitude of 1 GHz input data (51 frequency samples); red denotes 3X extrapolated data using signal model obtained from 1D signal poles. (b) Blue shows the Fourier-based compressed pulse using 1 GHz input data; red denotes the Fourier-based compressed pulse using 3X frequency extrapolation frequency data; red circles are the dB (Awt) from MATLAB code positioned versus range estimates computed from1D HRSE technique [$R = sn\_poles/(4\pi df/c)$].

B.2.2   Two-Dimensional MATLAB Code

```
function [sn_poles,pn_poles]=cmimo3(data,order);

%****************input********************
% data is the 2-D (time, frequency) data matrix
% order:number of scattering centers associated with the
data
%*****************************************************************
***

%****************output****************************************
**
%(sn_poles,pn_poles)
%  ************************************************************
**
```

```
[k1,n1]=size(data);
J=fix(n1/2);
L=fix(k1/2);
A1=mimo3_appB(data,J,L,order);
eigenv1=eig(A1);
A2=mimo3_appB(data.',L,J,order);
eigenv2=eig(A2);
f1=-angle(eigenv1);
f2=-angle(eigenv2);
f2b=sort(f2);
D1=ones(order,1)*f2.';
D2=f2b*ones(1,order);
[q,p]=find(abs(D1-D2)==0);
[M1,R1]=eig(A1);
R2=inv(M1)*A2*M1;
f22=-angle(diag(R2));
f22b=sort(f22);
f22b=f22b(q);
D1=ones(order,1)*f22.';
D2=f22b*ones(1,order);
[q,p]=find(abs(D1-D2)==0);
eigenv2=eigenv2(q);

ceigenvls=[eigenv1./abs(eigenv1) eigenv2./abs(eigenv2)];

sn_poles=-imag(log(ceigenvls(:,2)));
pn_poles=-imag(log(ceigenvls(:,1)));

function A=mimo3_appB(data,J,L,order);

%*********************input*******************
% data = input (Q by K)
%J=frequency correlation average = [K/2]
%L=time correlation average = [Q/2]
%***********************************************

%********************output**************************
%**********
% A = state matrix used to compute eigenvalues for MIMO sys-
tems 1 and 2
```

```
%*************************************************************
********

Hb=block_H_appB(data,J,L);
[k1,n1]=size(data);
[k,n]=size(Hb);
[U,S,V]=svd(Hb,0);
V=V';
Obs=U(:,1:order)*sqrt(S(1:order,1:order));% observability
matrix
A=Obs(1:k-(n1-J+1),:)\Obs(n1-J+2:k,:);

function Hv=block_H_appB(data,J,L);

%**********************input*******************
% data = input data samples
% L= correlation window in horizontal direction
% J= correlation window in vertical direction

[m,n]=size(data);
H1=zeros(m*(n-J+1),J);
for k=1:m
H1((k-1)*(n-J+1)+1:k*(n-J+1),:)=get_dat_appB(data(k,:).',J);
end
for l=1:L
Hv(:,(l-1)*J+1:l*J)=H1((l-1)*(n-J+1)+1:(m-L+l)*(n-J+1),:);
end

function A=get_dat_appB(y,p);

N=length(y);
  A=zeros(N-p+1,p);
  y=y.';
  for n=1:N-p+1
    A(n,:)=y(n:n+p-1);
  end
```

2D HRSE Example

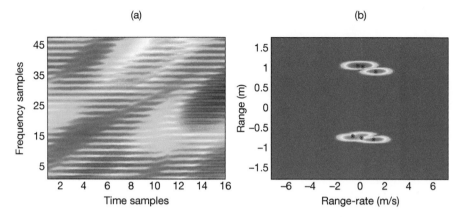

Figure B.2   2D HRSE MATLAB code applied to a typical frequency-time data block from simulations in chapter 2 using the cylindrical target geometry illustrated in figure 2.7. (a) Amplitude of input data matrix where $K = 48$ frequencies, $Q = 16$ pulses. (b) Fourier-based range, range-rate image using input data matrix. Range, range-rate spectral estimates (order $= 6$) extracted from input data block using 2D HRSE MATLAB code superimposed on image ($R_n = sn\_poles/(4\pi df/c), \dot{R}_n = pn\_poles/(4\pi dt/\lambda)$).

Appendix C: 2D Spectral Estimation: An ESPRIT Approach

C.1 Introduction

The two-dimensional (2D) estimation of signal parameters via rotational invariance technique (ESPRIT)-based high resolution spectral estimation(SE) technique described in this appendix is adapted from [125] and modified using the frequency-time (k, q) data block indexing notation and scattering center amplitude, range, and range-rate $(R_n, \dot{R}_n, D_n)$ notation described earlier in the text. Otherwise, the development is directly extracted from [125]. The technique is presented in sufficient detail for comparison to the state-space HRSE technique development in appendix B. Comparison of the two HRSE technique developments provides much insight, not only into the similarities of the techniques, but to the contrasting underlying assumptions that lead to each formulation. The interested reader should refer to [125] for a more detailed development and applications of the ESPRIT technique, as well as a comprehensive discussion of the historical evolution of ESPRIT-based spectral estimation. The reference also describes an enhancement to the technique that allows the scattering center observables to be individually tracked as a sequence of data blocks is processed.

The 2D all-pole signal model introduced in chapter 2 is characterized by equation (2.13). The development in [125] is based on the same signal model, but is developed using the complex transpose of the signal model matrix used throughout this text. Adapting the signal model defined by equation (2.13) to the notation used in [125], the transpose of the equation (2.13) can be expressed in the form

$$E(q, k) = \Sigma_n D_n \exp(jq\alpha_n + jk\beta_n), \qquad (C.1)$$

where $\alpha_n = -4\pi \dot{R}_n \Delta t / \lambda$, $\beta_n = -4\pi R_n \Delta f / c$, D_n is the signal model complex amplitude, R_n and $\dot{R}_n$ are the range and range-rate of the nth scattering center, Δt is the pulse time sampling interval, and Δf is the frequency sampling interval. Contrasted with the HRSE development in appendix B, the pulse time index q in equation (C.1) ranges over the rows of E and the frequency sampling index k ranges over the columns of E, both assumed to be centered on zero.

Analogous to the state-space HRSE technique developed in appendix B, the ESPRIT technique estimates the signal model poles $p_n = \exp(j\alpha_n)$ and $s_n = \exp(j\beta_n)$ for each scattering center contributing to the data by manipulating the data array $E(q, k)$ in such a way that (s_n, p_n) become the solutions of two coupled eigenvalue problems. Once (s_n, p_n) are estimated, the extraction of the range and range-rate estimates $(R_n, \dot{R}_n)$ are given by equations (2.18) and (2.19).

C.2   Overview

Two-dimensional ESPRIT exploits the property of the signal model that the contributions of spectral component n to any two equally sized subarrays of the data array E differ only by a shift factor determined by the number of rows and columns by which the subarrays are offset from one another. For an offset of p rows and m columns, that shift factor is $\exp(jp\alpha_n + jm\beta_n)$, and because it depends on n, the two subarrays are, in total, two different linear combinations of all the individual spectral contributions. Of special significance are the values $\exp(j\alpha_n)$ and $\exp(j\beta_n)$ taken by the shift factor when one of the pair p and m is one and the other zero.

This feature makes it possible to rearrange and select the elements of E, in the manner described in the next section, to create the shift matrices U_0, U_t, and U_f of block-Hankel structure with the following three properties:

1. Each column is a different linear combination of all the individual spectral contributions.

2. The rank, in the absence of noise, is equal to the total number N of spectral components in the data (which assumes that the smaller dimension of each matrix is no less than N).

3. The contribution of spectral component n to U_t or U_f differs only by the shift factor $\exp(j\alpha_n)$ or $\exp(j\beta_n)$ from the contribution of the same spectral component to U_0.

These properties imply that in the absence of noise there exists a vector x_n that can recombine the columns of U_0, U_t, and U_f to extract the particular spectral component n and also that the result of the extraction on the three matrices differs only by the factors $\exp(j\alpha_n)$ and $\exp(j\beta_n)$. Specifically, they imply $U_t x_n = U_0 x_n \exp(j\alpha_n)$ and $U_f x_n = U_0 x_n \exp(j\beta_n)$, two coupled generalized eigenvalue problems. The exponents of their eigenvalues, according to the derivation in the last section, are proportional to the range and range-rate of scatterer n.

The details of the procedure described previously are developed in the next section, including an initial singular-value decomposition of the data to strip away noise, evaluation of the three shift matrices U_0, U_p, and U_f, a procedure for correctly pairing the N eigenvalues of each of the two eigenvalue problems, and a final gradient-following coupled least-squares fine tuning of the paired solutions, which in the presence of noise are not independently determined.

C.3 The ESPRIT Details

C.3.1 The Coupled Eigenvalue Problems

In the absence of noise, every two-dimensional subarray of the data array E of a given size will be a different linear combination of the same N spectral components of the form $\exp(jq\alpha_n + jk\beta_n)$. Therefore, if the elements of each subarray are rearranged systematically into a single column vector, and these column vectors are then concatenated to form a matrix H, each column of H will be a different linear combination of the same N spectral components. In addition, it follows that no two spectral components can have the same vector contribution to any column unless their corresponding scattering centers lie in the same locations. This implies that there will exist a set of vectors x_n, one for each scattering center, such that the column vector Hx_n will differ by only a scalar factor from the vector contribution of scattering center n to any one of the columns of H.

Conventionally, H is a rearrangement of E into block-Hankel form. First a Hankel matrix is created from each column of E [note that Hankel matrix element (p, m) is column element $(p+m-1)$]. Then the block-Hankel matrix is assembled from the Hankel matrices [where block (p, m) (p, m) is the Hankel matrix from column $(p+m-1)$ of E).

Next, the singular-value decomposition $H = U\Sigma V'$ is used to reduce the second dimension of H to match its essential rank N by retaining only the columns of U corresponding to the significant singular values (following MATLAB notation, A' is the conjugate transpose of A). One test for significance is the size of the singular value relative to the maximum singular value; another is its absolute size. U retains the property of H that there exists a vector x_n for which the column vector Ux_n will differ by only a scalar factor from the vector contribution of point scattering center n to any one of the columns of H.

The block-Hankel rearrangement of E implies that the contribution of scattering center n to any one row of U differs from its contribution to the previous row in the same block by the factor $\exp(j\alpha_n)$, and from its contribution to the corresponding row in the previous block by the factor $\exp(j\beta_n)$. Thus by deleting specific sets of rows[1] of U, three versions (U_0, U_p, U_f) of U can be created such that, in the absence of noise, $U_t x_n = \exp(j\alpha_n)\,U_0 x_n$ and $\tilde{U}_f x_n = \exp(j\beta_n)\,U_0 x_n$. Multiplying both sides of both equations by $(U_0'U_0)^{-1}U_0'$ recasts them into the two coupled eigenvalue problems

$$(R_t - \tau_n\,I)x_n = 0, \tag{C.2}$$

$$(R_f - v_n\,I)x_n = 0, \tag{C.3}$$

where the square $N \times N$ matrices R_t and R_f are given by

$$R_h = (U_0'U_0)^{-1}U_0'U_h, \, (h = t, f), \tag{C.4}$$

where $\tau_n = \exp(j\alpha_n)$, $v_n = \exp(j\beta_n)$, and I is the identity matrix.

The range R_n and range-rate $\dot{R}_n$ of the scattering centers can then be evaluated from the eigenvalues $\exp(j\alpha_n)$ and $\exp(j\beta_n)$ using $\alpha_n = -4\pi\dot{R}\Delta t/\lambda$ and $\beta_s = -4\pi R_n\Delta f/c$.

C.3.2   Eigenvalue Pairing

Solving the eigenvalue problems equations (C.2) and (C.3) separately would deliver two lists of eigenvalues. But the elements in the two lists would not in general be in the same order and so the eigenvalues could not be matched properly in pairs. However, because the eigenvectors of the two eigenvalue problems are the same, the eigenvectors can be evaluated just once, as eigenvectors of the weighted matrix sum $c_1 R_t + c_2 R_f$, where the scalar weights c_1 and c_2 can be chosen essentially arbitrarily. Then the properly paired eigenvalues corresponding to each eigenvector x_n are given by $\tau_n = x_n'R_t x_n$ and $v_n = x_n'R_f x_n$ [126]. One choice for the weights is $c_1 = 1$ and $c_2 = 0$, but in the presence of noise a more accurate estimate of the eigenvalues is obtained if both matrices are brought to bear by making both weights non-zero. This pairing technique fails in the rare situation in which the weighted eigenvalue sum $c_1\tau_t + c_2 v_f$ for one scattering center happens to be the same as that for another. If that happens, it can be avoided simply by changing the scalar weights in the weighted matrix sum $c_1 R_t + c_2 R_f$.

Another pairing technique is described by [127].

---

1.   Deleting from U the last row of each block of rows, and also the whole last block of rows creates U_0. Deleting the first row of each block and the whole last block of rows creates U_t. Deleting the last row of each block and the whole first block of rows creates U_f.

C.3.3   Effect of Noise

In the presence of noise, the right sides of equations (C.2) and (C.3) will be error vectors rather than zero. Now the problem becomes one of finding the eigenvalue pair (τ_n, v_n) and corresponding eigenvector x_n that together minimize the sum of the squared two-norms of the two error vectors $\|(R_t - \tau_n I)x_n\|_2^2 + \|(R_f - v_n I)x_n\|_2^2$.

The calculus of variations applied to this mean-square minimization problem leads to the following set of simultaneous equations for the x_n, τ_n and v_n:

$$[(R_t' - \tau_n * I)(R_t - \tau_n I) + (R_f' - v_n * I)(R_f - v_n I)]x_n = 0, \tag{C.5}$$

$$\tau_n = x_n' R_t x_n, \tag{C.6}$$

$$v_n = x_n' R_f x_n, \tag{C.7}$$

where x_n is construed to be of unit norm.

Their solution can be found using a gradient method starting from initial estimates of the τ_n and v_n evaluated by solving equations (C.2) and (C.3) in the manner described. From them, the matrix in square brackets in equation (C.5) is evaluated and then factored by singular-value decomposition to determine the normalized vector x_n that minimizes the norm of the left side of equation (C.5). Substituting this vector in equations (C.6) and (C.7) yields the updated values of τ_n and v_n that then become the estimates for the next iteration. It is found in practice that the process converges quickly, making it necessary to use only a few iterations. The inclusion of this gradient method as a final step in the technique has been shown in tests to significantly improve the precision of the final values of the τ_n and v_n.

A potential noise-reduction technique that should be mentioned is that of so-called total least-squares (TLS) processing [128, 129, 130]. The idea is that all the columns of all three of the mutually shifted matrices U_0, U_p, U_f are closely related. Specifically, each of these columns, apart from the additive noise it contains, is a different linear combination of the same N column vectors, one for each of the N scattering centers. Accordingly, cleaner versions of them can be evaluated by using singular-value decomposition to factor the block matrix $[U_0, U_p, U_f]$. However, tests of radar imaging have shown the inclusion of TLS processing to produce no solid improvement. This is probably attributable to the fact that column-relatedness is already a key property in the processing, and is therefore essentially fully exploited without TLS processing.

C.4   MATLAB Code and Data Examples

C.4.1   Two-Dimensional MATLAB Code

```
function [sn_poles,pn_poles]=ESPRIT(data,order)

%******************input********************
% data is the 2-D (time, frequency) data matrix
% order:number of scattering centers associated with the
data
%***************************************************************
***

SVDR=100;    % dynamic range defining the significant singular
values
E=data;
[nt,nf]=size(E);    % evaluate the dimensions of E
St=length(E(:,1))/2;
Sf=length(E(1,:))/2;
EI=reshape(1:nt*nf,nt,nf);    % establish a unique numeric
label for each element of E
HI=hankel2(EI,St,Sf);    % rearrange the numeric labels into
block-Hankel form HI
[H1,H2]=size(HI);    % evaluate the dimensions of the block-
Hankel matrix

H=reshape(E(HI(:)),H1,H2); % use HI to place the elements of
E in block-Hankel form H
[U,G,V]=svd(H,0);    % evaluate the SVD of H
Sp=min(order,ones(1,St*Sf)*(diag(G)>max(diag(G))/SVDR));
% number Sp of significant
% singular values found, up to order

Rt=U(1:(nt-St)*(nf-Sf+1),1:Sp)\U((1:(nt-St)*(nf-Sf+1))
+nf-Sf+1,1:Sp);    % square cross-
% range master matrix of dimension Sp

Ir=reshape(1:H1,nf-Sf+1,nt-St+1);
Ir=reshape(Ir(1:nf-Sf,1:nt-St+1),1,(nf-Sf)*(nt-St+1));
% row indices of shifted submatrices
```

```
Rf=U(Ir,1:Sp)\U(Ir+1,1:Sp);    % square slant-range master
matrix of dimension Sp
[X,dummy]=eig(Rt+Rf);Lt=diag(X'*Rt*X);Lf=diag(X'*Rf*X);
% pole estimates
sn_poles=-angle(Lf);    % slant-range poles
pn_poles=-angle(Lt);    % cross-range poles
function H=hankel2(x,St,Sf)
%
% Generates a block-Hankel matrix of dimension (nt-
Sf+1)*(nf-Sf+1) by St*Sf from
% the nt-by-nf data array x.
% The rows of H are the concatenated and transposed columns
of every possible
% different St by Sf submatrix that can be defined in x.
%
[nt,nf]=size(x);
H=zeros((nt-St+1)*(nf-Sf+1),St*Sf);
for It=1:nt-St+1
  for If=1:nf-Sf+1
    H(If+(It-1)*(nf-Sf+1),:)=reshape(x(It:It+St-1,If:If+
    Sf-1),1,St*Sf);
  end
end
return
```

2D HRSE Example

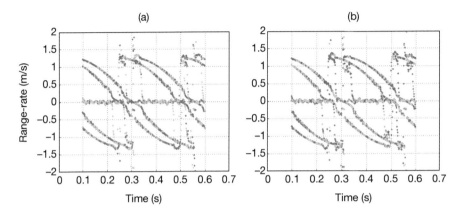

Figure C.1   Comparison of the sequential range-rate estimates using the 2D-HRSE state-space MATLAB routine listed in appendix B to the 2D HRSE ESPRIT MATLAB routine listed in appendix C. Both use the spinning canonical cylinder simulation discussed in figure 2.8(a) where $N = 6$, $Q_0 = 16$ pulses in block, $n_f = 48$ frequency samples, and 2 GHz bandwidth at X-band. For a given data block in the sequence, the data matrix inputs to both routines are identical. The conversion to range, and range-rate is given by $(R_n = sn\_poles/(4\pi df/c), \dot{R}_n = pn\_poles/(4\pi dt/\lambda))$. (a) Range-rate sequence over the (0.1–0.6 s) time interval using the state-space–based MATLAB routine. (b) Range-rate sequence over the (0.1–0.6 s) time interval using the ESPRIT-based MATLAB routine.

Appendix D: Location Estimation Using Target Space Filters for $\{R_n, \dot{R}_n\}$ Observables

Application of the target space persistence filter developed in section 5.6, along with the incorporation of the acceleration observable estimate $\ddot{R}_n$, eliminate the need for sequential correlation in range, range-rate observable space in order to obtain a three-dimensional (3D) scattering center location estimate. For a given frequency-time data block, the resulting triad of observable estimates $(R_n, \dot{R}_n, \ddot{R}_n)$ provides unique location estimates from a single frequency-time data block, allowing for a relatively simple filter in target space to eliminate erroneous location estimates. However, depending on the time sampling interval of the data sequence, particularly if the radar pulse repetition frequency relative to the target motion is limited, estimates of $\ddot{R}_n$ might become problematic. In this case only range, range-rate observables can be used to obtain the scattering center location estimates.

When only range, range-rate observables are used, observables extracted from multiple data blocks must be correlated before equation (5.7) can be applied to obtain the scattering center location estimates. For example, in section 5.5, a nearest-neighbor correlation scheme was used to generate a correlated sequence of range and range-rate observables for each scattering center. Given this association, a number of sequential observables can be concatenated in the form of equation (5.7) to yield unambiguous solutions the scattering center location estimates. The range, range-rate observable association requirement can be circumvented by introducing techniques that take an alternative approach: they hypothesize all possible associations, applying equation (5.7) to each hypothesized association, and apply a nearest-neighbor correlation filter on the resulting scattering center estimates in target space. The techniques also exploit the notion of *scattering center persistence* introduced in section 5.6.1 by assuming that each true scattering center location estimate accumulates in the 3D target space bins corresponding to the true scattering center.

D.1 The Nearest-Neighbor Target Space Persistence Filter

The technique begins by assuming all possible associations between the range, range-rate observables in two or more sequential data blocks. For example, for N scattering

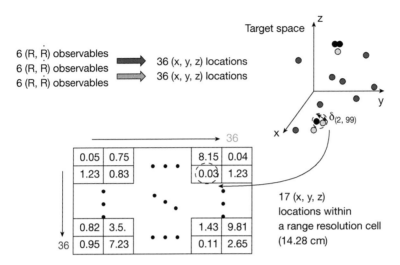

Figure D.1   An iteration of the target-space nearest-neighbor technique using scattering center location estimation and gating.

centers and two independent data blocks, this leads to N^2 possible associations and N^2 location estimates. To reduce the number of potential scattering center location estimates that need to be tallied in target space, apply a nearest-neighbor filter in target space (vs. range, range-rate observable space) to filter out those location estimates that are widely separated. The resulting location estimates in target space are then additionally filtered using the persistence filter developed in section 5.6.

As an example, applying the nearest-neighbor target space persistence filter, consider processing the range, range-rate observables from three independent frequency-time data blocks. Assume the three data blocks are centered at $t = t_b$, $t = t_{b+L}$, and $t = t_{b+2L}$, where L denotes the data block incremental block jump index. Using the observables from the first two data blocks, first form all possible scattering center location estimates using equation (5.6). Then form all possible scattering center location estimates using observables from the second and third data blocks. An example is depicted in figure D.1 for $N = 6$. Next, a distance matrix is formed where the (ij)th element in the array is the distance, in target space, between the ith scattering center position estimate from the first pair of data blocks and the jth scattering center position estimate from the second pair of data blocks. Those elements in the distance array that are less than some other desired threshold (e.g., the radar range resolution) are then selected, and their corresponding scattering center position estimates from the first observation data block are plotted on a persistence grid. Figure D.1 contains an example of a distance matrix. Thus, a scattering center location estimate is counted in the persistence grid only if there exists a scattering center detection in another set of data blocks at some specified later time whose position estimate is within a given gating distance from the current estimate.

Potentially better performance in the nearest-neighbor approach is achieved by also forming possible scattering center location estimates using observables from the first and third data blocks and applying the nearest-neighbor method to three sets of possible scattering center location estimates, not just two. Each final scattering center location estimate is then the average of three position estimates (one from each set), and its amplitude is the average of the three associated amplitudes. This approach is taken to obtain the results shown in this section. The gating distance used in the subsequent examples is that of the sensor's range resolution.

D.2 PHD Persistence Particle-Filter Approach

An alternative technique for reducing the number of potential scattering center location estimates tallied in the persistence grid is to use a probability hypothesis density (PHD)-based particle filter. The PHD filter [131], which has typically been applied to multitarget, multisensor tracking problems, is a Bayesian-based filter that propagates a density in target space and indicates the expected number of targets present[1]. The peaks in the density indicate the target locations. For more information on the PHD particle filter and its derivation, the reader is referred to [132–135].

Consider processing a sequence of frequency-time data blocks. For each iteration in the sequence, the PHD-based particle filter fuses range, range-rate observables from multiple data blocks to determine the location of the scattering center position estimates, and it reduces the number of scattering center detections to be plotted in the persistence grid to be roughly equal to the number of observables at the current time step. This method is tempting to use because of its capability for association-free data fusion of the multiple observables and its ability to remove the ghost scattering centers caused by persisting scattering center estimates from incorrectly associated observables. The nearest-neighbor filter technique is incapable of removing all such ghost scattering centers.

In the 3D scattering center location estimation application considered here, the *target locations* will be the locations of the scattering centers on our target body. For

---

1. Integrating any area of the PHD gives the expected number of objects located in that area.

computational efficiency, a particle filter implementation of the PHD is used. The particle filter consists of particles, such as

$$\xi_i = \begin{bmatrix} x_i \\ y_i \\ z_i \end{bmatrix},$$

(D.1)

with associated weights to represent the propagated PHD. Because target space typically presents a large volume to cover, the particles are placed in areas of high importance, so that computational resources are efficiently allocated. The PHD, similar to target-state parameters in other tracking filters, is propagated according to a target motion model and updated according to the observable at each iteration. When applied to the 3D scattering center location problem, a stationary target motion model is used because the target body and scattering centers are fixed in target space. Only *process noise* is added to provide some particle diversity at each time step. The birth particles injected into the filter correspond to the potential scattering center locations generated by equation (5.7).

Figures D.2 and D.3 show block diagrams of the flow of the PHD particle-filter persistence technique. This technique forms all possible scattering center location estimates using equation (5.6), but the PHD particle filter then uses the observables from the current data block to determine how likely each of the potential scatter locations is. Those that are less likely are less likely to pass through the filter. The expected locations of the scattering centers at the current iteration are then determined, given the state of the filter from the previous iteration. The detections from the PHD particle filter, that is, the expected scattering center locations for the current iteration, are accumulated in a persistence grid. After all the radar data has been processed, the persistence grid counts are thresholded to reveal the discrete-point scattering center locations of the target.

The PHD particle filter uses the current observable at each time step to filter the true scattering center estimates from the incorrect ones. The ghost scattering centers mentioned earlier result from incorrectly associated observables in equation (5.6) and are thus less likely to have produced the observables seen in a single data block of data. Therefore, the likelihood is that $f(z_n|\xi_i)$ will be small compared to the true scattering centers' likelihoods, and the ghost scattering centers will be filtered out by the PHD filter.

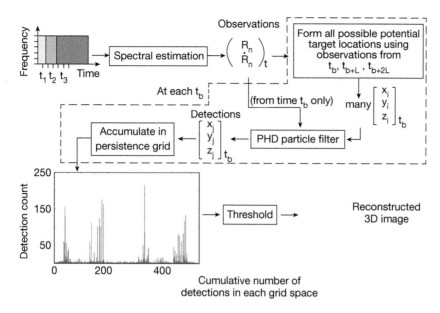

Figure D.2 Block diagram of the PHD particle-filter persistence approach, where *L* is the data block look-ahead parameter.

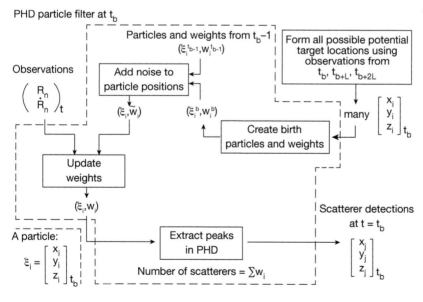

Figure D.3 Block diagram of a single cycle of the PHD particle filter.

PHD particle weights

The PHD particle filter contains a weight-update step whose purpose is to adjust the particle weights, and hence the PHD that the particles represent, to account for current observables. Details of the general PHD particle-filter update step can be found in [131, 134]. The technique described in this section assumes that the probability of detection (p_D) of a scattering center is near one, and the probability of false alarm (p_{FA}) is near zero[2]. The weight update is thus based primarily on the single-object observable likelihood $f(z_n|\xi_i)$, which is the likelihood that an observable z_n was generated by an object located at the position specified by particle ξ_i.

Figure D.4 contains a diagram of the weight-update process. For each particle in the filter, a hypothetical observable is generated from the particle and the current motion matrix. This hypothetical observable is the observable that one would expect to see were a scattering center to be located at the position specified by the particle. Next, each of these hypothetical observables is compared to each of the M actual observables at the current time to generate the likelihoods $f(z_n|\xi_i)$. Each likelihood is a multidimensional Gaussian distance in observable space and is given by the following equation, which assumes that range and range-rate are independent:

$$f(z_n|\xi_i) = N(R_n - (\hat{k}\cdot\xi_i), \sigma_R)\cdot N(\dot{R}_n - (\dot{\hat{k}}\cdot\xi_i), \sigma_{\dot{R}})$$
$$= \frac{1}{\sqrt{2\pi\sigma_R^2}}e^{\frac{1}{2\sigma_R^2}(R_n - (\hat{k}\cdot\xi_i))^2} \cdot \frac{1}{\sqrt{2\pi\sigma_{\dot{R}}^2}}e^{\frac{1}{2\sigma_{\dot{R}}^2}(\dot{R}_n - (\dot{\hat{k}}\cdot\xi_i))^2}. \tag{D.2}$$

where σ_R^2 and $\sigma_{\dot{R}}^2$ are the range and range-rate observable variances, respectively. The weights, ω_i, are then updated according to

$$\omega_i = (1 - p_D)\tilde{\omega}_i + \sum_{n=1}^{M} \frac{p_D(\xi_i)f(z_n|\xi_i)\tilde{\omega}_i}{C(z_n) + \Sigma_{j=1}^{N}p_D(\xi_j)f(z_n|\xi_j)\tilde{\omega}_j}, \tag{D.3}$$

where

$$C(z_n) = \frac{p_{FA}}{R(\text{resolution})\dot{R}(\text{resolution})}. \tag{D.4}$$

---

2. Actual values used are $p_D(\xi) = 0.99$ and $p_{FA} = 0.01$.

and $\tilde{\omega}_i$ are the prior particle weights. Setting $p_D(\xi) \approx 1$ and $p_{FA} \approx 0$ results in:

$$\omega_i \approx \sum_{n=1}^{M} \left[\frac{f(z_n|\xi_i)}{\sum_{j=1}^{N} f(z_n|\xi_j)\tilde{\omega}_j} \right] \tilde{\omega}_i. \tag{D.5}$$

The interpretation of equation (D.5) is illustrated in figure D.5. The likelihoods associated with each observable are normalized, and the weight of each particle is taken to be the sum of the likelihoods over all the observables associated with that particle. Hence, the total updated weight of all the particles, which is what specifies the expected number of scattering centers to extract from the PHD at each iteration, is roughly equal to the number of observable pairs, $z_n = (R_n, \dot{R}_n)$ at the current iteration.

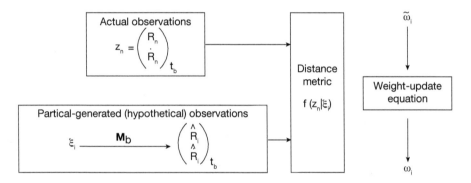

Figure D.4 Block diagram of the weight-update process at iteration t_b.

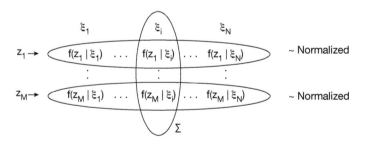

Figure D.5 Block diagram of the weight-update process at iteration t_b.

D.3   A Hybrid Nearest-Neighbor, PHD-Based Persistence Filter

The technique presented in this section attempts to exploit both the computationally simple efficacy of the nearest-neighbor approach presented in section D.1, and the filtering performance of the PHD particle-filter technique presented in section D.2. This is achieved by discarding the particle filter and using only the data-update portion of the PHD approach. Instead of determining which particles are more likely to have been produced by the observables extracted from a radar data block, look directly at the output of the nearest-neighbor filter and determine the likelihood that each potential scattering center location found by the nearest-neighbor technique was produced by the observable from the current data block.

Thus, the hybrid technique consists of applying the nearest-neighbor technique of section D.1, and then computing the PHD at each potential scattering center, assuming $p_D = 1$, according to the following equation:

$$v_i = \sum_{n=1}^{M} \frac{f(z_n | \xi_i)}{C(z_n) + \sum_{j=1}^{N} f(z_n | \xi_j)}, \tag{D.6}$$

where the parameters are the same as in section D.2, except that ξ_i is no longer a particle, but one of N potential scattering center locations produced by the nearest-neighbor filter. Those potential scattering center locations with the highest non-zero v_i are plotted in the persistence grid, and the rest are discarded. The number of potential scattering center locations plotted in the persistence grid is the lesser of either the number of expected scattering centers, as given by the rounded sum of the v_i,[3] or the number of potential scattering centers with non-zero v_i. Thus, the power of the nearest-neighbor filter has been combined with the data-fusion capability of the PHD filter to filter out all but the most realistically probable scattering center locations.

The previous process is repeated for each subsequent collection of data blocks. Because a particle filter is not used, this technique is less computationally intensive than the PHD particle filter approach. Also, unlike in the PHD particle-filter approach, the results include amplitude information for each scattering center. The nearest-neighbor portion of the technique averages the amplitudes of the observables used to generate each potential scattering center location. These average amplitude values simply accompany the scattering center if it is plotted in the persistence grid.

---

3. Recall that the PHD integrates to the expected number of objects present, and was shown in section D.2 to integrate to roughly the number of observables (M) in the current snapshot.

An example illustrating the scattering center location estimates using the hybrid neighbor, PHD-based persistence filter applied to the canonical spinning cylindrical target introduced in chapter 2, section 2.5, is shown in figure D.6. For this example, the motion parameters used in section 2.5 having 0° precession (pure spin) are modified to add 5° of precession, with the other values remaining the same. A comparison of the two motions is illustrated in figure D.7. A threshold was used in the target space persistence filter corresponding to 20 percent of the maximum count observed in the persistence grid. Note that there are scattering centers that appear on the z-axis. These are caused by the projection onto the z-axis of the slipping scattering centers that occurs when they are processed by this discrete-scattering center processing technique as discussed in section 5.5.3.

The extracted $(R_n, \dot{R}_n)_b$ observables versus block time t_b are illustrated in figure D.8. A Fourier-based spectral estimator was used to extract the observables from each data block. Note that in comparing the range, range-rate observables in figure D.8 resulting from 5° precession to those of figure 2.11 resulting from 0° precession, even a small amount of precession produces a marked change in the range-rate observable.

The next section introduces an approach for estimating the radius of the two slipping scattering centers.

D.3.1   Persistence Filtering for Slipping Scattering Centers in Target Space

The persistence filter techniques presented so far are applicable only for correctly estimating discrete-point scattering center locations. A technique that handles axially symmetric slipping scattering centers, provided precession exists in the object's motion, is developed in this section. When the slipping-scattering center estimation technique is applied in conjunction with the discrete-scattering center estimation approach, a more complete picture of an object is obtained. Unfortunately, for objects with spin-only motion, a unique solution for the location of the slipping scattering centers is theoretically unobtainable.

The slipping-scattering center scattering center location technique developed in this section is also a persistence-based technique, whereby it assumes that a slipping scattering center will persist for some time along a ring centered about the spin axis of the target. In other words, a slipping scattering center persists at a fixed two-dimensional cylindrical coordinate location in target space, which provides an estimate of the slipping scattering center radius. The slipping-scattering center scattering center location technique is first applied to all the observable data separately from the discrete-scattering center scattering center location techniques. Note that, because slipping scattering

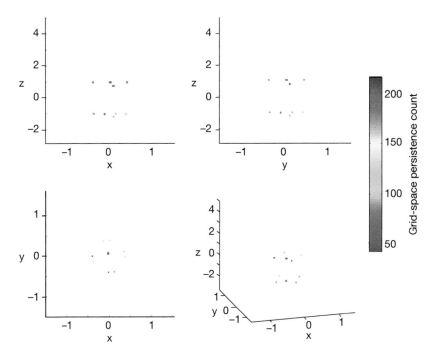

Figure D.6 Results of applying the discrete-scattering center location technique to the precessing cylinder data set with 5° precession. Note that the slipping scattering centers project onto the z-axis.

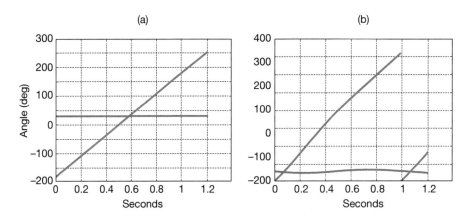

Figure D.7 Motion assumptions. (a) Motion 1, spin with no precession where $\kappa = 30°$, $T_p = 10{,}000s$, $T_s = 1$ s, $\theta_p = 0$, $a_p = -180°$, $a_s = 0$. (b) Motion 2, spin with 5° precession where $\kappa = 30°$, $T_p = 1s$, $T_s = 100$ s, $\theta_p = 5°$, $a_p = 0$, $a_s = 10°$.

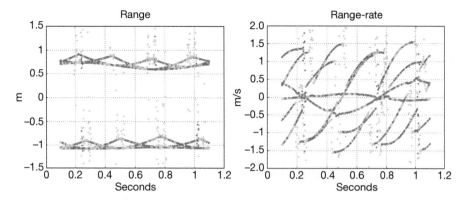

Figure D.8 Extracted range, range-rate observable sequence for motion assumption 2. Spin with 5° precession.

center location estimation is a two-dimensional problem, an estimate can be made from each range, range-rate observable pair and association is not required. That is, for each observable pair, a system of two equations and two unknowns characterizes the estimate of the slipping scattering center location [see equation (5.13) for which $\dot{\phi} = 0$]:

$$
\begin{bmatrix} \sin(\theta) & \cos(\theta) \\ \dot{\theta}\cos(\theta) & -\dot{\theta}\sin(\theta) \end{bmatrix}_{t_b} \cdot \begin{bmatrix} \rho_n \\ z_n \end{bmatrix} = \begin{bmatrix} R_n \\ \dot{R}_n \end{bmatrix}_{t_b} . \tag{D.7}
$$

where $\theta(t)$ is the aspect angle motion, and (ρ_n, z_n) are the radius and z-axis locations of the slipping scattering center estimates. Assuming the motion of the target is known, each range, range-rate observable pair can be inserted into equation (D.7) to obtain its corresponding polar coordinate location (ρ_n, z_n). Similar to the discrete-scattering center scattering center location techniques, the two-dimensional target space is discretized into a grid. This time, the grid is in polar coordinates, and all that is needed is a two-dimensional (2D) array to maintain counts of the number of observables whose scattering center's polar coordinates fall inside each grid space. All observable pairs are input into equation (D.7). If the observable pair corresponds to a slipping scattering center, estimates of (ρ_n, z_n) will persist in the two-dimensional target space grid. Other observable pairs will be spread randomly over this space and be removed by the target space filter. To extract the polar coordinate positions of the persistent slippery scattering centers, a threshold is applied to the counts in the array. The polar coordinate positions

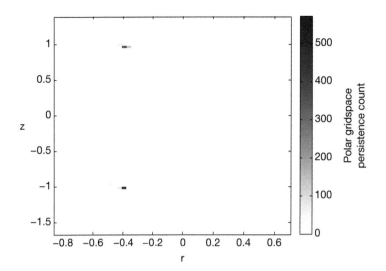

Figure D.9 Persistence grid for finding polar coordinates of slipping scattering centers in data set of the spinning and processing cylinder. Note that two distinct slipping scattering centers are evident.

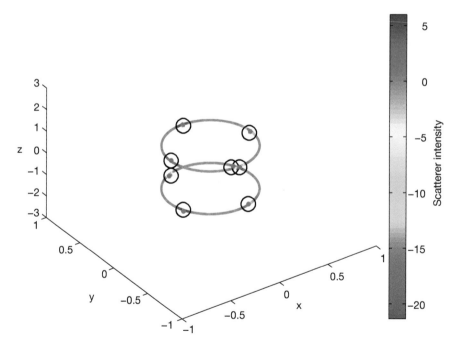

Figure D.10 Hybrid scattering center location estimates including both discrete and slipping scattering centers. Canonical spinning cylinder with 5° precession. Circular rings are added around discrete scattering center locations for clarity.

whose counts exceed the threshold are then plotted in the target space. An example of the results of this technique is shown in figure D.9. A threshold was used corresponding to 25 percent of the maximum count observed in the persistence grid.

The technique correctly predicts the presence of two slipping scattering centers having radius 0.4 m located at $z = -1, +1$ m.

The fixed and slipping scattering center techniques presented in this appendix can be combined into an integrated technique for estimating 3D scattering center locations including both the slipping and fixed-point scattering centers. First, the slipping-scattering center filter of section D.3.1 is applied to the sequential observable data to determine the locations of the slipping scattering centers. Next, the range, range-rate observables that correspond to the slipping scattering center locations are removed. The hybrid filter technique of section D.3 is applied to the remaining observables to extract the discrete point scattering centers. The results using this integrated technique are illustrated in figure D.10. Both the slipping scattering center and discrete scattering center locations are correctly estimated.

Appendix E: Acceleration Estimation MATLAB Code and Input Parameters

E.1   MATLAB Acceleration Code

The input parameters to the acceleration code listed here are determined from the linear signal model characterized by the initial data block illustrated in figure 3.1, as well as the iteration parameters defined in figure 3.7. Determination of the linear signal model parameters are consistent with the development and code of appendixes B and C.

```
Input Parameters:

%linear signal model input data
  % sn_poles,pn_poles
  % data2d_in: frequency,time input data over expanded data
  block
  % tdata,freq: frequency,time data vectors over expanded
  data block
% iteration parameters
  % Q0 =number of pulses in linear signal model
  % 2N0 =number of pulses used for each iteration
  % Nit =number of iterations
  % order1=model order
%----------------------------------------------------------------
--------------------------------------------------------------------
-----------------------------------------------------------------
------

[R,Rd,Rdd,Awt]=Acceleration_text(sn_poles, pn_poles, data2d_
in, tdata, Q0, N0, Nit, order1, freq);

% %initialize parameters
c=3.e8;
dt=tdata(2)-tdata(1);
```

```
df=freq(2)-freq(1);
f0=freq(fix(length(freq)/2));
lambda0=c/f0;

% transpose input data to conform with appendices B and C
% data2d is time-frequency matrix
data2d=data2d_in.';
%
% %data over initial data block
data2d_1=data2d(length(data2d(:,1))/2-Q0/2+1:1:length
(data2d(:,1))/2+Q0/2,:);

% signal model to data match over initial data block
  [Awt]=weight_appB(data2d_1,sn_poles,pn_poles,freq,order1);

% initialize Rdd=0
Rdd=zeros(order1,1);

% increment Q0 data block over Nit iterations
for m=1:Nit

% set up data for block incremetss
N0_inc=N0+(m-1)*N0;
Ndata2=length(data2d(:,1))/2-Q0/2-N0_inc+1:1:length(data2d
(:,1))/2+Q0/2+N0_inc;
data2d_2=data2d(Ndata2,:);
tsect2=tdata(Ndata2);

   % extract nonlinear signal model parameters from each
   expanded data block
[sn_poles,pn_poles,Rdd,Awt]=R_Rd_Rdd_text(data2d_1, . . .
   data2d_2,freq,tsect2,order1,sn_poles,pn_poles,Rdd);

end

R=sn_poles/(4*pi*df/c);
Rd=pn_poles/(4*pi*dt/lambda0);

function
[Awt]=weight_appB(data2d,sn_poles,pn_poles,freq,order1)

% data is time-frequency matrix
[Q,K]=size(data2d);
```

```
Kf=[-K/2+1:K/2];
Qt=[-Q/2:Q/2-1];
df=freq(2)-freq(1);
f0=freq(fix(length(freq)/2));

F=[];F1=[];F11=[];
for n=1:order1
   F1(:,:,n)=(exp(-j*sn_poles(n)*Kf')*exp(-j*pn_
   poles(n)*Qt))...;
      .*exp(-j*Kf'*pn_poles(n)*Qt*df/f0);
   F11(:,n)=reshape(F1(:,:,n),K*Q,1);
   F=[F,F11(:,n)];
end

Edata2d=reshape(data2d.',K*Q,1);

Awt=inv(F'*F+1.e-5*eye(order1))*F'*Edata2d;

function [sn_poles,pn_poles,Rdd,Awt]= . . .
   R_Rd_Rdd_text(data2d_1,data2d_2,freq,tsect2,order1
   ,sn_poles,pn_poles,Rdd)

% iterate within each expanded data block
Nit_inblock=2;
   for k=1:Nit_inblock

[sn_poles,pn_poles,Rdd,Awt]=Rdditerate_
text(data2d_2,data2d_1,sn_poles,pn_poles, . . .
   freq,tsect2,order1,Rdd);

end

function
[sn_poles,pn_poles,Rdd,Awt]=Rdditerate_
text(data2d,data2d_1,sn_poles,pn_poles, . . .
freq,tsect,order1,Rdd)

c=3.e8;
dt=tsect(2)-tsect(1);
df=freq(2)-freq(1);
f0=freq(fix(length(freq)/2));
lambda0=c/f0;
```

```
[Q,K]=size(data2d);
Kf=[-K/2+1:K/2];
Qt=[-Q/2:Q/2-1];

Rdd0=Rdd;
Rdd_scaled0=Rdd0*(2*pi*dt^2)/lambda0;

[Awt]=weight_appB(data2d_1,sn_poles,pn_poles,freq,order1);

% nonlinear signal model over expanded data block
[E0]=nonlinear_model(sn_poles,pn_poles,Rdd_
scaled0,Awt,Kf,Qt,order1,freq);

diff=data2d.'-E0;

T1=ones(size(Kf));T2=Qt.*Qt;
[pn_prime]=weight_acc_text(diff.',pn_poles,sn_poles,
freq,order1,Rdd_scaled0,T1,T2);

% update Rdd
del_Rdd=-lambda0/(2*pi*dt^2)*imag(pn_prime./Awt);
Rdd1=Rdd0+del_Rdd;
Rdd_scaled1=Rdd1*(2*pi*dt^2)/lambda0;

[E0]=nonlinear_model(sn_poles,pn_poles,Rdd_scaled1,
Awt,Kf,Qt,order1,freq);
diff=data2d.'-E0;

T1=ones(size(Kf));T2=Qt;
[pn_prime]=weight_acc_text(diff.',pn_poles,sn_poles,
freq,order1,Rdd_scaled0,T1,T2);

% update pn_poles
pn_poles=pn_poles-imag(pn_prime./Awt);

[Awt]=weight_appB(data2d_1,sn_poles,pn_poles,freq,order1);
[E0]=nonlinear_model(sn_poles,pn_poles,Rdd_scaled1,Awt,
Kf,Qt,order1,freq);
diff=data2d.'-E0;

T1=Kf;T2=ones(size(Qt));
[pn_prime]=weight_acc_text(diff.',pn_poles,sn_poles,
freq,order1,Rdd_scaled0,T1,T2);
```

```
% update sn_poles
sn_poles=sn_poles-imag(pn_prime./Awt);

[Awt]=weight_appB(data2d_1,sn_poles,pn_poles,freq,order1);
[E0]=nonlinear_model(sn_poles,pn_poles,Rdd_
scaled1,Awt,Kf,Qt,order1,freq);
diff=data2d.'-E0;

T1=ones(size(Kf));T2=Qt.*Qt;
[pn_prime]=weight_acc_text(diff.',pn_poles,sn_
poles,freq, . . .
  order1,Rdd_scaled1,T1,T2);

del_Rdd=-lambda0/(2*pi*dt^2)*imag(pn_prime./Awt);
Rdd=Rdd1+del_Rdd;

function
[E0]=nonlinear_model(sn_poles,pn_poles,Rdd_
scaled,Awt,Kf,Qt, . . .
  order1,freq)

df=freq(2)-freq(1);
f0=freq(fix(length(freq)/2));

E0=zeros(length(Kf),length(Qt));
for n=1:order1
  E0=E0+Awt(n)*(exp(-j*sn_poles(n)*Kf'))* . . .
(exp(-j*pn_poles(n)*Qt).*exp(-j*Rdd_scaled(n)*Qt.*Qt)) . . .
.*exp(-j*Kf'*pn_poles(n)*Qt*df/f0);
end
function [Awt]=weight_acc_text(data2d,pn_poles,sn_
poles,freq,
order1,Rdd_scaled,T1,T2)

[Q,K]=size(data2d);
Kf=[-K/2+1:K/2];
Qt=[-Q/2:Q/2-1];

df=freq(2)-freq(1);
f0=freq(fix(length(freq)/2));
```

```
F=[];F1=[];F11=[];
for n=1:order1
F1(:,:,n)=(T1'.*exp(-j*sn_poles(n)*Kf')*(exp(-j*pn_
poles(n)*Qt).*exp(-j*Rdd_scaled(n)*Qt.*Qt).*T2))...
.*exp(-j*Kf'*pn_poles(n)*Qt*df/f0);
F11(:,n)=reshape(F1(:,:,n),K*Q,1);
F=[F,F11(:,n)];
end

Edata2d=reshape(data2d.',K*Q,1);

Awt=inv(F'*F+1.e-6*eye(order1))*F'*Edata2d;
```

E.2   Example Using the Slotted Cylindrical Target

The slotted cylindrical target was first introduced in chapter 6 to illustrate the applicability of the narrowband imaging techniques developed there. It will be used in this appendix to illustrate the application of the autocorrelation filter applied to the choice of input parameters essential to enhancing the accuracy of the acceleration observable estimate using the nonlinear signal model iteration scheme developed in chapter 3. The wideband autocorrelation filter will be used to set the input parameters for the iteration portion of the acceleration estimate. The results are obtained using the MATLAB code listed previously.

Chapter 9 used 1 GHz synthetic wideband data for the slotted cylindrical target to illustrate the joint solution technique for extracting the motion and target scattering center locations. In this appendix the simulated X-band data set using the slotted cylindrical target is extended to 2 GHz bandwidth and the synthetic data are generated using a rotation rate $T = 2\pi$ s corresponding to $\dot{\varphi} = 1$. Proper choice of input parameters to the signal model iteration scheme is essential to enhancing the accuracy of the $(R_n, \dot{R}_n, \ddot{R}_n)_b$ observable sequence. As described in chapter 3, the inputs to the acceleration estimation code are partially determined by the size of the initial data block characterized by the linear signal model. The number of pulses, Q_0, contained in the initial data block is determined by the coherence of the data over the data block. Choice of the data block time duration for various data block coherence times has been discussed in detail in chapter 4 (see figures 4.7 and 4.8). It is seen there that optimum choice of Q_0 corresponds to choosing a data block time extent corresponding to the coherence time T_c as defined in chapter 4. The optimum choice for Q_0 can be determined from the average of each range gate autocorrelation function, averaged over the

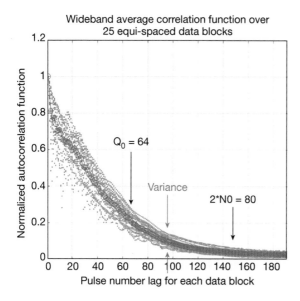

Figure E.1   Superimposed autocorrelation functions (red) averaged across the range gates for each of 25 data blocks equally spaced over the (0, 360) degree rotation angle. The blue circles indicate the average of the 25 autocorrelation functions.

data block range gates illustrated in figure 9.13a. The red dots in figure E.1 illustrate the average of the autocorrelation function across the range gates for each data block, superimposed on the plot for 25 equi-spaced sequential data blocks defined over the (0, 360) degree data set. The compressed pulses are obtained using a 2 GHz bandwidth. The blue circles illustrate the average of the 25 superimposed autocorrelation functions. The relatively small variance in the red curves indicates that the coherence of the data is nearly the same for each data block, as one might expect for a target having a uniform rotation rate. On examination of figure E.1, using a coherence time estimate of T_c, scaled to pulse lag, of 64 pulses results in a value of $Q_0 = 64$ pulses for the size of the initial data block.

Once the initial data set size is optimized, one is assured that increments in data block size beyond the initial data block of Q_0 pulses place the signal model in the nonlinear region, which is required to properly model the acceleration observable. Choice of N_0 for this example is indicated in figure E.1 as $N_0 = 40$, and the iteration parameter *Nit* is set to three for this example. Figure E.2 illustrates the range-time intensity plot, Doppler- time intensity plot, (scaled to range-rate) and acceleration-time intensity plot

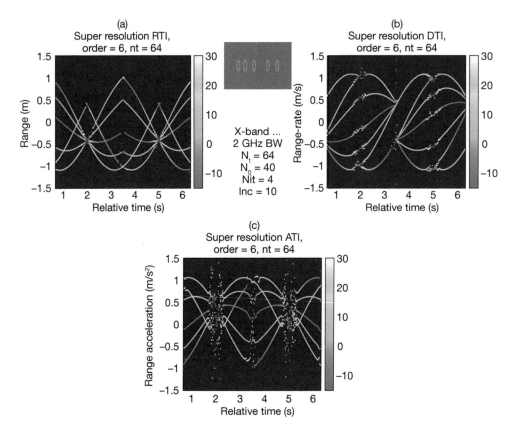

Figure E.2 Range, range-rate, range-acceleration observable sequences extracted from the 2 GHz X-band simulation data using the following parameters: nt = Q_0 = 64 initial pulses, N_0 = 40 pulse block increments and three iterations as described in chapter 3. Each data block was incremented sequentially by 10 pulses. Color scale in decibels (dB) (a) Range-time intensity plot;(b) Doppler- time intensity plot (scaled to range-rate), and (c) acceleration-time intensity plot

using the synthetic data as extended data blocks of 304 pulses are incremented by 10 pulses (0.5°) and the target rotates through the (0, 360) degree angular sector. The color of the traces in the time intensity plots is indicative of the relative radar cross section of the assumed diffraction coefficients for each of the components located on the target.

As a check on the accuracy of the extracted acceleration observables, because the target is in pure tumble motion as defined in chapter 9, and the look angle to the target lies in the tumble plane, one can use equation (9.17).

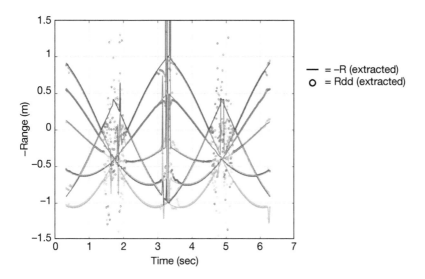

Figure E.3   Check on the estimated range-acceleration sequence $\ddot{R}_n = -R_n\dot{\varphi}^2$ for the slotted cylindrical target. $\ddot{R}_n = -R_n$ for the assumed motion $\dot{\varphi} = 1$. Estimates for each range observable $-R_n$ superimposed on the estimate for the acceleration observable $\ddot{R}_n$.

$$(\dot{\varphi})^2 = -\ddot{R}_n/R_n. \tag{E.1}$$

Because $\dot{\varphi} = 1$, it follows that ideally, $\ddot{R}_n = -R_n$ for the observable sequences extracted for each of the scattering centers. Thus, the estimates for $-R_n$ can be superimposed over the estimates for $\ddot{R}_n$ and the results plotted as a function of rotation over the complete rotation period. The comparison is illustrated in figure E.3. Note by examining the color code for each observable, the estimates are properly correlated to each scattering center, and the agreement $\ddot{R}_n = -R_n$ is quite good, except in those localized regions where the acceleration estimate is poor as indicated in figure E.2(c).

Appendix F: Integrating Static Range and Field Test Measurements into a Computational, Measurements-Based Signature Model

F.1 Static Range and Field Test Coordinate System Alignments

As stated at the outset of this text, the objective of the processing framework is to produce a high-fidelity radar signature simulation model. Typically, the computational model is represented in a target-centered coordinate system having its origin located at a position on the target chosen at the convenience of the analyst, usually dictated by the symmetry properties of the target, for example, the z-axis colocated with the target's longitudinal axis. Given a model of the target the simulation realizations can introduce any mass distributions and/or moments of inertia imbalances as the user desires by selecting a different location for the center of rotation (CoR) on the target defined relative to the reference CoR of the computational model. Characterization of the computational model can be obtained from a variety of sources. Three primary sources considered in this text are knowledge of the target topology, as well as radar static range and field measurements. Integrating the measurement data into the target-fixed coordinate system associated with the computational model is dependent on associating the CoR location and axes orientation introduced specifically to process the measurement data with that associated with the computational model. In the static range context precise measurements can be made to transform data from the measurement framework to the model framework. The field measurement scenario presents a more formidable task as effects of target imbalances and translational motion must be removed based on the information contained in the measurements themselves.

Static range measurements are taken on a standard radar cross section (RCS) range having controlled, known motion, with the target's CoR set by the position of the target on the measurement pylon. Thus, the observation angle to the target, and the distance to its scattering centers relative to the CoR are precisely known. In addition, the static range allows precise control of the rotational motion of the target.

Conversely, field measurements are conducted during an actual flight test where the motion and CoR of the target are dependent on the mass distribution of the target and must be inferred from measurements. The radar's viewing angle coverage is typi-cally more robust relative to the static range because it allows for Nyquist angle sam-ples as a function of both aspect and roll angles over a motion not realizable on the static range. However, in the context of field measurements the inability to precisely remove target translational and rotational errors, as well as the signal-to-noise ratio of the measurements and bandwidth of the sensor limits the accuracy of results. Some measurement configuration comparisons between static range and field measurements are discussed in chapter 5, figure 5.2.

In analyzing field data, it is important to recognize that four different coordinate systems come into play, and it is critical to be cognizant in which system the analyst is working at any given time. First, there is coordinate system 1 (CS1), the radar measurement system. In this context, each frequency-time data block processed using the high resolution spectral estimation (HRSE) or Fourier technique is projected into a two-dimensional range, cross-range coordinate system defined relative to the radar line of sight (LoS) to the target. This results in what is generally described as a range, Doppler (range-rate) image. Next is coordinate system 2 (CS2), the (x_0, y_0, z_0) inertial space coordinate system, which is dynamic in nature and is driven by the target's rota-tional motion. The analyst generally prefers to orient this coordinate system with the z_0-axis aligned with $\underline{J}$, the target's angular momentum vector, and the origin placed at the target's CoR. This may be difficult to accomplish, and errors in the orientation here lead to errors in the parameters characterizing the target model as one applies the motion-to-target space mapping referenced to a target-centered coordinate system. Coordinate system 3 (CS3) is the (x', y', z') target-fixed system introduced to process the measurement data. It has origin at the target CoR, but perhaps contains some resid-ual effects due to inaccuracies of compensating for the effects of the target's transla-tional motion. Coordinate system 4 (CS4) is the (x, y, z) target-fixed computational system that has origin at a CoR introduced by the analyst used for the computational model. Association of the (x', y', z') measurement system with the (x, y, z) computa-tional system is required for integrating the measurement data into the computational model. This is where all data from every measurement venue is combined for incorpo-ration into a high-fidelity radar signature model for the target.

The objective of the measurements-based modeling (MBM) technique is to inte-grate both the static range and field measurement data into a single signature model, referenced to a specific origin centered on the target, and referenced to an axis orienta-tion fixed to the target. Simulated signature data (using electromagnetic codes applied to the target model) compared to the measurement data are an important means for assessing the fidelity of the target model, as well as interpreting the signature measurement.

This appendix addresses the problem of fusing field and static range measurement data into the computational signature model characterizing the target.

For simplicity, assume the computational signature model is represented by equation (15.1) with $E_{PTD} = 0$. Denote this geometrical theory of diffraction–based signature model as

$$E(f, \theta, \varphi) \sim \sum D_n(f, \theta, \varphi) e^{-j\frac{4\pi f}{c} \hat{k}(\theta, \varphi) \cdot \underline{r}_n} \tag{F.1}$$

where the scattering center locations $\underline{r}_n$ are defined in the target-centered coordinate system having axes (x, y, z) and origin O_t as illustrated in figure F.1. The location of O_t is set by the target's geometric and symmetry properties, for example, typically on the z-axis for azimuthally symmetric targets. The unit vector $\hat{k}(\theta, \varphi)$ given by equation (2.4) defines the LoS between O_t and the transmit/receiver. The model defined by equation (F.1) is analytical (computational) in nature and defined over the 4π steradian observation space, where (θ, φ) cover the regions $0 \leq \theta \leq \pi$ and $0 \leq \varphi \leq 2\pi$. As discussed in chapter 15, for the MBM model, the diffraction coefficients are parameter-based, and estimated from the measurement data to characterize $D_n(f, \theta, \varphi)$ for each scattering component. This is accomplished using a parametric model $D_n(f, \theta, \varphi, c_1, c_2, \ldots, c_m)$ to approximate the measurement of $D_n(f, \theta, \varphi)$ as discussed in section 15.1. The parameters $(c_1, c_2, \ldots, c_m)$ are determined by fitting a functional model to the estimated diffraction coefficients over the data observation regions.

Denote the surface topology of the target model by the surface constraint $S(x, y, z) = 0$ referenced to origin O_t. Because the model of the target is specified, the scattering center locations used in the computational model are known, and the N fixed-point scattering centers $\underline{r}_n = (x_n, y_n, z_n)$ are defined to lie on the surface of the target model at the locations $S(x_n, y_n, z_n) = 0$, $n = 1, \ldots, N$. The computational signature model characterized by equation (F.1) is defined relative to the target-centered system.

However, incorporating static range and/or field data into the into the computational model will require some additional corrections in order to bring their respective coordinate systems into alignment with that of the target-centered computational framework. Because the computational model is phase and observation angle referenced to a particular coordinate system, and the static range and field measurement data sets are typically registered to different, but known coordinate systems, the estimated diffraction coefficients must be phase and viewing angle adjusted to align with the computational model.

Denote the origin of the target-centered static range or field measurement system as O_r, having target-centered coordinate system axes (x', y', z'), and the corresponding polar coordinate angles (θ', φ') to describe the motion. For static range measurements,

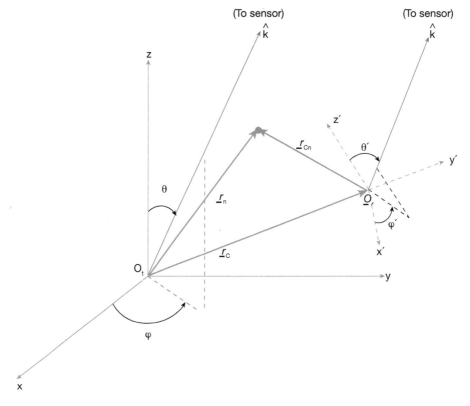

Figure F.1   The relationship of the computational signature model (x, y, z) coordinate system defined at origin O_t to the field/static range (x', y', z') coordinate system having origin O_r located at the target CoR. Defined in (x, y, z) target space, the origin O_r is related to O_t by the translation vector $\underline{r}_c$.

O_r is determined by the intersection of the pylon's axis of rotation with the target as placed on the pylon. It is important to recognize that for static range data collected over multiple roll cuts, coherent fusion of roll-cut data into the model defined by equation (F.1) requires that the target's repositioning for each roll-cut measurement sequence maintain the same origin O_r. Otherwise, the data from each roll-cut measurement sequence will have to be adjusted using techniques similar to those described in this appendix. For field measurements, O_r must be inferred from the measurements such as discussed in section F.4. Assume a static range and a field test experiment is conducted using identical targets. As a first step it is necessary to bring the two measurement scenes into a common geometric framework. The static range data are collected as the target rotates on the measurement pylon, and is tagged with the precise viewing angle to the target. To apply the target space mapping equations, the static range rotation is assigned a convenient rotation rate and the data are represented as a function of time. The motion to rotational angle mapping is known, thus providing a motion solution for the target space mapping equations.

In the case of the field measurement, the motion solution must be estimated from the data. For Newtonian motion referenced to inertial space, the motion of the target as viewed by the radar can be represented by the translation of its center of mass, and its rotation about its center of mass. Typically, the radar data will not have all the effects of the translational motion removed, and additional compensation procedures are required to properly register the measurement data in the rotational system. The location of the center of mass is determined by the mass distribution of the target; the rotational motion is characterized by Euler's equations of motion and is dependent on the target moments of inertia I_x, I_y, I_z that govern the solutions to Euler's equations of motion for a given torque applied to the target [120]. Techniques for removing the effects of translational motion from the received data are discussed in section F.2.

Denote $E_R(f, t)$ as the measured frequency-time pulse train data which contain only the effects of the rotational motion after compensating for the target's translational motion. $E_R(f, t)$ is registered to a CoR location on the target corresponding to either the static range or field measurement scenario. For the measurement system, the time evolution of the field scattered from the rotating target defined relative to the origin O_r, takes the form

$$E_R(f,t) \sim \sum_n D'_n(f,t) e^{-j\frac{4\pi f}{c}\hat{k}(t)\cdot \underline{r}_{cn}} \tag{F.2}$$

where the scattering center locations $\underline{r}_{cn}$ are registered to the center of rotation at O_r as indicated in figure F.1. Referenced to the target-centered space, the location of O_r is related to that of O_t by a constant vector $\underline{r}_c$. Consequently, the scattering center locations referenced to the measurement system O_r are related to those referenced to the model system O_t by the translation

$$\underline{r}_{cn} = \underline{r}_n - \underline{r}_c \tag{F.3}$$

The estimated diffraction coefficients are extracted from the measurement data using the estimated target model poles as described in section 15.4.3 using equations (15.3)–(15.6). The target model poles used in equation (15.4) are defined relative to the measurement system scattering center locations $\underline{r}_{cn} = (x_{cn}, y_{cn}, z_{cn})$, $1, \ldots, N$ referenced to O_r. The computational signature model diffraction coefficients and scattering center locations are defined relative to the origin located at O_t. Consequently, an estimate of $\underline{r}_c$ is required to fuse the estimated diffraction coefficients into the computational model.

The LoS viewing angle to the target is defined in the (x', y', z') system, so that $\hat{k}(t) = \hat{k}(\theta', \varphi')$, $D'_n(f, t) = D'_n(f, \theta', \varphi')$ and the spherical coordinate angles (θ', φ') change with time as a function of rotational motion referenced to the (x', y', z') system centered at O_r. The (x', y', z') axes associated with the origin O_r will in general not be aligned to the (x, y, z) axes associated with the computational model. The relationship of the computational signature model coordinate system defined at origin O_t to the field/ static range coordinate system having origin O_r located at the target CoR is illustrated in figure F.1.

The alignment of the static range and/or field measurement coordinate systems with the computational model defined relative to the origin O_t is critical. The alignment process for the static range data is straightforward. Although the location of the origin of the target as placed on the pylon might differ from that used for the computational model, the orientation of the (x', y', z') coordinate axes of the static range measurement system are defined by the observer, and can be chosen in a manner easily related to the (x, y, z) system associated with the computational model. The alignment process for the field measurement is inherently more difficult because the target motion is referenced to inertial space, and the coordinate system used to characterize the target motion might differ considerably from that of the computational system. Both viewing angle alignment and target offset adjustment to the computational system must be performed.

F.1.1   A Simulation Example

In order to illustrate the procedure for aligning the static range or field measurement coordinate systems to the computational model, a simulation example is used to represent both the static range measurement data and the field measurement data. The simulation uses the canonical cylindrical target described in chapter 2, and is illustrated in figure F.2. The data are generated using the computational signature model defined in equation (F.1), having scattering center locations defined relative to origin O_{r1} for the static range simulation and O_{r2} for the field measurement simulation. For simplicity, the edges of the cylinder are modeled as slipping rings, and the fixed-point scattering centers are omitted from the simulation. The side and end specular scattering from the cylinder are not included in the simulation model. The radar parameters are at X-band (10 Ghz) having a bandwidth of 1 Ghz. The motion parameters for the computational model used for each simulation data set are defined in the (x, y, z) coordinate system relative to figure 9.1(a) having the z-axis of the target aligned along the cylinder axis. The simulation motion parameters for each measurement scenario are given by

Static range simulation: $\kappa = \pi/2$, $\theta_p = \pi/2$, $T_p = 1\text{s}$, $\alpha_p = 0°$, $f_s = 0$, $\alpha_s = 0$.

Field measurement simulation: $\kappa = 70°$, $\theta_p = \pi/2$, $T_p = 1\text{s}$, $\alpha_p = 60°$, $f_s = 0$, $\alpha_s = 0$.

For the static range measurement simulation the CoR, O_{r1}, is located at the midpoint of the cylinder z-axis. The target can be viewed as in pure tumble motion, with the sensor viewing the target in the plane of tumble, typical of a static range measurement environment. For comparison to the field data, the static range data is assumed to have been collected as a function of time, having a rotation period of 1s, so that the pylon rotates 360 degrees over one rotational period. This also provides a range-rate scale to the static range data in order to provide both a range and range-rate observable sequence for application of the target space composite mapping equations.

For the field data experiment the target is also in pure tumble motion with tumble period $T_p = 1$ s, but the assumed mass distribution of the target is such that the CoR, O_{r2}, of the target is offset from O_{r1} with a displacement $\underline{r}_c$ relative to O_{r1} given by $\underline{r}_c = -0.2\hat{x} - 0.4\hat{z}$ m as illustrated in figure F.2. The orientation of the target relative to the radar LoS in inertial space is such that the viewing angle of the radar is elevated $20°$ out of the tumble plane. An arbitrary rotational phase of $60°$ is added to the simulation data. The field simulation measurement data are defined relative to the CoR at origin O_{r2}.

The target geometry, cylinder end slipping rings and the locations of the CoR associated with each respective measurement simulation are illustrated in figure F.2. Shadowing of each cylinder end not visible to the sensor LoS is included in the simulation.

As discussed previously, a separate coordinate system, using the primed notation (x', y', z') illlustrated in figure 9-1(b) is introduced to process the measurement data described as follows:

For the static range scenario, the target is positioned on the measurement pylon such that the cylinder axis (z-axis of the target-centered system) is orthogonal to the pylon rotation axis, and the sensor LoS to the target lies in the plane of rotation. The target is positioned such that the target-centered origin O_{r1} is located on the pylon rotation axis midway on the cylinder axis. The (x', y', z') axes are chosen such that the target rotates in the $(x', y', z' = 0)$ plane. In this coordinate system, the radar LoS is $\theta' = 90°$, and the target appears to be in tumble motion. The static range measurement data is processed assuming a motion $\varphi'(t) = 2\pi t / T_p$, $T_p = 1s$ such that the measurement rotational angle φ' satisfies $0 \leq \varphi' < 2\pi$.

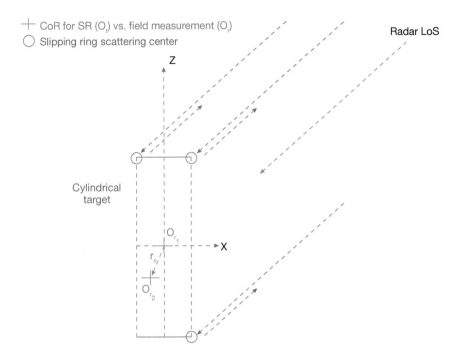

Figure F.2　Cylindrical target model illustrating the center of rotations assumed for the static range and field measurement scenarios. The edges of the cylinder are modeled as slipping rings. The side and end specular scattering from the cylinder are not included in the simulation model. The locations of the CoR associated with each respective measurement simulation are illustrated.

For the field measurement example, the target tumbles in the (x', y') plane so that z' is perpendicular to the tumble plane and aligned with the angular momentum vector. Consequently, the (x', y', z') axes associated with the field measurement are such that the target rotates in the $(x', y', z' = 0)$ plane and the radar LoS to the target is such that $\theta' = \kappa = 70°$.

Figure F.3 illustrates wideband two-dimensional (2D) composite target space mappings projected on the (x', y') plane obtained by processing the static range and field measurement data sets over a full tumble cycle. The composite mappings are determined by extracting the range, range-rate observable sequences from each simulated measurement data set over the time span of the data collection and applying the wideband composite mapping techniques developed in chapter 6, equation (6.14), over the complete rotation

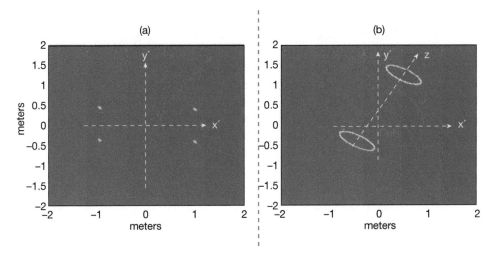

Figure F.3   Two-dimensional composite target space mappings obtained using each simulated measurement data set assuming the cylindrical target illustrated in figure F.2. (a) The composite mapping is formed using the 2D wideband motion to target mapping equations (6.14) using static range data collected by placing the cylindrical target on the measurement pylon positioned such that the cylinder axis is horizontal to the vertical pylon and rotated 360°. (b) The composite mapping is formed using the 2D wideband motion to target mapping equations (6.14) using field simulation measurements on the target assuming the target is in tumble motion and a motion estimate characterized by $0 \leq \varphi' < 2\pi$ such that $\Phi'(t=0) = 0$, and the radar viewing angle to the target is elevated 20° from the tumble plane.

cycle defined by $0 \leq \varphi' < 2\pi$. For the static range composite mapping, the motion is known; for the field experiment composite mapping, the rotational motion is estimated from the simulation data. The range, range-rate observable sequences for each measurement scenario are discussed later in this appendix and are illustrated in figure F.6.

Figure F.3(a) illustrates the 2D composite mapping associated with the static range data. The (x', y') axes associated with the target and pylon rotational angle φ' are referenced to the origin O_{r1} defined relative to the target position on the pylon. The rotation angle φ' is defined relative to the x'-axis. Observe that, since the side and end specular scattering are not included in the simulation model, only the two edges of each cylinder end lying in the plane of rotation exhibit scattering from the target.

Figure F.3(b) illustrates the composite mapping associated with the field measurement data. In order to develop the composite mapping, the motion must be estimated. The motion estimation can be carried out by first extracting the range, range-rate observable sequence over the time span of the data collection and using one of the techniques developed in chapters 8–10 in the text. As discussed in chapter 9, for a target in pure tumble motion, motion estimation is accomplished by introducing a parameter-based motion function $\varphi'(t)$ defined in (x', y') space, and the resulting composite mapping is projected into this (x', y') coordinate system. Because the field data received by the radar, after compensating for the translational motion, has rotational motion inherently registered to the CoR defined at the location of the origin O_{r2}, the motion estimate is referenced to this origin. Assume the motion estimate $\varphi'(t)$ is characterized by the functional form $\varphi'(t) = 2\pi t / T_p + \alpha_p$. Because the orientation of the target relative to the radar LoS is unknown at reference time $t = 0$, it is convenient to set $\alpha_p = 0$ and estimate the tumble period using the range observable sequence in figure F.6(b) such that $T_p = 1$s. Noting the 60° phase offset used in the field measurement data set simulation, and the assumption $\alpha_p = 0$, the resulting composite mapping appears rotated in (x', y') space. Observe that because the radar line of sight is elevated out of the tumble plane, and the wideband range, range-rate observables are used to form the composite mapping, the composite mapping is actually a slightly distorted projection of the target onto the (x', y') plane. Because the radar LoS is elevated above the tumble plane, the entire circumference of each cylinder edge (and the top portion of the target) becomes visible to the radar as the target rotates, as indicated by the elliptical shape of each end of the target in the composite mapping. The out-of-plane angle is given by $\pi/2 - \theta'$ and can be estimated from the composite mapping using the knowledge of the target cylindrical radius at each edge. The distorted projection of each ring (cylinder edge) onto the (x', y') plane is consistent with the results discussed in chapter 9, section 9.2 when using the wideband range, range-rate composite imaging technique for which the viewing angle to the target is elevated out of the rotational plane. The (x', y') axes associated with the target and the rotational angle φ' are illustrated in figure F.3b. A close examination of the two composite mappings shows that the field measurement mapping is rotated 60° relative to the static range mapping and the CoR, O_{r2}, of the field measurement is displaced from O_{r1} consistent with the CoR offset $\underline{r}_c = -0.2\hat{x} - 0.4\hat{z}$ m. The actual scaling of the displacement of O_{r2} relative to O_{r1} must take account of the composite mapping dimensional scaling caused by the 20° out-of-tumble plane viewing angle.

The resulting field measurement composite mapping is inherently registered to the CoR located at the assumed center of mass of the target. This is made evident by noting that the location of CoR O_{r2} occurs at the $(0, 0)$ origin of the composite target space mapping. Referencing the location of the field measurement CoR relative to the target geometry provides a high-resolution estimate of the location of the center of mass on the target, and is discussed later in section F.4. A useful analogy to note is that once the translational motion is appropriately compensated, the field data are registered as if the target is rotating about a synthetic pylon located in inertial space.

Treated separately, both composite mappings provide a valid representation of the target. However, in fusing the diffraction coefficients extracted from each measurement data set into a computational signature model, the extracted coefficients must each be adjusted appropriately.

In the general case of arbitrary motion, joint solutions for motion and target scattering center locations using the field data referenced to the origin O_r will provide a three-dimensional view of the target topology and an estimate of motion (θ', φ') defined relative to the (x', y', z') axes characterizing the target relative to the origin O_r. The field measurement viewing angle observation history must then be associated to the computational system for integration into the computational model. For the field data simulation example considered in figure F.3(b), the association of (θ', φ') corresponding to the field measurement to (θ, φ) used for the simulation data computational model can be developed as follows.

The field data are represented using the (θ', φ') measurement coordinate system illustrated in figure 9.2(b), while the computational signature model is referenced to the (θ, φ) coordinate system illustrated in figure 9.2(a). The field measurement derived motion and viewing angle history must be associated to the computational model coordinate system. This is accomplished by applying the Euler motion parameters estimated from the field measurement to the Euler-to-motion viewing angle transformation developed in appendix A, for which the aspect and roll angles (θ, φ) are registered to the z-axis of the computational model referenced to figure 9.1(a). The complete set of estimated motion parameters are given by

$$\theta_p = 90°, \ T_p = 1\text{s}, \ \alpha_p = 60°, \ \kappa = 70°, \ f_s = \alpha_s = 0.$$

and are, in fact, the motion parameters used to simulate the field measurement data set. Figure F.4 illustrates the motion (θ, φ) computed using the Euler-to-motion transformation using these parameters.

Observe that the field measurement viewing angles to the target, referenced to the computational model, present a wide variety of look angles, and are considerably different than those occurring for the typical roll-cut motion associated with static range measurements. Visualizing the target as observed by the radar having LoS out of the tumble plane, the target appears to roll (φ), and changes in the aspect angle θ allow a top view of the target as it rotates in the tumble plane. All locations along the

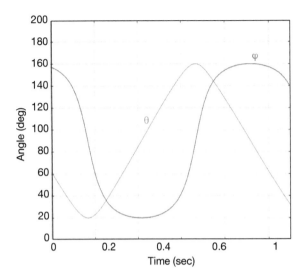

Figure F.4 The motion (θ, φ) determined using the Euler-to-motion transformation characterized by the complete set of motion parameters estimated from the simulation field data: $\theta_p = 90°$, $T_p = 1$s, $a_p = 60°$, $\kappa = 70°$, $f_s = a_s = 0$.

circumference of each cylinder end become visible to the sensor. For the example considered, each cylinder end modeled by a ring-scattering center exhibits no change in RCS versus roll. However, for a more realistic target, roll behavior of the estimated diffraction coefficient of the cylinder end would be observed. It should be noted that the same roll and aspect viewing angle to the target can be realized on the static range by elevating the sensor LoS out of the rotational plane, as discussed in chapter 15.

Two static range measurement types are complimentary when incorporating static range measurements into the computational model: roll-cut measurements versus sting platform measurements (see figure 15.9). For the roll-cut measurement having radar LoS in the rotational plane, the composite mapping will be two-dimensional having out of plane scattering components projected into the plane of the mapping; for a sting pylon measurement, the composite mapping will be three-dimensional. Knowledge of the target geometry becomes essential to associate the observation angles associated with the two measurement data sets. For the corresponding field data measurement, the case for a target in pure spin motion is analogous to a sting pylon static range measurement and the viewing angle association is straightforward. The case of a target in pure tumble and radar viewing angle elevated out of the tumble plane results in a more complex viewing angle association to the computational model.

F.1.2   Integration of the Measurements into the Computational Model

The phase correction required to register the diffraction coefficients extracted from the static range or field measurement for incorporation into the computational signature model is most easily developed assuming the observation angles for each measurement have been properly associated to the computational model. Assuming the viewing angle association has been made, the radar viewing angles to the target referenced to the (x', y', z') measurement system can be registered in the (x, y, z) computational system. The association is illustrated in figure F.1. Since the target is in the far field of the radar, the viewing angle to the target relative to both the origin of the computational system and the measurement system is the same.

The target scattering center locations defined relative to O_r are defined by $\underline{r}_{cn}$. Referenced to target space, the origin O_r is simply related to O_t by the translation of a constant vector $\underline{r}_c$, and the scattering center locations $\underline{r}_{cn}$ are defined by equation (F.3) relative to the target centered (x, y, z) axes, and are independent of the viewing angle to the target. Using the motion solution for (θ', φ'), the time variation of the scattered field in equation (F.2), defined in the (x', y', z') measurement system, takes the form

$$E_R(f,\theta',\varphi') \sim \sum D'_n(f,\theta',\varphi') e^{-j\frac{4\pi f}{c}\hat{k}(\theta',\varphi')\cdot \underline{r}_{cn}} \tag{F.4}$$

The spherical coordinate angles (θ', φ') in equation (F.4) are defined relative to the origin O_r of the target center of rotation so that $\hat{k}(t) = \hat{k}(\theta', \varphi')$, $D'_n(f,t) = D'_n(f\,\theta',\varphi')$ and the spherical coordinate angles (θ', φ') change with time as a function of rotational motion. The spherical coordinate angles (θ, φ) for the computational model in equation (F.1) are defined relative to the origin O_t having axes (x, y, z). Using equation (F.3) in equation (F.4), the signature model characterizing the field experiment can be expressed relative to the (x_n, y_n, z_n) coordinates of the scattering center locations $\underline{r}_n$ defined relative to O_t for the computational model.

$$E_R(f,\theta',\varphi') \sim \sum D'_n(f,\theta',\varphi') e^{-j\frac{4\pi f}{c}\hat{k}(\theta',\varphi')\cdot(\underline{r}_n - \underline{r}_c)} \tag{F.5}$$

Express equation (F.5) in the form

$$E_R(f,\theta',\varphi') \sim \sum \left\{ D'_n(f,\theta',\varphi') e^{j\frac{4\pi f}{c}\hat{k}\cdot(\theta',\varphi')\cdot \underline{r}_c} \right\} e^{-j\frac{4\pi f}{c}\hat{k}(\theta',\varphi')\cdot \underline{r}_n} \tag{F.6}$$

Note that in equation (F.6), the observation angle (θ', φ') is defined relative to the measurement observation space and the association of (θ, φ) to (θ', φ') is required in order to apply the estimated diffraction coefficients to the computational model. Denote the viewing angle association of the (x, y, z) axes to the (x', y', z') axes as $(\theta, \varphi) = T_a(\theta', \varphi')$, where the notation $T_a(\theta', \varphi')$ denotes the association of (θ', φ') to (θ, φ). Comparing equation (F.6) to the computational model defined in equation (F.1), the diffraction coefficients $D_n'(f, \theta, \varphi)$ extracted from the field experiment must be modified by

$$D_n(f, \theta, \varphi) = D_n'(f, T_a(\theta', \varphi'))e^{j\frac{4\pi f}{c}\hat{k}(T_a(\theta', \varphi'))\cdot \underline{r}_c}, \tag{F.7a}$$

for incorporation into the computational model, as well as the radar LoS viewing angle association

$$\hat{k}(\theta, \varphi) = \hat{k}(T_a(\theta', \varphi')). \tag{F.7b}$$

An example of the association mapping is illustrated in figure F.4.

The impact of equation (F.7a) is significant: integration of the field test data into the MBM computational signature model defined by equation (F.1) requires an estimate of the location, $\underline{r}_c$, of the center of mass relative to the origin O_p, as well as the association of the field data measurement angle observation space to the computational model viewing angle space. As discussed previously, this association is straightforward for the two complimentary field measurement cases of tumble and spin, but more complex for arbitrary motion.

F.2   Data Registration Relative to the Target Rotational Motion

For the field measurement, the data received by the sensor is not inherently registered to the rotational motion of the target and additional compensation procedures are required to properly register the observable sequence in the rotational system. For Newtonian motion referenced to inertial space, the motion of the target as viewed by the radar consists of two fundamental types: translational, and rotational about the target's center of rotation. The observable sequence extracted directly from the received pulse train contains information on both the translational and rotational motion. Figure F.5 illustrates a notional scenario for a sensor viewing a target on a Newtonian-like trajectory, illustrating the combination of translational and rotational motion.

Assuming the motion is torque-free, the direction of the angular momentum vector $\underline{J}$ remains constant relative to inertial space and the LoS of the radar to $\underline{J}$, denoted as

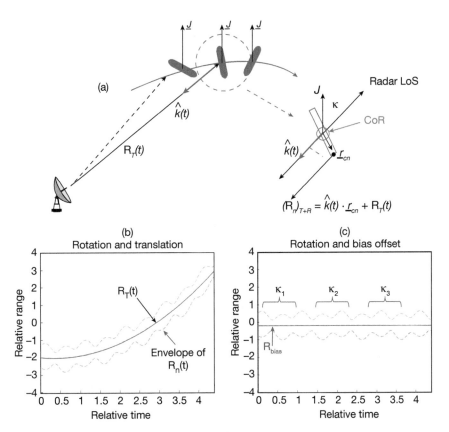

Figure F.5 (a) Combined translation and rotational motion for a target undergoing torque-free Newtonian motion. (b) Example range observable sequence including both rotational and translational motion. (c) Range observable sequence envelope after the temporal variation of the translation motion has been removed, resulting in a constant range bias offset. Independent piecewise segments of data, each having a different κ, are notionally indicated on the figure.

the angle κ, changes as a function of time. The time duration along the trajectory illustrated corresponds to the most general Condition C discussed in section 5.1 during which the look angle κ to the angular momentum vector J changes over the time span of the data collect. Denote the radar range to the target trajectory defined relative to the center of mass as $R_T(t)$. The long-term trend of $R_T(t)$ characterizes the translational motion of the target center of mass, while the short-term variations of the observable sequence, referenced to $R_T(t)$, characterize the rotational motion defined relative to the CoR. Of prime interest in this text is Condition B defined in section 5.1 that there exist piecewise segments of data over which κ is constant and over this time span the rotational motion can be decomposed from the translational motion. The following development will discuss techniques required to register the observable sequence relative to the target CoR.

Upon examination of the inset illustrated in figure F.5(a) the range to the nth scattering center on the target resulting from the combination of the translational and rotational motion as viewed by the sensor takes the form

$$(R_n)_{T+R} = \hat{k}(t) \cdot \underline{r}_{cn} + R_T(t), \, n = 1, \ldots, N, \tag{F.8}$$

where $\underline{r}_{cn}$ denotes the location of the nth scattering center on the target defined relative to the target center of rotation. The range $R_T(t)$ changes with the trajectory motion of the target, and because it follows the trajectory of the center of mass, it is independent of the location of the scattering centers on the target. The term $\hat{k}(t) \cdot \underline{r}_{cn} \equiv R_n(t)$ characterizes the rotational motion of the nth scattering center relative to the look angle to the sensor as it rotates about the target center of rotation, and defines the observable sequence of interest for characterizing the target signature.

The extracted range observable sequence resulting from both translation and rotation at block times $t = t_b$, $b = 1, \ldots, B$ is given by

$$(R_{n,b})_{T+R} = R_n(t_b) + R_T(t_b), \, n = 1, \ldots, N, b = 1, \ldots, B. \tag{F.9}$$

where $R_T(t_b)$ denotes the translational motion as a function of block time, and $R_n(t_b) = \hat{k}(t_b) \cdot \underline{r}_{cn}$ the range observable sequence resulting from the rotational motion of the target. Figure F.5(b) illustrates the envelope of an example range observable sequence characterized by equation (F.9) for a spinning/precessing target extracted from sliding blocks of pulse train data defined relative to the sensor's look angle to the target. The example motion illustrates the combined translational and rotational motion for the case where the rotational motion dominates the translational motion.

As depicted in figure F.5(a) the look angle κ changes over the time span of the data collect. However, as discussed in chapter 5, section 5.1, of primary interest is Condition B such that the data are processed over a data interval that allows for significant scattering center migration, but κ, the viewing angle between the radar and the angular momentum vector, J, is constant. The objective here is to solve for the motion parameters appropriate to this time interval and use them to unwrap the scattering center time histories back to a target-fixed coordinate system using the composite target space mapping equations. Techniques to accomplish this unwrapping form the basis for the techniques introduced in part 2 of the text. However, these techniques assume the translational motion has been removed. Thus, when processing the observables extracted from a sequence of data blocks covering the time interval of constant κ, alignment of the range, range-rate and range-acceleration observable sequences to the target CoR becomes essential. This is accomplished by incorporating an estimate of $R_T(t)$ into the framework.

If the track time on target is sufficiently long, an estimate of the temporal variation of $R_T(t)$, denoted as $\hat{R}_T(t)$, can be determined and modeled. A range alignment of the range observable sequence compensating for $R_T(t_b)$ effectively registers the observable sequence to the target CoR. This range alignment is applied to equation (F.9), which results in the estimate of the rotational component of the range observable sequence in the form

$$(R_{n,b})_R = R_n(t_b) + R_T(t_b) - \hat{R}_T(t_b), n = 1, \ldots, N, b = 1, \ldots, B. \quad (F.10)$$

Various factors primarily determined by the track data signal-to-noise ratio result in errors in the estimate of the translational motion $R_T(t)$. For long track times and closely spaced data blocks, the estimate becomes increasing more accurate. When applied to the observable sequence over the shorter time frame of constant κ, the range alignment in equation (F.10) typically reduces to a constant range bias, denoted as R_{bias}. The resulting range observable sequence characterizes the behavior of the rotational motion of the target, relative to a fixed bias offset.

$$(R_{n,b})_R = R_n(t_b) + R_{bias}, n = 1, \ldots, N, b = 1, \ldots, B. \quad (F.11)$$

Figure F.5(c) illustrates a portion of the range observable sequence envelope corresponding to a fixed $\kappa = \kappa_1$ after the temporal variation of the translational motion has been removed, resulting in the constant range bias, unique to each κ, indicated in equation (F.11).

In practice, the correction to remove the translational motion from the observable sequence is first applied directly to the received radar frequency-time pulse train data. Denote the received pulse train data including both translational and rotational motion as $E_{R+T}(f, t)$. Using the estimate of $\hat{R}_T(t)$, the pulse train data corresponding to only the rotational motion, $E_R(f, t)$ is given by

$$E_R(f, t) = E_{R+T}(f, t) e^{+j\frac{4\pi f}{c}\hat{R}_T(t)}. \quad (F.12)$$

Applying the sequential processing techniques introduced in chapters 2 and 3 to $E_R(f, t)$ results in range, range-rate and range-acceleration sequences, apart from a range bias, directly aligned to the target CoR.

When the radar, observes the target over a much larger segment of the trajectory, individual observation intervals can have different κ, and different values of R_{bias}. In such a case the larger observation interval can be divided into smaller intervals for which κ is constant or nearly so. Piecewise segments of the range observable sequence over which κ is nearly constant are indicated in figure F.5(c). By processing piecewise segments of data independently, each having a different κ, the time variation of κ can be estimated. Fusing multiple estimates of the target scattering center locations from multiple values of κ is made possible by noting that the location of the CoR on the target is independent of the observation interval. Thus, the estimates of the target scattering center locations referenced to target space, apart from a possible different orientation in (x', y', z') space, must be the same for each individual observation interval. The noncoherent combination of the scattering center estimates from the individual observation intervals, after rotational adjustment, provides not only more robust location estimates, but also a more broader viewing angle characterization of the diffraction coefficients relative to the larger segment of the viewing angle space.

In combining data blocks widely spaced in time for which a smooth estimate of $\hat{R}_T(t)$ is not available, compensating for the temporal variation of $R_t(t)$ presents a more complex estimation problem and is not considered in this text.

As discussed throughout the text, the time evolution of the range, range-rate and range-acceleration sequences contain much information relative to the motion and locations of the target scattering centers. The character of this time evolution sequence relative to these attributes can change considerably depending on the motion and look angle to the target. The next sections use the static range and field data simulation examples discussed in the first section of this appendix to contrast the nature of the observable sequence for each scenario for two contrasting motion solutions.

F.3   Static Range versus Field Data: Characteristics of the Observable Sequence

In order to contrast the different behavior of the temporal evolution of the observable sequences resulting from static range measurements versus field measurements, consider the observable sequence characterizing each respective scenario using the simulated measurement data discussed in section F.1.1. Recall that both the static range and field simulated measurement data were simulated as functions of time, using a rotational

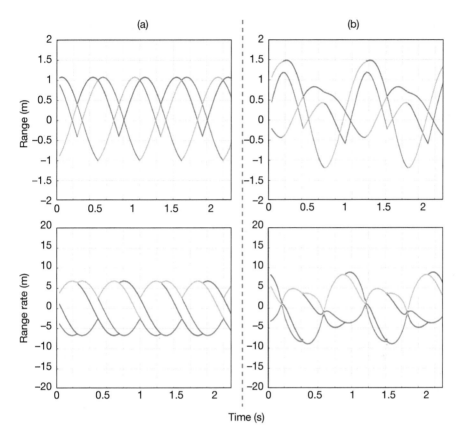

Figure F.6   (a) Range and range-rate observable sequences extracted from simulation data for static range data collected on the cylindrical target illustrated in figure F.1 having CoR located symmetrically on the target *z*-axis. (b) Range and range-rate observable sequences extracted from simulation data for the field test data collected on the same cylindrical target, but having CoR offset $\underline{r}_c$ to the static range origin O_{r1} and viewing angle to the tumble plane elevated 20°.

motion having a rotation period of 1s. Thus, the range and range-rate observation sequences extracted from each data set are properly scaled for direct comparison. The range, range-rate observable sequences extracted from the static range wideband data are illustrated in figure F.6(a). The slope discontinuities in the range observable sequence are due to the simulated effects of shadowing of the cylinder edges as discussed previously.

Note that because of the symmetry and axial positioning of the target on the pylon, the temporal variation of the static range observable sequence referenced to the origin O_{r1} provides a clear picture of the harmonic rotational motion of the target as well as the correlation of the rotational motion to the scattering center locations on the target.

For the field experiment simulation, the mass distribution on the target is such that the center of mass is offset in both the x- and z-directions, resulting in the CoR referenced to the origin O_{r2} defined by $\underline{r}_c = -0.2\hat{x} - 0.4\hat{z}$ m indicted in figure F.2. The location of $\underline{r}_c$ for a real target in the field experiment would be determined by the mass distribution of the target as discussed previously. For the field measurement simulation, the radar LoS was set to lie at 20° out of the tumble plane. This corresponds to decreasing κ to 70° relative to $\kappa = 90°$ for the static range example. The corresponding range and range-rate observable sequences extracted from the simulation field data and used to form the composite mapping illustrated in figure F.3(b), are illustrated in figure F.6(b). Comparing figure F.6(a) to figure F.6(b) it becomes clear that, although measured on the same target but having different motions, the character of the observable sequences for the static range scenario versus the field data scenario can appear quite different. In particular, it becomes more difficult to visually correlate the time evolution of the field data derived observable sequence to the motion and scattering locations on the target.

F.4   Estimating the Location of the CoR

Using the target space composite mapping techniques discussed in the text, one can process the measurement data to develop an estimate of the target scattering center locations $\underline{r}_{cn}$, $n = 1, \ldots, N$ defined relative to the CoR of the target measurement scenario. The location of the CoR relative to the location of the scattering center location estimates is defined by the (0,0) coordinates of the composite target space mapping. These estimates can be correlated to an estimate of the target model topology and computational model scattering center locations defined by $S(x_n, y_n, z_n)$, $n = 1, \ldots, N$ to provide an estimate of $\underline{r}_c$. An example of this correlation is illustrated in figure F.3(b). For field data, referencing the location of the CoR relative to estimates of the target surface topology provides a high-resolution estimate of the location of the center of mass on the target.

Appendix G: A Polynomial Filter Estimate of Scattering Center Acceleration

G.1   Introduction

In chapter 2 a linear all-pole signal model was introduced and used to develop the high resolution spectral estimate (HRSE) technique for estimating the observable sequence $(R_n, \dot{R}_n, D_n)_b, b = 1, \ldots, B$, for the range, range-rate and amplitude observable associated with each scattering center. In chapter 3 several approaches that might be used to estimate the range-acceleration associated with each scattering center, $\ddot{R}_n$, were discussed. The HRSE technique developed in chapter 3 provides an estimate of $\ddot{R}_n$ by extending the linear signal model to a nonlinear signal model incorporating the range-acceleration variable $\ddot{R}_n$. This technique for estimating $\ddot{R}_n$ has three basic attributes:

1. The technique is applicable to targets having closely spaced scattering centers relative to the radar bandwidth because the side-lobe reduction weighting required for conventional Fourier transform processing is not used, and the HRSE spectral estimation processing techniques inherently provide better resolution.

2. The complex amplitudes (D_n), $n = 1, \ldots, N$ determined using the nonlinear signal model are phase matched to the data block. This property exploits the dependence of the data block phase on $\ddot{R}_n$ as the data block size is extended beyond the limits of the linear signal model.

3. The range and range-rate observable sequence resulting from iteratively applying the nonlinear signal model to a sequence of expanding data blocks provides a smoothing effect to the initial observable sequence obtained using the linear signal model.

Nevertheless, for some applications, in particular where the optimal resolution of the HRSE processing technique is not required, it is useful to examine an alternate, computationally faster approach to estimating the acceleration observable. This appendix introduces a polynomial smoothing technique used to estimate $\ddot{R}_n$. The technique uses the range, range-rate observable sequence obtained using either the linear or nonlinear signal model as input, and provides a filtered version of each sequence as output as well as an estimate of the associated range-acceleration sequence. It is based on the assumption that each scattering center implicit in the observable sequence is well-resolved and can be tracked over short time intervals using a simple range-sorting correlation filter. The technique is particularly applicable to targets in near pure tumble motion having scattering centers that tend to persist over large angular changes of the radar viewing angle to the target.

G.2   Development

Assume the range and range-rate observable sequence is extracted from a sequence of frequency-time data blocks as described in chapters 2 and 3 using either the Fourier or HRSE spectral estimation techniques. Consider a segment of the range observable sequence consisting of S observable samples covering a short time interval T_{fit}. The time interval is assumed short enough that the S range, range-rate observable samples can be correlated as a function of time and associated with the same scattering center over the time interval defined by T_{fit}, and can be modeled by a second-order polynomial time variation over this interval. Figure G.1 illustrates a typical example of a range observable sequence assuming $(N=3)$ scattering centers on the target, where the label on the graph illustrates the observable sequence samples as a function of time and the red circles denote the second-order polynomial fit to the range observable samples over T_{fit} referenced to the time stamp $t=t_c$ centered about the data samples.

The data are assumed noise free for this example. The rationale for using a second-order polynomial fit is given by the discussion relative to equation (3.1) incorporating the scattering center acceleration phase into the nonlinear signal model. Define the local time variation t' referenced to t_c as $t' = t - t_c$. Using equation (3.1), the time variation of the range associated with the nth scattering center, $R_n(t')$ defined relative to $t_0 = t_c$, is given by

$$R_n(t') = \left(R_{n0} + \dot{R}_{n0}(t - t_c) + \frac{1}{2}\ddot{R}_{n0}(t - t_c)^2 \right), \tag{G.1}$$

Assume equation (G.1) is applied to S range observable sequence samples, $R_n(t'_s)$, $s = 1, \ldots S$, covering the region T_{fit} centered about t_c. The coefficients R_{n0}, $\dot{R}_{n0}$ and $\ddot{R}_{n0}$ characterize the polynomial fit to the S range observable samples. So determined, the coefficients R_{n0}, $\dot{R}_{n0}$, and $\ddot{R}_{n0}$ provide filtered estimates of the range, range-rate, and range-acceleration observables defined relative to time $t=t_c$ based on weighting the values of the S range samples defined over T_{fit}.

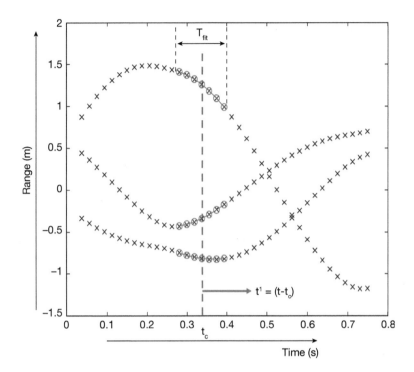

Figure G.1 Segment of the range observable sequence consisting of S observable samples covering a short time interval T_{fit}.

The corresponding polynomial fit to the range-rate observable sequence associated with the (paired) range-rate observable sequence is given by the time derivative of equation (G.1):

$$\dot{R}_n(t') = (\dot{R}_{n0} + \ddot{R}_{n0}(t - t_c)). \tag{G.2}$$

To evaluate R_{n0}, $\dot{R}_{n0}$, and $\ddot{R}_{n0}$ over a given time interval T_{fit}, assume equations (G.1) and (G.2) are evaluated at the times $t = t_s$, $s = 1,\ldots,$ S for each of the S range and range-rate observable samples centered about $t = t_c$. This leads to the system of $2S$ equations defined by

$$R_n(t'_s) = \left(R_{n0} + \dot{R}_{n0}(t_s - t_c) + \frac{1}{2}\ddot{R}_{n0}(t_s - t_c)^2 \right), s = 1,\ldots,S \tag{G.3}$$

$$\dot{R}_n(t_s') = (\dot{R}_{n0} + \ddot{R}_{n0}(t_s - t_c)),\, s = 1,\ldots,S. \tag{G.4}$$

The $R_n(t_s')$ and $\dot{R}_n(t_s')$ are the samples of the original observable sequence over T_{fit}, and R_{n0}, $\dot{R}_{n0}$, and $\ddot{R}_{n0}$ represent the filtered estimates of these observables defined relative to the center time t_c of the S samples. The $2S$ equations defined by equations (G.3) and (G.4) can be solved using a best mean square fit to the observable samples resulting in the $3N$ estimates of R_{n0}, $\dot{R}_{n0}$, and $\ddot{R}_{n0}$, $n = 1,\ldots,N$.

By extracting the polynomial fit to the HRSE or Fourier observable sequence as the sample region defined by T_{fit} is sequentially incremented through the measured observable

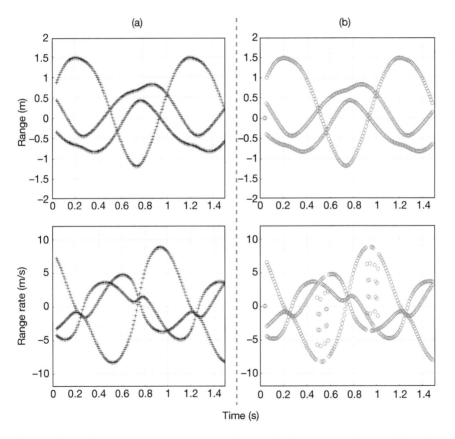

Figure G.2 Filtered range, range-rate estimates (red) obtained after applying the polynomial filter technique using the observable sequence data of figure G.2(a) (black) as input.

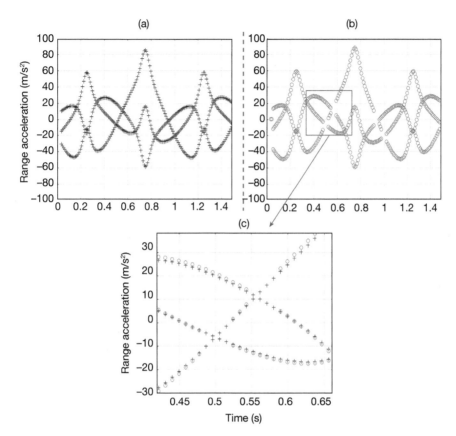

Figure G.3 Range-acceleration observable sequence estimates using two techniques. (a) The nonlinear signal model technique developed in chapter 3 versus (b) using the polynomial filter technique.

sequence, the resultant observable sequence $(R_{n0}, \dot{R}_{n0}, \ddot{R}_{n0})_b\ n = 1, \ldots, N; b = 1, \ldots, B$ defines a filtered version of the original range, range-rate observable sequence, and additionally includes an estimate of $\ddot{R}_n$ as a function block time.

Example 1: Simulation Data

Figure G.2(a) illustrates an example range and range-rate observable sequence extracted from wideband simulation data. The range, range-rate observables in figure G.2(a) are used as input to the polynomial fit technique. The observable sequence covers roughly one and a half rotation cycles. Figure G.2(b) illustrates the polynomial filtered range, range-rate estimates obtained after applying the polynomial filter technique using the observable sequence data of figure G.2(a) as input to the filter.

Figures G.3(a) and G.3(b) compare the range-acceleration observable sequence estimates using two techniques: the nonlinear signal model technique developed in chapter 3 applied to the simulation data [figure G.3(a)] versus using the polynomial filter technique [figure G.3(b)].

Observe that in general, because the input observable sequence is mostly resolved in range and range-rate for this example, the agreement using the two techniques to estimate the range, range-rate and range-acceleration observable sequences is generally good. However, in some isolated cases the agreement is poor, particularly over the time intervals $0.5 < t < 0.6$ and $0.9 < t < 1.0$ s. This occurs in regions where the joint range and range-rate resolution for the input data are poor. As an example, figure G.3(c) illustrates an expanded view of the comparison of the range-acceleration estimates using the two techniques over the region $0.5 < t < 0.6$s. The red circles correspond to the polynomial filter technique, and occasionally give poor agreement (indicated by estimates exceeding the graphical limits and missing from the figures) to the nonlinear signal model in some regions over this time interval.

References

1. R. J. Sullivan, *Radar Foundations for Imaging and Advanced Concepts*, SciTech Publishing, Raleigh, NC, 2000.

2. W. W. Camp, J. T. Mayhan, and R. M. O'Donnell, "Wideband Radar for Ballistic Missile Defense and Range-Doppler Imaging of Satellites," *Lincoln Lab. J.* 12, no. 2 (2000).

3. D. L. Mensa, *High Resolution Radar Imaging*, Artech House, Inc., Norwood, MA, 1981.

4. D. R. Wehner, *High Resolution Radar*, 2nd ed., Artech House, Norwood, MA, 1994.

5. M. A. Stuff, "Three-Dimensional Analysis of Moving Target Radar Signals: Methods and Implications for ATR and Feature Aided Tracking," *Proc. SPIE Int. Soc. Opt. Eng.* 3721 (April 1999): 485–496.

6. R. L. Fante, *Signal Analysis and Estimation*, Wiley and Sons, Hoboken, NJ, 1988.

7. K. M. Cuomo, J. E. Piou, and J. T. Mayhan, "Ultrawideband Coherent Processing," *IEEE Trans. Antennas Propag.* 47 (1999).

8. L. D. Vann, K. M. Cuomo, J. E. Piou, and J. T. Mayhan, *Multi-Sensor Fusion Processing for Enhanced Radar Imaging*, Lincoln Lab Technical Report TR-1056 (April 2000).

9. R. F. Harrington, *Field Computation by Moment Methods*, Macmillan, New York, NY, 1968.

10. C. A. Balanis, *Advanced Engineering Electromagnetics*, 2nd ed., Wiley and Sons, Hoboken, NJ, 2012.

11. R. F. Harrington, *Time-Harmonic Electromagnetic Fields*, McGraw-Hill, New York, NY, 1961.

12. J. B. Keller, "Geometrical Theory of Diffraction," *J. Opt. Soc. Am.* 52, no. 2 (February 1962): 116–130.

13. R. G. Kouyoumjian, "Asymptotic High-Frequency Methods," *Proc. IEEE* 53 (August 1965).

14. J. E. Piou, K. M. Cuomo, and J. T. Mayhan, *A State-Space Technique for Ultrawide-Bandwidth Coherent Processing*, Lincoln Lab Technical Report 1054 (July 1999).

15. M. L. Burrows, "Two-Dimensional ESPRIT with Tracking for Radar Imaging and Feature Extraction," *IEEE Trans. Antennas Propag.* 52 (February 2004).

16. K. Suwa, T. Wakayama, and M. Iwamoto, "Three-Dimensional Target Geometry and Target Motion Estimation Method Using Multistatic ISAR Movies and Its Performance," *IEEE Trans. Geosci. Remote Sens.* 49 (June 2011).

17. C. Yang and E. Blasch, "Estimating Target Range-Doppler Slope for Maneuver Indication," *Proc. SPIE Int. Soc. Opt. Eng.* 6968 (2008).

18. M. Iwamoto and T. Kirimoto, "A Novel Technique for Reconstructing Three-Dimensional Target Shapes Using Sequential Radar Images," *Proc. IEEE IGARSS* 4 (July 2001).

19. M. Ferrara, G. Arnold, and M. Stuff, "Shape and Motion Reconstruction from 3D-to-1D Orthographically Projected Data via Object-Image Relations," *IEEE Trans. Pattern Anal. Mach. Intell.* 31 (October 2009).

20. J. T. Mayhan, M. L. Burrows, K. M. Cuomo, and J. E. Piou, "High Resolution 3D 'Snapshot' Imaging and Feature Extraction," *IEEE Trans. Aerosp. Electron Syst.* 37 (April 2001).

21. K. M. Cuomo, private communication.

22. S. L. Borison, S. B. Bowling, and K. M. Cuomo, "Super-Resolution Methods for Wideband Radar," *Lincoln Lab. J.* 5, no. 3 (1992).

23. D. L. Mensa, *High Resolution Radar Imaging*, Artech House, Inc., Norwood, MA, 1981, p. 104.

24. D. A. Ausherman, A. Kozma, J. L. Walker, H. M. Jones, and E. C. Poggio, "Developments in Radar Imaging," *IEEE Trans. Aerosp. Electron. Syst.* AES-20 (July 1984).

25. J. T. Mayhan, *"Phase-Enhanced" 3D Snapshot ISAR Imaging and Interferometric SAR*, Lincoln Lab Technical Report 1135 (December 2009).

26. J. E. Piou, "Balanced Realization for 2-D Data Fusion," AIAA Guidance, Navigation and Control Conference, San Francisco, CA, August 2005.

27. W. W. Camp, J. T. Mayhan, and R. M. O'Donnell, "Wideband Radar for Ballistic Missile Defense and Range-Doppler Imaging of Satellites," *Lincoln Lab. J.* 12, no. 2 (2000).

28. M. L. Skolnik, *Radar Handbook*, McGraw-Hill, New York, NY, 1990.

29. M. L. Skolnik, *Introduction to Radar Systems*, McGraw-Hill, New York, NY, 1980, pp. 369–370.

30. R. A. Becker, *Introduction to Theoretical Mechanics*, McGraw-Hill, New York, NY, 1954.

31. P. R. Escobal, *Methods of Orbit Determination*, Wiley and Sons, Hoboken, NJ, 1976.

32. R. F. Harrington, *Time-Harmonic Electromagnetic Fields*, McGraw-Hill, New York, NY, 1961, p. 280.

33. H. Palmeri, "Jointly Estimating 3D Target Shape and Motion from Radar Data," PhD thesis, Rensselaer Polytechnic Institute, Troy, NY, 2012.

34. W. W. Camp, J. T. Mayhan, and R. M. O'Donnell, "Wideband Radar for Ballistic Missile Defense and Range-Doppler Imaging of Satellites," *Lincoln Lab. J.* 12, no. 2 (2000).

35. S. L. Borison, S. B. Bowling, and K. M. Cuomo, "Super-Resolution Methods for Wideband Radar," *Lincoln Lab. J.* 5, no. 3 (1992).

36. M. L. Skolnik, *Radar Handbook*, McGraw-Hill, New York, NY, 1990.

37. M. L. Skolnik, *Introduction to Radar Systems*, McGraw-Hill, New York, NY, 1980, pp. 369–370.

38. R. A. Becker, *Introduction to Theoretical Mechanics*, McGraw-Hill, New York, NY, 1954, pp. 296–302.

39. P. R. Escobal, *Methods of Orbit Determination*, Wiley and Sons, Hoboken, NJ, 1976.

40. K. M. Cuomo, J. E. Piou, and J. T. Mayhan, "Ultrawideband Coherent Processing," *IEEE Trans. Antennas Propag.* 47 (1999).

41. R. F. Harrington, *Time-Harmonic Electromagnetic Fields*, McGraw-Hill, New York, NY, 1961, p. 280.

42. H. Mott, *Polarization in Antennas and Radar*, Wiley-Interscience, Hoboken, NJ, 1986.

43. E. M. Kennaugh and D. L. Moffett, "Transient and Impulse Response Approximations," *Proc. IEEE* 53, no. 8 (August 1965).

44. D. G. Dudley, "A State-Space Formulation of Transient Electromagnetic Scattering," *IEEE Trans. Antennas Propag.* 33, no. 10 (October 1985).

45. E. Rothwell, K. M. Chen, D. P. Nyquist, P. Ilavarasan, J. Ross, R. Bebermeyer, and Q. Li, "Radar Target Identification and Detection Using Short EM Pulses and the E-Pulse Technique," in *Ultra-Wideband, Short-Pulse Electromagnetics*, edited by H. L. Bertoni, L. Carin, and L. B. Felsen, Springer, New York, NY, 1993, pp. 475–482.

46. H. L. Bertoni, L. Carin, and L. B. Felsen, eds., *Ultra-Wideband, Short-Pulse Electromagnetics*, Springer, New York, NY, 1993.

47. J. M. McCorkle, V. Sabio, R. Kapoor, and N. Nandhakumar, "Transient Synthetic Aperture Radar and the Extraction of Anisotropic and Natural Frequency Information," in *Detection and Identification of Visually Obscured Targets*, edited by C. E. Baum, Taylor and Francis, New York, NY, 1999.

48. C. E. Baum, *Detection and Identification of Visually Obscured Targets*, Taylor and Francis, New York, NY, 1999.

49. D. G. Dudley, P. A. Nielsen, and D. F. Marshall, "Ultra-Wideband Electromagnetic Target Identification," in *Ultra-Wideband, Short-Pulse Electromagnetics*, edited by H. L. Bertoni, L. Carin, and L. B. Felsen, Springer, New York, NY, 1993, pp. 457–474.

50. H. L. Bertoni, L. Carin, and L. B. Felsen, eds., *Ultra-Wideband, Short-Pulse Electromagnetics,* Springer, New York, NY, 1993.

51. K. M. Chen and D. Westmoreland, "Impulse Response of a Conducting Sphere Based on Singularity Expansion Method," *Proc. IEEE* 69, no. 6 (June 1981).

52. K. M. Cuomo, J. E. Piou, and J. T. Mayhan, "Ultrawideband Coherent Processing," *IEEE Trans. Antennas Propag.* 47 (1999).

53. J. E. Piou, "Balanced Realization for 2-D Data Fusion," AIAA Guidance, Navigation and Control Conference, San Francisco, CA, August 2005.

54. M. L. Burrows, "Two-Dimensional ESPRIT with Tracking for Radar Imaging and Feature Extraction," *IEEE Trans. Antennas Propag.* 52 (February 2004).

55. R. F. Harrington, *Field Computation by Moment Methods,* Macmillan, New York, NY, 1993, pp. 107–125.

56. J. B. Keller, "Geometrical Theory of Diffraction," *J. Opt. Soc. Am.* 52 (February 1962): 116–130.

57. J. T. Mayhan, M. L. Burrows, K. M. Cuomo, and J. E. Piou, "High Resolution 3D 'Snapshot' Imaging and Feature Extraction," *IEEE Trans. Aerosp. Electron. Syst.* 37 (April 2001).

58. C. A. Balanis, *Advanced Engineering Electromagnetics*, 2nd ed., Wiley and Sons, 2012.

59. R. G. Kouyoumjian, "Asymptotic High-Frequency Methods," *Proc. IEEE* 53 (August 1965).

60. P. H. Pathak, G. Carluccio, and M. Alabani, "The Uniform Geometrical Theory of Diffraction and Some of Its Applications," *IEEE Antennas Propag. Mag.* 55, no. 4 (August 2013).

61. P. H. Pathak, "High-Frequency Techniques for Antenna Analysis," *Proc. IEEE* 80 (January 1992): 44–65.

62. K. Suwa, T. Wakayama, and M. Iwamoto, "Three-Dimensional Target Geometry and Target Motion Estimation Method Using Multistatic ISAR Movies and Its Performance," *IEEE Trans. Geosci. Remote Sens.* 49 (June 2011).

63. C. Yang and E. Blasch, "Estimating Target Range-Doppler Slope for Maneuver Indication," *Proc. SPIE Int. Soc. Opt. Eng.* 6968 (2008).

64. M. Iwamoto and T. Kirimoto, "A Novel Technique for Reconstructing Three-Dimensional Target Shapes Using Sequential Radar Images," *Proc. IEEE IGARSS* 4 (July 2001).

65. C. Stiller and J. Konrad, "Estimating Motion in Image Sequences," *IEEE Signal Process Mag.* (July 1999).

66. A. Papoulis, *Systems and Transforms with Applications in Optics*, McGraw-Hill, New York, NY, 1968, p. 66.

67. A. Papoulis, *Systems and Transforms with Applications in Optics*, McGraw-Hill, New York, NY, 1968, pp. 70–73.

68. P. Swerling, *The "Double Threshold" Method of Detection*, Rand Corp. Report RM-1008 (December 1952).

69. M. L. Skolnik, *Radar Handbook,* McGraw-Hill, New York, NY, 1990.

70. R. L. Fante, *Signal Analysis and Estimation*, Wiley and Sons, Hoboken, NJ, 1988.

71. M. Ferrara, G. Arnold, and M. Stuff, "Shape and Motion Reconstruction from 3D-to-1D Orthographically Projected Data via Object-Image Relations," *IEEE Trans. Pattern Anal. Mach. Intell.* 31 (October 2009).

72. H. Palmeri, "Jointly Estimating 3D Target Shape and Motion from Radar Data," PhD thesis, Rensselaer Polytechnic Institute, Troy, NY, 2012.

73. D. A. Ausherman, A. Kozma, J. L. Walker, H. M. Jones, and E. C. Poggio, "Developments in Radar Imaging," *IEEE Trans. Aerosp. Electron. Syst.* AES-20 (July 1984).

74. M. Iwamoto and T. Kirimoto, "A Novel Technique for Reconstructing Three-Dimensional Target Shapes Using Sequential Radar Images," *Proc. IEEE IGARSS* 4 (July 2001).

75. J. T. Mayhan, M. L. Burrows, K. M. Cuomo, and J. E. Piou, "High Resolution 3D 'Snapshot' Imaging and Feature Extraction," *IEEE Trans. Aerosp. Electron. Syst.* 37 (April 2001).

76. S. I. Borison, S. B. Bowling, and K. M. Cuomo, "Super-Resolution Methods for Wideband Radar," *Lincoln Lab. J.* 5, no. 3 (1992).

77. J. T. Mayhan, *RCS Scatterer Extraction Using A Priori Target Information*, Lincoln Lab Technical Report 1093 (March 2004).

78. D. L. Mensa, *High Resolution Radar Imaging*, Artech House, Inc., Norwood, MA, 1981, p. 104.

79. R. F. Harrington, *Time-Harmonic Electromagnetic Fields*, McGraw-Hill, New York, NY, 1961, p. 311.

80. S. Kusiak, *Advanced Narrowband Electromagnetic Size and Shape Determination*, Lincoln Lab Technical Report 1116 (February 2008).

81. R. F. Harrington, *Field Computation by Moment Methods*, Macmillan Company, New York, NY, 1993, pp. 75–78.

82. L. D. Vann, K. M. Cuomo, J. E. Piou, and J. T. Mayhan, *Multisensor Fusion Processing for Enhanced Radar Imaging*, Lincoln Lab Technical Report 1056 (April 2000).

83. A. Dumanian, C. Burt, and B. Knapp, "A Component Model Approach for the RCS Validation of an Electrically Large Open-Ended Cylindrical Cavity," IEEE AP-S International Symposium 2007, Honolulu, HI, June 2007.

84. R. J. Sullivan, *Radar Foundations for Imaging and Advanced Concepts*, SciTech Publishing, Raleigh, NC, 2000.

85. B. D. Rigling and R. L. Moses, "Three-Dimensional Surface Reconstruction from Multistatic SAR Images," *IEEE Trans. Image Process.* 14, no. 8 (August 2005).

86. W. G. Carrara, et. al., *Spotlight Synthetic Aperture Radar*, Artech House, Norwood, MA, 1995, pp. 369–374.

87. J. T. Mayhan, *"Phase-Enhanced" 3D Snapshot ISAR Imaging and Interferometric SAR*, Lincoln Lab Technical Report 1135 (December 2009).

88. R. A. Becker, *Introduction to Theoretical Mechanics*, McGraw-Hill, New York, NY, 1954, pp. 296–302.

89. P. R. Escobal, *Methods of Orbit Determination*, Wiley and Sons, Hoboken, NJ, 1976.

90. Y. Masutani, T. Iwatsu, and F. Miyazaki, "Motion Estimation of Unknown Rigid Body Under No External Forces and Moments," IEEE International Conference on Robotics and Automation, San Diego, CA, May 1994.

91. T. J. Broida and R. Chellappa, "Estimation of Object Motion Parameters from Noisy Images," *IEEE Trans. Pattern Anal. Mach. Intell.* PAMI-8 (January 1986).

92. A. N. Netravali and J. Salz, "Algorithms for Estimation of Three-Dimensional Motion," *AT&T Tech. J.* 64, no. 2 (February 1985).

93. T. S. Huang and A. N. Netravali, "Motion and Structure from Feature Correspondences: A Review," *Proc. IEEE* 82 (February 1994).

94. M. D. Lichter and S. Dubowsky, "State, Shape and Parameter Estimation of Space Objects from Range Images," IEEE International Conference on Robotics and Automation, New Orleans, LA, April 2004.

95. J. F. Kinkel and M. Thomas, "Estimation of Vehicle Dynamics and Static Parameters from Magnetometer Data," *J. Guid. Control Dyn.* 20 (January–February 1997).

96. S. Kidera, H. Yamada, and T. Kirimoto, "Accurate 3-Dimensional Imaging Method Based on Extended RPM for Rotating Target," *IEICE Trans. Commun.* E95-B (October 2012).

97. J. F. Kinkel and M. Thomas, "Estimation of Vehicle Dynamics and Static Parameters from Magnetometer Data," *J. Guid. Control Dyn.* 20 (January–February 1997).

98. C. Tomasi and T. Kanade, "Shape and Motion from Image Streams under Orthography: A Factorization Method," *Int. J. Comput. Vision* 9, no. 2 (November 1992), pp. 137–154.

99. M. Ferrara, G. Arnold, and M. Stuff, "Shape and Motion Reconstruction from 3D-to-1D Orthographically Projected Data via Object-Image Relations," *IEEE Trans. Pattern Anal. Mach. Intell.* 31 (October 2009).

100. C. Tomasi and T. Kanade, "Shape and Motion from Image Streams under Orthography: A Factorization Method," *Int. J. Comput. Vision* 9, no. 2 (November 1992), pp. 137–154.

101. L. L. Kontsevich, M. L. Kontsevich, and A. K. Shen, "Two Algorithms for Reconstructing Shapes," *Avtometriya* 5 (1987): 72–77. [Translation in *Optoelec. Instrum. Data Proc.* 5 (1987): 76–81].

102. M. Ferrara, G. Arnold, and M. Stuff, "Shape and Motion Reconstruction from 3D-to-1D Orthographically Projected Data via Object-Image Relations," *IEEE Trans. Pattern Anal. Mach. Intell.* 31 (October 2009).

103. L. D. Vann, K. M. Cuomo, J. E. Piou, and J. T. Mayhan, *Multi-Sensor Fusion Processing for Enhanced Radar Imaging*, Lincoln Lab Technical Report 1056 (April 2000).

104. K. M. Cuomo, J. E. Piou, and J. T. Mayhan, "Ultra-Wideband Coherent Processing," *IEEE Trans. Antennas Propag.* 47 (1999).

105. S. L. Borison, S. B. Bowling, and K. M. Cuomo, "Super-Resolution Methods for Wideband Radar," *Lincoln Lab. J.* 5, no. 3 (1992).

106. K. M. Cuomo, J. E. Piou, and J. T. Mayhan, "Ultra-Wideband Coherent Processing," *IEEE Trans. Antennas Propag.* 47 (1999).

107. S. Y. Kung, K. S. Arun, and D. V. Bhaskar Rao, "State-Space and Singular-Value Decomposition-Based Approximation Methods for the Harmonic Retrieval Problem," *J. Opt. Soc. Am.* 73, no. 12 (December 1983).

108. H. Akaike, "A New Look at the Statistical Model Identification," *IEEE Trans. Autom. Control* AC (1974): 716–723.

109. M. Wax and T. Kailath, "Detection of Signals by Information Theoretic Criteria," *IEEE Trans. Acoust. Speech Signal Process* ASSP-33, no. 2 (1985): 387–392.

110. J. Rissanen, "Modeling by Shortest Data Description," *Automatica* 14 (1978): 465–471.

111. M. Wax and I. Ziskind, "Detection of the Number of Coherent Signals by the MDL Principle," *IEEE Trans. Acoust. Speech Signal Process.* ASSP-37, no. 8 (1989): 1190–1196.

112. W. H. Press, S. A. Teukolsky, W. T. Vetterling, and B. P. Flannery, *Numerical Recipes in C: The Art of Scientific Computing*, 2nd ed. Cambridge University Press, Cambridge, UK, 1992.

113. J. B. Keller, "Geometrical Theory of Diffraction," *J. Opt. Soc. Am.* 52 (February 1962): 116–130.

114. L. D. Vann, K. M. Cuomo, J. E. Piou, and J. T. Mayhan, *Multi-Sensor Fusion Processing for Enhanced Radar Imaging*, Lincoln Lab Technical Report 1056 (April 2000).

115. R. O'Donnell, "Lecture 7, Radar Cross Section," AESS UNH Radar Lectures on Radar Systems, 2013.

116. E. F. Knott and T. B. A. Senior, "Comparison of Three High-Frequency Diffraction Techniques," *Proc. IEEE* 62, no. 11 (November 1974).

117. T. B. A. Senior and J. L. Volakis, *Approximate Boundary Conditions in Electromagnetics*, Institution of Electrical Engineers, 1995.

118. J. T. Mayhan, M. L. Burrows, K. M. Cuomo, and J. E. Piou, "High Resolution 3D 'Snapshot' Imaging and Feature Extraction," *IEEE Trans. Aerosp. Electron. Syst.* 37 (April 2001).

119. M. L. Burrows, private communication.

120. R. A. Becker, *Introduction to Theoretical Mechanics,* McGraw-Hill, 1954, New York, NY, pp. 296–302.

121. P. R. Escobal, *Methods of Orbit Determination,* Wiley and Sons, Hoboken, NJ, 1976.

122. J. E. Piou, "System Realization Using 2-D Output Measurements," IEEE American Control Conference, Boston, MA, June–July 2004.

123. S. Y. Kung, K. S. Arun, and D. V. B. Rao, "State-Space and Singular Value Decomposition-Based Approximation Methods for the Harmonic Retrieval Problem," *J. Opt. Soc. Am.* 73 (December 1983): 1799–1811.

124. J. E. Piou, J. T. Mayhan, and K. M. Cuomo, *Technique Development and Performance Bounds for Sparse-Band, Sparse-Angle Processing*, Lincoln Lab Technical Report NTP-4 (June 2001).

125. M. L. Burrows, "Two-Dimensional ESPRIT with Tracking for Radar Imaging and Feature Extraction," *IEEE Trans. Antennas Propag.* 52 (February 2004).

126. F. Vanpoucke, M. Moonen, and Y. Berthoumieu, "An Efficient Subspace Algorithm for 2-D Harmonic Retrieval," Proceedings of ICASSP, IEEE International Conference on Acoustics, Speech, and Signal Processing, New York, NY, April 1994.

127. J. E. Piou, this text, appendix B.

128. B. Ottersten, M. Viberg, and T. Kailath, "Performance Analysis of the Total Least Squares ESPRIT Technique," *IEEE Trans. Signal Process.* 39, no. 5 (May 1991): 1122–1135.

129. G. H. Golub and C. F. Van Loan, *Matrix Computations*, Johns Hopkins, Baltimore, MD, 1996, p. 395.

130. P. Strobach, "Total Least Squares Phased Averaging and 3-D ESPRIT for Joint Azimuth-Elevation-Carrier Estimation," *IEEE Trans. Signal Process.* 49, no. 1 (January 2001): 54–62.

131. R. P. S. Mahler, "Multitarget Bayes Filtering via First-Order Multitarget Moments," *IEEE Trans. Aerosp. Electron. Syst.* 39, no. 4 (2003): 1152–1178.

132. T. Zajic and R. P. S. Mahler, "A Particle-Systems Implementation of the PHD Multitarget-Tracking Filter," in *Proceedings, Signal Processing, Sensor Fusion, and Target Recognition XII*, edited by I. Kadar, SPIE, 2003, pp. 291–299.

133. B.-N. Vo, S. Singh, and A. Doucet, "Sequential Monte Carlo Implementation of the PHD Filter for Multi-Target Tracking," in *Proceedings, Sixth International Conference of Information Fusion*, International Society of Information Fusion, 2003, pp. 792–799.

134. B.-N. Vo, S. Singh, and A. Doucet, "Sequential Monte Carlo Methods for Multi-Target Filtering with Random Finite Sets," *IEEE Trans. Aerosp. Electron. Syst.* 41, no. 4 (October 2005): 1224–1245.

135. M. Tobias and A. D. Lanterman, "Techniques for Birth Particle Placement in the Probability Hypothesis Density Particle Filter Applied to Passive Radar," *IET Radar Sonar Navig.* 2, no. 5 (October 2008): 351–365.

Contributors

Much of the material in this book derives from work conducted at Lincoln Laboratory over the past 25 years and has been published in the open literature. In some cases, the authors have asked the original contributors to provide content adapted for inclusion into this book. This section attempts to provide acknowledgment of their efforts.

Chapter 2

Many of the concepts presented in this chapter were formulated over a period of time with collaboration from Drs. Jean Piou, Kevin Cuomo, and Michael Burrows. Development of the two-dimensional spectral estimation techniques, essential to the application of the sequential estimation concept and described in appendixes B and C, is the result of work by Dr. Piou and Dr. Burrows, respectively. The all-pole signal model development and its connection to the geometrical theory of diffraction were extracted from [20]. Figure 2.5 is used courtesy of Dr. Audrey Dumanian.

Chapter 5

The basis for most of the material in this chapter was adapted from [20], coauthored by Drs. Jean Piou, Kevin Cuomo, and Michael Burrows. Development of the target space filter concept was in collaboration with Dr. Martin Tobias, and extended in concept by Dr. Tobias in appendix D.

Chapter 6

Figures 6.1 and 6.2 are used courtesy of Drs. Michael Burrows and Chuck Burt. The narrowband cross-range pole integration technique was developed in concert with Shirin Kubat, as well as the narrowband image analysis of sections 6.4.2 and 6.4.3 using this technique. Simulation data was developed by Dr. Audrey Dumanian.

Chapter 10

The analysis in this chapter was carried out independently by Dr. Michael Burrows, and adapted for incorporation into this text to complement the joint solutions technique developed in chapter 9.

Chapter 11

Considerable material in this chapter was adapted from [8] coauthored by Drs. Laura Vann, Kevin Cuomo, and Jean Piou.

Chapter 12

The material in this chapter was developed in concert with Dr. Audrey Dumanian.

Chapter 13

The material in this chapter was adapted from [7], coauthored by Drs. Kevin Cuomo and Jean Piou. The technique for joining the two sparse bands, essential to the fusion technique, was developed by Dr. Cuomo.

Chapter 14

The basis for most of the material in this chapter was adapted from [8].

Chapters 15 and 16

The material in these two chapters was developed in concert with Dr. Audrey Dumanian and Dr. Michael Burrows.

Appendixes

Appendix B is based on a development by Dr. Jean Piou, adapted from [26]. Appendix C is based on a development by Dr. Michael Burrows, adapted from [15]. Appendix D is based on a development by Dr. Martin Tobias.

About the Authors

Dr. John A. Tabaczynski began his career at the Massachusetts Institute of Technology Lincoln Laboratory (LL) in 1962 as a summer intern. In November 1966 he came to the Laboratory as a full-time staff member. During his career he served as Assistant, Associate, and in February 1978 he became full Group Leader of the Ballistic Missile Defense Analysis and Systems Group. In October 1984 he was promoted to Associate Division Head of the BMD Division. His primary technical interests evolved over the years. Initial tasks involved the application of Kalman filtering techniques to the problem of radar tracking and discrimination. Through the 70s he participated in a number of major defense system studies that shaped the evolving Ballistic Missile Defense System architecture. In the late 80s he was responsible for managing the Laboratory's Kwajalein Missile Range Program. Through the 90s he was active in promoting advanced radar technology for BMD applications. In 2000 he worked with staff at the Missile Defense Agency to lead a national effort in the area of ballistic missile technology development and testing. In 2004 he was appointed Principal Laboratory Researcher where he continued to work on areas of advanced radar technology. He held a BSEE from MIT (1960), and earned an MS (1962) and PhD (1965) in Electrical Engineering from Purdue University. Before joining Lincoln, he was a staff member at NASA's Jet Propulsion Laboratory in Pasadena, CA. He was an avid hiker, backpacker, and photographer, having hiked the trails and mountains of New England, the Adirondacks, and South America.

Dr. Joseph T. Mayhan received his B.S degree from Purdue University, Lafayette, Indiana, in 1963 and his MS and PhD degrees from The Ohio State University, Columbus, Ohio, in 1964 and 1967, respectively. From 1968 to 1969, he was employed at Avco Corporation, Wilmington, MA, working in the areas of re-entry systems analysis, microwave breakdown, and nonlinear interactions of electromagnetic waves in plasmas. In 1969, he joined the faculty of the University of Akron in Ohio, serving as an Associate Professor in the Department of Electrical Engineering. Since June 1973 he has been employed at MIT's LL in Lexington, Massachusetts. During this time, he has worked in the areas of satellite communications antennae, adaptive antenna design, spatial spectral estimation using multiple beam antennas, electromagnetic scattering from actively loaded targets, and radar data analysis. He served two four-year tours at the laboratory's field site at Kwajalein in the Marshall Islands, one tour as Leader of the ALTAIR deep space tracking radar and a second as Site Manager of the Lincoln Technical Program at Kwajalein. On return he served as Leader of the Sensor Systems and Measurements Group. Since his retirement from MIT, he continues working on a part-time basis as a senior member of the technical staff. He has published extensively in the areas of non-linear interactions of electromagnetic waves in plasma media, adaptive antenna design, spatial spectral estimation techniques, electromagnetic scattering, and radar signal processing. He twice received the RWP King Award from the Institute of Electrical and Electronics Engineers (IEEE) Antennas and Propagation Society (AP-S) for paper published in the journal *IEEE Transactions on Antennas and Propagation*, and served as a Distinguished Lecturer for AP-S.

Index

Acceleration estimation, 73
  Fourier vs. HRSE, 87, 91
  nonlinear signal model, 85
  scattering center location, 153
Admittance matrix, 182
  vs. mutual coupling, 162
  vs. operational frequency band, 163
All-pole model, 55
Aspect and roll angles, 21
  Euler motion, 65
  time evolution, 65
Autocorrelation, 101
  autocorrelation contour, 120
  generalized autocorrelation matrix, 102
Autocorrelation contour, vs. range
    range-rate image, 123

Bandwidth, 14
  baseband waveform, 26
  signal model, 27, 28
  vs. transmitted frequency, 22, 26
Bandwidth extrapolation (BWE), 15,
    306
Bistatic
  definition, 16
  imaging, 197
Block processing
  block time, 57
  data blocks, 34
  sequential estimation processing, 60, 69
Block time, 18, 57

Bounding volume, 160
  definition, 175
  estimation, 176

Center of mass, 24, 50, 61
Center of rotation, 21, 128, 234, 280
Chirp waveform, 14, 18, 27
Coherence, definition, 8, 100, 280,
    330
Coherence time, vs. linear imaging block
    size, 112, 121, 107
Coherency metrics, 100, 106
  coherent, 105
  noncoherent, 103
Coherent integration, 30, 103, 104
Component modeling, 352, 399
Composite target space mapping, 158,
    241, 324
Computational coordinate system, 50,
    236
Coordinate systems, 458
  alignment, 460
  center of rotation, 21
  computational model, 236
  measurements, 238
Correlation, 59
  nearest neighbor technique, 148
  range, range-rate pairs, 147
Correlation lag function, 209
  monostatic, 210
Creeping waves, 47, 163, 165

Cross range
  definition, 51
  poles, scaling to target dimensions, 176
Cross-range line integration, 179
Cylindrical cavity, 180

Data extrapolation, 285, 306
Data fusion scenarios, 288
  signal models, 282
  sparse band vs. sparse angle, 283
Delayed returns, scattering center typing,
  160, 163
DFT processing gain, vs. HRSE, 30
Diffraction Coefficient, 10, 53, 134
  GTD component modeling, 383, 401,
  403
Direction cosine space, 284, 267
Doppler processing, 3, 34, 104
  Range-rate scaling, 35

Enhanced BWE, 285, 306
Euler motion, 25
  characterization, 64, 407
  Euler parameters, 64, 407
  rotational motion, 21, 24, 470
Extended coherent processing, 140, 320
Extracted Observables, 4

Far field, 280, 345, 348
Fast time, 18, 27
Field data, 24
  vs. static range data, 32
Fusion, 39, 283, 363
  sparse angle, 280, 346
  sparse band, 42, 346

Generalized autocorrelation function, 102
Geometrical theory of diffraction, 43, 76,
  127

Global signature model, 281, 331
GTD signature model, 53, 344

Hankel matrix, 339
High resolution spectral estimation
  (HRSE), 4
  comparison to Fourier, 34

Imaging, 34
  composite mapping, 185, 324
  range-Doppler, 3, 34
  range-rate, 3, 36
Impulse response, 44, 46
  examples, 47, 48
  linear system, 46
Inertial coordinates, 129, 407
  Euler motion, 407
  vs. body fixed coordinates, 409
Integration of pulses, 103
  coherent, 104
  noncoherent, 103
Interferometric ISAR
  bistatic, 198
  monostatic, 207
  static range data, 387

Joint motion-target framework,
  motion-target decoupling, 129,
  139

Linear independence, composite mapping
  equations, 204
Linear signal model, 31, 111
  one-dimensional, 312
  two-dimensional, 55

Matched filter, definition, 22
Measurement coordinate system, 128,
  182

Model fidelity, 372, 398
  narrowband, 381
  wideband, 382
Motion, rotational vs. translational, 21,
  23, 129, 407
Motion estimation, 25, 127, 131
  known target, 217, 224
  joint solutions, 231, 256, 265
  motion solution, 52, 61
Motion observation space, field data vs.
  static range data, 129, 136
Motion solution, definition, 25, 61
Multiple targets, pulse compression, 26, 27
Multisensor fusion, 286
Multistatic sensors, ID encoding, 17
Mutual coherence, 279
  concept, 280
  duality, 302
Mutual coupling, delay mechanisms, 163

Narrowband imaging, 132, 179
Near field, 345
Newton's laws of motion
  torque free, 471
  translational vs. rotational motion, 472
Noncoherent integration, 103
Nonlinear signal model, scattering center
  acceleration, 85, 93
Nyquist sampling, range and angle, 27,
  183, 369

Observability conditions, 128
Observables, extracted from measurement
  data, 56, 57
Operational frequency, vs. baseband
  frequency, 26

Persistence filter, 153, 434
Phase derived range, 14

Phase-enhanced grid sampling, 205
Physical Theory of Diffraction, vs. GTD,
  367
Polar formatting, 284
Polarization, 21
Pulse compression
  range scaling, 29
  range-rate scaling, 35
Pulse repetition frequency, 52
Pulse repetition interval, 52

Radar calibration, 20
Radar cross section, signature modeling,
  23, 25, 38
Radar design architecture, functions,
  17
Radar fixed coordinate system, 50
Radar imaging techniques, narrowband
  vs. wideband, 165, 169, 174
Radar observables, 58, 50
Radar signature modeling, 49, 53, 367
Radar technology evolution, 14
Range bias, 219, 234, 473
Range Doppler coupling, 18
Range-Doppler imaging, 35, 37
Range gates, 35
  compressed pulse, 30
  definition, 29, 35
Range-rate image phase difference,
  201
Range-rate imaging
  technique, 35
  vs. the autocorrelation contour, 123
Range-rate scaling, 35
Range scaling, 35. See also Range gates
Range windowing, for enhanced SNR, 29
Receiver processing chain
  frequency domain, 22
  time domain, 33

Resolution, 13, 40, 132
  Fourier vs. HRSE, 68, 69
  Range vs. angle, 131
Resonating gaps and cavities, examples, 180
Roll angle, 21, 135
Rotational motion, vs. translational
    motion, 24, 50, 128

Scattering center acceleration, 73, 84, 96
Scattering center dispersion metric, 234,
    258
Scattering centers
  delay mechanisms, 163
  fixed, 53
  shadowing, 63
  slipping, 62
  specular, 164
Scattering phenomenology, vs.
    operational frequency band, 40
Sensor calibration, 20
Sensor fusion scenarios, sparse band vs.
    sparse angle, 286
Sequential estimation processing, 60
Shadowing, 62, 63, 70, 269
Sharpness, vs. resolution, 308
Signal model, 41, 49, 305
Signal processing
  back-end, 18
  front-end, 18
  signal flow models, 26
Signal-to-noise ratio
  received signal, 22
  receiver processing chain, 22
Signature modeling
  using GTD, 53
  using other methods, 366
Singular value decomposition, 266
Slipping scattering center, 53
  characterization, 62
  modeling with GTD, 62

Slotted cylindrical target, 180
  characterization, 180
  simulation examples, 184
Slow time, 18, 32, 33
Specular reflection, all pole modeling,
    164
Spherical coordinates, 50
  rotational motion, 51
  target fixed coordinates, 50, 51,
    129
Static range data
  roll cut measurement, 135
  sting platform, 135
  3D measurements, 385
  vs. field data, 129, 135
Sting measurement platform, 135, 385,
    468
Synthetic aperture radar (SAR), 195
Synthetic data, frequency angle space, 41,
    286, 307

Target bounding volume, 160, 175
Target center of rotation, range bias, 217,
    234, 472
Target fixed coordinates, 50
  aspect and roll angles, 65
  center of rotation, 128, 217, 234
Target frequency response, 44
  linear wire, 48
  spherical target, 47
Target model fidelity, 374, 380, 399
Target model observables, vs. HRSE
    model observables, 373, 374
Target motion, translational vs.
    rotational, 24, 50, 128
Target motion coupling, 139
  characterization, 138
  singular value decomposition, 266
Target motion mapping, special cases,
    141

Target space mapping equations, 138
  narrowband, 174
  wideband, 169
Three-dimensional imaging, 129, 139
  bistatic, 195
  monostatic, 138
  static range data, 385
Time-bandwidth product, baseband
    waveform characterization, 32
Time delay, wideband scattering
    mechanisms, 160, 168
Time domain signal flow
  receiver output signal model, 26
  window filter, 30
Time scales, 17, 18
Translational motion, center of mass, 24,
    50
Transmit waveform, 26
Traveling and surface waves, examples,
    163, 165
Tumble period, 229, 242
Two-dimensional strip
  diffraction coefficients, 353
  scattering from, 353
  simulation, 355

Ultra-wideband, synthetic vs. measured
    data, 42, 330

Waveform generator, 21, 25
Wideband delay mechanisms, 163